$107-40

50

DIETARY FAT AND CANCER
Genetic and Molecular Interactions

ADVANCES IN EXPERIMENTAL MEDICINE AND BIOLOGY

Recent Volumes in this Series

Volume 413
OPTICAL IMAGING OF BRAIN FUNCTION AND METABOLISM 2: Physiological
Basis and Comparison to Other Functional Neuroimaging Methods
Edited by Arno Villringer and Ulrich Dirnagl

Volume 414
ENZYMOLOGY AND MOLECULAR BIOLOGY OF CARBONYL METABOLISM 6
Edited by Henry Weiner, Ronald Lindahl, David W. Crabb, and T. Geoffrey Flynn

Volume 415
FOOD PROTEINS AND LIPIDS
Edited by Srinivasan Damodaran

Volume 416
PLATELET-ACTIVATING FACTOR AND RELATED LIPID MEDIATORS 2: Roles in
Health and Disease
Edited by Santosh Nigam, Gert Kunkel, and Stephen M. Prescott

Volume 417
DENDRITIC CELLS IN FUNDAMENTAL AND CLINICAL IMMUNOLOGY, Volume 3
Edited by Paola Ricciardi-Castagnoli

Volume 418
STREPTOCOCCI AND THE HOST
Edited by Thea Horaud, Anne Bouvet, Roland Leclerq, Henri de Montclos,
and Michel Sicard

Volume 419
ADP-RIBOSYLATION IN ANIMAL TISSUES: Structure, Function, and Biology of Mono
(ADP-ribosyl) Transferases and Related Enzymes
Edited by Friedrich Haag and Friedrich Koch-Nolte

Volume 420
ADVANCES IN CIRRHOSIS, HYPERAMMONEMIA, AND HEPATIC
ENCEPHALOPATHY
Edited by Vicente Felipo

Volume 421
CELLULAR PEPTIDASES IN IMMUNE FUNCTIONS AND DISEASES
Edited by Siegfried Ansorge and Jürgen Langner

Volume 422
DIETARY FAT AND CANCER: Genetic and Molecular Interactions
Edited under the auspices of the American Institute for Cancer Research

DIETARY FAT AND CANCER

Genetic and Molecular Interactions

Edited under the auspices of the

American Institute for Cancer Research
Washington, D.C.

PLENUM PRESS • NEW YORK AND LONDON

Library of Congress Cataloging-in-Publication Data

Dietary fat and cancer : genetic and molecular interactions / edited
   under the auspices of the American Institute for Cancer Research.
      p.   cm. -- (Advances in experimental medicine and biology ; v.
   422)
      "Proceedings of the American Institutefor Cancer Research's
   Seventh Annual Conference on Dietary Fat and Cancer: Genetic and
   Molecular Interactions, held August 28-30, 1996 in Washington,
   D.C."--T.p. verso.
      Includes bibliographical references and index.
      ISBN 0-306-45683-4
      1. Cancer--Nutritional aspects--Congresses. 2. Lipids in human
   nutrition--Congresses. 3. Lipids--Pathophysiology--Congresses.
   4. Cancer--Molecular aspects--Congresses. 5. Cancer--Genetic
   aspects--Congresses.   I. American Institute for Cancer Research.
   II. Conference on Dietary Fat and Cancer: Genetic and Molecular
   Interactions (1996 : Washington, D.C.) III. Series.
      [DNLM: 1. Neoplasms--etiology--congresses. 2. Neoplasms-
   -genetics--congresses. 3. Dietary Fats--adverse effects-
   -congresses. 4. Dietary Fats--metabolism--congresses. 5. Fatty
   Acids--genetics.   W1 AD559 v.422 1997 / QZ 202 D5646 1997]
   RC268.45.D5626   1997
   616.99'4071--dc21
   DNLM/DLC
   for Library of Congress   97-4389
                                                               CIP

PO 01628
18.2.98

Proceedings of the American Institute for Cancer Research's Seventh Annual Conference on Dietary Fat
and Cancer: Genetic and Molecular Interactions, held August 28 – 30, 1996, in Washington, D.C.

ISBN 0-306-45683-4

© 1997 Plenum Press, New York
A Division of Plenum Publishing Corporation
233 Spring Street, New York, N. Y. 10013

http://www.plenum.com

10 9 8 7 6 5 4 3 2 1

Printed in the United States of America

PREFACE

The annual research conference for 1996 of the American Institute for Cancer Research was again held at the Loews L'Enfant Plaza Hotel in Washington, DC, August 29 and 30. The topic for this, the seventh in the series, was "Dietary Fat and Cancer: Genetic and Molecular Mechanisms." Two separate presentations were given as the conference overview. "Fat and Cancer: The Epidemiologic Evidence in Perspective" noted that dietary fat can be saturated, largely from animal or dairy sources, or mono- or polyunsaturated, mostly from plant sources. Unlike animal fats, fish contain relatively high levels of protective omega-3 fatty acids. Although the hypothesis that dietary fat is associated with cancer is plausible, the mechanisms involved are reasonable, and many animal studies support the hypothesis, there are many obstacles in any direct extrapolation to humans, including imprecise measures of dietary fat intake, variability in individual diets, and species variations. Despite these limitations, there is a weak positive correlation between colon cancer and dietary fat intake, but with substantial differences for various ethnic groups. In the case of breast cancer, there is substantial variation among countries and ethnic groups, but the overall evidence indicated an association with fat in the diet. Epidemiologic studies of dietary fat and prostate cancer are more consistent and most show a positive relationship. However, it was not clear which types of dietary fat were implicated in the effect. Surprisingly, although the major etiologic factor for lung cancer is smoking, there are some fairly good correlations between high fat or cholesterol intake and this type of cancer. For other organ sites such as pancreas, endometrium, ovary, kidney, bladder, and gallbladder, there is no compelling evidence for an effect of dietary fat.

The second overview paper "Dietary Lipids and the Cancer Cascade" mentioned that although animal models implicate dietary lipid as a factor in development of malignancies, testing this hypothesis in human intervention studies has been very slow. Problems include cost, controlling for changes in foods and nutrients, and quantitating levels of lipids and any changes accurately. Better elucidation of the molecular, biochemical, and cellular processes modulated by lipids will eventually allow novel means to assess accurately the intake of dietary fats.

The cancer process, from its beginning with a single mutated cell, requires many years to evolve to a life-threatening malignancy, thus affording opportunity for diet to influence both the rate of progression and the biological behavior of the neoplasm. Genetics has an important role in cancer initiation and progression, either by virtue of familial or germ line genes that are inherited and influence cancer risk, or by acquired defects in critical genes comprising the evolving tumor. Examples are the BRCA1 and BRCA2 tumor suppressor genes; if a woman inherits a mutation in these genes, her risk of breast cancer increases measurably. Genes that determine capacity to either activate or detoxify chemical carcinogens also influence the pattern of acquired mutations in critical tumor suppressor genes. Persons who inherit two noneffective copies of the GSTM1 (for glutathione-S-transferase) gene have a significant increase in bladder cancer if they smoke since GST detoxifies a carcinogen in tobacco smoke. Inheriting the deletion carries little risk for those who do not smoke. These polymorphisms that affect risk of cancer may potentially interact with dietary factors. Since options for intervention are limited for those with defective genes, prevention strategies involving diet and nutrition are more practical.

Extrapolation from studies in mice with a genetic background leading to increased risk for mammary cancer indicates that delaying the onset of tumor development by dietary means is feasible. In mice lacking the critical p53 gene (knockout mouse) moderate calorie restriction delayed the onset of spontaneous tumors. Further, p53 may serve as an activator of apoptosis for damaged cells before they progress on to neoplasia. Angiogenesis, the means by which tumors develop vascular systems, also is important. There are preliminary data that these processes can be controlled to a certain extent by dietary means. Reducing various cancer rates through dietary intervention is possible.

The first of the subsequent sessions was entitled "Dietary Fat and Tumor Growth Regulation," with the first paper on specific studies of gene expression in chemically induced mammary carcinogenesis. In female rats given the mammary carcinogen methylnitrosourea and fed diets high in omega-6 polyunsaturated fatty acids (promoters), there was higher expression of cyclooxygenase 2 (COX-2) transcripts than in rats fed omega-3 fatty acids or control diets. Additional experiments suggested that a certain level of dietary cholesterol may inhibit de novo cholesterol synthesis in preneoplastic mammary epithelial cells, thus decreasing their proliferative rate and neoplastic development. Likewise, in an estrogen-independent and highly metastatic human breast cancer cell line (MDA-MB-231) there was expression of a high level of COX-2. Exposure to the promoting agent TPA caused further sustained enzyme expression. In contrast, the estrogen receptor positive human breast cancer cell line MCF-7, which has low invasive capacity, had a high level of COX-1 expression and only a transient response to TPA. Reducing dietary intake of the omega-6 fatty acids, increasing the omega-3, developing a specific drug for inhibition of COX-2, and developing an inhibitor of lipoxygenase, another enzyme system involved in oxidative activation of fatty acids, were suggested as future chemopreventive approaches.

Increased breast cancer cell proliferation was implicated as an important adverse factor in the survival of cancer patients. In cell cultures long chain saturated fatty acids such as stearate inhibited cancer cell proliferation but there was no effect on membrane fluidity. Inhibition by stearate paralleled a decrease in association of the transducer molecules guanine nucleotide binding proteins (G-proteins) with the epidermal growth factor receptor (EGFR) but tyrosine phosphorylation of the EGFR was not affected. The unsaturated long chain fatty acid oleate increased cell proliferation and G-protein association with EGFR. It was suggested that stearate may merit a specific controlled place in the diet.

The second session "Dietary Fat and Signal Transduction" explored the mechanisms by which unsaturated fatty acids influence the processes involved in neoplasia. The eicosanoids, derived from oxygenated arachidonic acid metabolites and the hydroxyeicosatetraenoic acids (HETEs) are involved in multiple steps of neoplasia and metastasis. HETEs affect tumor cell motility, interactions with the extracellular matrix, the release of lysosomal enzymes, and they act as signaling molecules. However, the unsaturated fatty acids in fish oils, namely eicosapentaenoic acid and docosahexaenoic acid, tend to reduce the development of colon tumors in rats treated with a colon carcinogen. These fatty acids modulate favorably the balance between colonic cell proliferation and apoptosis and decrease carcinogen-induced ras mutation in the colon. Further, in azoxymethane-treated rats, dietary fish oils decreased colonic phospholipase C gamma 1 expression and diacyl-glycerol mass and blocked the azoxymethane-induced decrease in steady-state levels of select colonic protein kinase C (PKC) isozymes. The beneficial action of fish oils may reside in blocking the activation and downregulation of PKC isozymes.

The emphasis then shifted to a different dietary component-choline. A deficiency leads to hepatocellular carcinoma in male rats. Adapting rat hepatocytes to grow in a choline-deficient medium caused the cells to acquire resistance to apoptosis and to produce hepatocellular carcinomas when injected into nude mice. These adapted hepatocytes were resistant to the transforming growth factors (TGF-Beta 1), which also induce cell death. Thus choline deficiency may act through induction of alterations in growth factor signaling pathways. Although peroxisome proliferators have been implicated as inducers of hepatic tumors in rodents, both humans and subhuman primates apparently are minimally responsive or resistant to their action. The session on "Dietary Fat and Peroxisome Proliferator Activated Receptors" provided information on these cell constituents. Peroxisomes are cellular organelles that contain a variety of enzymes, but especially those involved in fatty acid beta-oxidation. Lack of these systems can lead to multiple functional deficiencies since oxidation of cholesterol, biosynthesis of cholesterol and other lipids and dolichols, all depend on peroxisome-localized enzymes. Hypolipidemic drugs, leukotriene D4 antagonists, some herbicides and pesticides, phthalate esters, various solvents, and the endogenous chemicals, phenylacetate and dehydroepiandosterone, are only some of the many peroxisome proliferators. A nuclear receptor was cloned from mice (peroxisome proliferator receptor or PPAR) and found to be a member of the nuclear receptor superfamily that includes estrogen, progesterone, thyroid hormone, and other receptors. Three members of the PPAR family, alpha, beta and gamma, have been cloned and appear to have different functions and tissue distribution. PPAR alpha appears to have a role in modulating lipid catabolism and perhaps lipid synthesis. PPAR alpha activates gene expression through binding to a response element (PPRE), and it interacts with other members of the nuclear receptor superfamily. Peroxisome proliferators may lead to hepatocellular carcinoma in rats by increasing intracellular hydrogen peroxide, by damaging cytoplasmic mitochondrial DNA, or by inhibiting apoptosis. Mice lacking the PPAR alpha receptor were produced and showed no apparent gross phenotypic abnormalities. In addition, they showed no response to peroxisome proliferators such as clofibrate or dehydroepiandrosterone sulfate, but they did have an altered lipid metabolism. These null mice are being used to investigate species and other differences in response to peroxisome proliferators.

The complexity of the PPAR activation pathway has been enhanced by finding that PPAR DNA binding is linked to heterodimerization with a member of the retinoid X (RXR) family of receptors. In addition, the PPAR/farnesol activation pathway converges at some points with another steroid receptor orphan family member, the farnesoid X activated receptor (FXR). The data suggest that farnesol is a common intermediate for endogenous activators of both PPAR alpha and FXR and that FXR can be integrated into the PPAR alpha-mediated oxidative stress hypothesis.

"Dietary Fat and Gene Expression," the next session, provided more information on the functions of PPAR and related receptors. Several genes expressed in adipose cells, such as that for adipocyte lipid-binding protein (aP2) and phosphoenolpyruvate carboxykinase have binding sites for PPAR/RXR heterodimers. These intracellular aP2s control accessibility of the substrates and are negative elements in polyunsaturated fatty acid control of gene expression.

In discussing dietary fatty acids, the short chain fatty acids are often disregarded, but acetate, propionate, and butyrate have important biological and clinical effects in the normal and diseased colon. Butyrate, especially, regulates expression of growth-related genes, such as p53, c-fos, thymidine kinase, and c-myc. Expression of c-myc is correlated with cellular proliferation and inversely with cellular differentiation. Butyrate acts rapidly on c-myc, presumably through induction of a protein that decreases c-myc. Butyrate also induces apoptosis in several colon and breast carcinoma cell lines. It inhibits the growth of the colon carcinoma cells in the early G1 phase of the cell cycle.

Although mammals are the organisms of interest with respect to PPAR, peroxisome proliferation also occurs in yeast, where levels of peroxisomal beta-oxidation enzymes are regulated by the available carbon source. Expression of the genes encoding the enzymes occurs in the presence of glucose and activation occurs when a fatty acid such as oleate is supplied for growth. The rate-limiting enzyme in the peroxisomal beta-oxidation cycle is POX1 (for peroxisomal acyl-CoA oxidase); an activating sequence required for oleate-specific activation of POX1 was identified. Disrupting this gene in the yeast led to cells which did not grow in the presence of oleic acid.

An overview on the metabolism of fats, derivation of energy and genetic control summarized the situation. Dietary fat regulates gene expression through multiple pathways, thus causing changes in carbohydrate and lipid metabolism. Polyunsaturated fatty acid (PUFA) regulatory factor (PUFA-RF) was not PPAR, although PPAR apparently is involved in one mechanism for fatty acid regulation. This appears to be transcription of S14 gene in liver where it acts by sequestering RXR. PUFA can have either a beneficial action as in the n-3 PUFA suppression of serum triglycerides or a detrimental one as in the saturated and n-6 PUFA promotion of insulin resistance.

The final session "Future Directions and Implications of Research on Dietary Fat and Genetics" afforded a panel of experts the opportunity to present their individual views on this matter. The role of body fat distribution and genetic predisposition in relation to breast cancer was mentioned. In addition the relative risk was fairly high (4.8) for a relationship between obesity and both breast and ovarian cancer, versus a risk of 2.1 for breast cancer alone and a relative risk of 1.1 for women with no family history of breast or ovarian cancer. In animal studies dietary fat definitely influences the outcome of various carcinogenicity studies, but molecular studies may not always provide a rationale for the observed effects.

Although the evidence for the role of specific genes in human cancer is growing, some experts consider that environmental factors are still the most important determinant of cancer risk. Studies of populations with certain cancer susceptibilities may provide more definitive human data. Integrative studies on nutrition and cancer, the time in life when exposure is important, and epidemiologic studies that assess cancer genetics using molecular biology techniques were all considered necessary. There is general interest in any relationship between body shape and risk of breast or ovarian cancer, as well as the difference in function of body fat from different anatomical sites.

Besides the scientific sessions, there was a special evening workshop on "Diet, Nutrition and Cancer Prevention: Research Opportunities, Approaches and Pitfalls," directed toward obtaining funding and the interactions with funding agencies.

There were 57 poster abstracts presented with participants from eight countries besides the United States, confirming the international interest in research in the diet/cancer field.

Thus, although the epidemiologic evidence is not always definitive, it does hint toward an association between dietary fat and cancer incidence. Fat has an important role in the diet, but the type and quantity consumed can interact with the genetic background of the individual to increase the risk of disease. There is a need for most people in the affluent world to reduce the amount of dietary fat. Educational efforts toward that end are most appropriate.

The Editors

CONTENTS

Chapter 1
Fat and Cancer: The Epidemiologic Evidence in Perspective 1
Laurence N. Kolonel

   Classification of Fat in Epidemiologic Research 1
   Plausibility of the Fat–Cancer Hypothesis 3
   Some Issues Related to Epidemiologic Studies of Fat and Cancer 3
   Epidemiologic Associations of Dietary Fat and Cancer 4
      Colorectal Cancer 4
      Breast Cancer 6
      Prostate Cancer 9
      Lung Cancer 11
      Other Cancers 12
   Conclusions and Implications for Public Health 13
   References 14

Chapter 2
Dietary Lipids and the Cancer Cascade 21
Steven K. Clinton

   Introduction 21
   The Genetic Basis of Cancer Predisposition and Susceptibility 23
   Cellular Proliferation, Apoptosis, and Immortality 25
   Angiogenesis and Metastasis 28
   The Relationship between Dietary Lipids, Energy Intake, and Cancer in
      Experimental Models 29
   Future Directions: Dietary Lipids and the Cancer Cascade 32
   References 34

Chapter 3
Molecular Studies on the Role of Dietary Fat and Cholesterol in Breast Cancer
Induction 39
Michael C. Archer, Ahmed El-Sohemy, Laurie L. Stephen, and Alaa F. Badawi

   Introduction 39

Dietary Fat and Rat Mammary Tumorigenesis   40
Dietary Cholesterol and Rat Mammary Tumorigenesis   43
    Acknowledgments   45
References   45

Chapter 4
Fatty Acid Regulation of Breast Cancer Cell Growth and Invasion   47
David P. Rose, Jeanne M. Connolly, and Xin-Hua Liu

Introduction   47
Effects of Omega-6 Fatty Acids on Normal Mammary Epithelial and Breast
    Cancer Cell Growth   48
    Stimulation *in Vitro*   48
    Effects *in Vivo*   48
Suppression of Breast Cancer Cell Growth by n-3 FAs   49
Fatty Acids, Eicosanoids, and the Metastatic Phenotype   49
    Cyclooxygenases and Lipoxygenases   49
    Cell Growth   50
Invasion and Metastasis   51
Estrogen Dependence and Arachidonic Acid Metabolism   52
Commentary   52
    Acknowledgments   53
References   53

Chapter 5
Fatty Acids and Breast Cancer Cell Proliferation   57
Robert W. Hardy, Nalinie S. M. D. Wickramasinghe, S. C. Ke, and Alan Wells

Introduction   57
    Dietary Fat and the Development of Breast Cancer   57
    Breast Cancer Cell Proliferation   58
Effects of Fatty Acids on EGF Induced Breast Cancer Cell Proliferation   60
    How Do Individual Long Chain Fatty Acids Alter EGF Induced Breast
        Cancer Cell Proliferation?   60
    Is Stearate Inhibition of EGF Induced Cell Proliferation a Cytotoxic or
        Membrane Fluidity Effect?   60
    What EGFR Induced Signaling Pathway Is Affected?   62
Future Studies   64
    Which Signaling Pathway Is Affected by Fatty Acids Altering
        EGFR/G-Protein Interaction?   64
    Do Fatty Acids Affect Other EGFR Related Signaling Mechanisms?   64
    Are Select Fatty Acids Safe Dietary Supplements?   65
Summary   65
    Acknowledgments   65
References   65

Chapter 6
Lipoxygenase Metabolites and Cancer Metastasis 71
Keqin Tang and Kenneth V. Honn

Introduction 71
    Metastasis 71
    Eicosanoids 72
12(S)-HETE: Involvement in Multiple Steps of Cancer Metastasis 73
    Effect of 12(S)-HETE on Tumor Cell Interactions with Extracellular
        Matrix (ECM) 74
    Effect of Exogenous 12(S)-HETE on Tumor Cell Motility 75
    Effect of Exogenous 12(S)-HETE on the Release of Lysosomal Enzymes ... 76
    Effect of Exogenous 12(S)-HETE on the Tumor Cell Infrastructure and
        Cystoskeleton 76
    12(S)-HETE as a Signaling Molecule 77
    Hypothesis for Role of 12(S)-HETE in Metastasis 79
    Acknowledgment 80
References 80

Chapter 7
**Modulation of Intracellular Second Messengers by Dietary Fat during Colonic
Tumor Development** 85
Robert S. Chapkin, Yi-Hai Jiang, Laurie A. Davidson, and Joanne R. Lupton

Dietary Factors and Colon Cancer 85
    Dietary Fat and Colon Cancer 86
    Dietary N-3 Polyunsaturated Fatty Acids and Colon Cancer 86
Experimental Carcinogenesis Model 87
The Role of Diacylglycerol and Phospholipase C-γ 1 in Colon Cancer 87
    Protein Kinase C Activation 89
    Role of Protein Kinase C in Tumor Promotion 89
    Role of Protein Kinase C in Colon Cancer 89
    Noninvasive Detection of Protein Kinase C Isozymes as a Biomarker for
        Colon Cancer Using Fecal Messenger RNA: Clinical Implications 91
Summary 91
    Acknowledgments 92
References 92

Chapter 8
Diet, Apoptosis, and Carcinogenesis 97
Craig D. Albright, Rong Liu, Mai-Heng Mar, Ok-Ho Shin, Angelica S. Vrablic,
Rudolf I. Salganik, and Steven H. Zeisel

Abstract 97
Introduction 98
    Diet and Carcinogenesis 98
    Choline Deficiency and Apoptosis 99
    Choline Deficiency and p53 Expression 100
    Choline Deficiency and TGFα/EGF Signaling 101

        Choline Deficiency and TGFß1 Signaling 102
    Summary 103
    Acknowledgment 104
    References 105

Chapter 9
The Role of Peroxisome Proliferator Activated Receptor α in Peroxisome
Proliferation, Physiological Homeostasis, and Chemical Carcinogenesis 109
Frank J. Gonzalez

    Introduction 109
        Peroxisomes and Peroxisome Proliferation 109
    PPARα 110
        The Peroxisome Proliferator-Activated Receptor 110
        Function of PPARs 111
        Mechanism of Gene Activation by PPARα 112
    Peroxisome Proliferator-Induced Hepatocarcinogenesis 114
    Species Differences and Human Risk Assessment 116
    The PPARα-Null Mouse 116
        Isolation and Characterization of the PPARα cDNA 116
        Production of the PPARα-Null Mouse 116
        Lipid Metabolism in the PPARα-Null Mice 117
        The Role of PPARα in DHEA Induction of Peroxisome Proliferation 117
    PPARα and Growth Control 118
    Conclusions and Future Studies 118
        PPARα and Peroxisome Proliferation 118
        PPARα and the Mechanism of Action of Non-Genotoxic Carcinogens 119
        PPARα and Species Differences in Response to Peroxisome Proliferators .. 119
    References 120

Chapter 10
A Hypothetical Mechanism for Fat-Induced Rodent Hepatocarcinogenesis 127
Daniel J. Noonan and Michelle L. O'Brien

    PPAR, a Family of Transcription Factors Regulating Fat Metabolism 127
    PPARα as a Mediator of the Oxidative Stress Hypothesis 128
    PPARα as a Regulator of Fat Catabolism 128
    Isoprenoids as Regulators of PPARα Activity 130
    Farnesol as an Activator of PPARα and FXR 131
    Integration of FXR into the Oxidative Stress Hypothesis 132
    Acknowledgments 132
    References 133

Chapter 11
Short Chain Fatty Acid Regulation of Intestinal Gene Expression 137
John A. Barnard, J. A. Delzell, and N. M. Bulus

    Introduction 137

Biology of Short Chain Fatty Acids in the Mammalian Colon 137
Molecular and Cell Biology of the Short Chain Fatty Acid, Butyrate, in Cells of
    Gastrointestinal Origin 138
    Effects of Butyrate on Intestinal Cell Growth and Differentiation 138
    Effects of Butyrate on Epithelial Cell Apoptosis 140
Summary of Studies on the Induction of Growth Inhibition and Differentiation
    by Butyrate in HT-29 Cells 140
    Butyrate Inhibits HT-29 Cell Growth 140
    Butyrate Inhibits Growth in Early G1 140
    Effect of Butyrate on Growth and Differentiation-Related Gene Expression
        in HT-29 Cells 141
    Mechanism of Down Regulation of *myc* Expression by Butyrate 141
    Current Studies 142
References 142

Chapter 12
Regulation of Gene Expression in Adipose Cells by Polyunsaturated Fatty Acids 145

David A. Bernlohr, Natalie Ribarik Coe, Melanie A. Simpson, and
Ann Vogel Hertzel

Summary 145
Introduction 146
Material and Methods 147
    Materials 147
    Lipid Binding Protein Purification 147
Separation of Adipose Cells 148
Extraction of Triglycerides and Fatty Acids from Fat Cells and Thin Layer
    Chromatography 148
    Determination of Total Free Fatty Acid and Triglyceride Levels 149
    Western Analysis 149
    Northern Analysis 149
    Ligand Binding Studies 149
    Competition Assays 150
Results 150
Discussion 153
Abbreviations 155
    Acknowledgments 155
References 155

Chapter 13
Regulation of Peroxisomal Fatty Acyl-CoA Oxidase in the Yeast *Saccharomyces cerevisiae* 157

Gillian M. Small, Igor V. Karpichev, and Yi Luo

Abstract 157
Introduction 157
    Peroxisome Proliferation in Mammals 158
    Peroxisome Proliferation in Yeast 158
Regulation of Peroxisomal Acyl-CoA Oxidase 159

        Identification of *cis*-Acting Elements in the Acyl-CoA Oxidase Gene 159
        Purification of a Protein Required for Oleate-Induction of *POX1* 159
    Current and Future Research 163
        Acknowledgments 164
    References 165

Chapter 14
Dietary Fat, Genes, and Human Health 167
Donald B. Jump, Steven D. Clarke, Annette Thelen, Marya Liimatta, Bing Ren, and
Maria V. Badin

    Introduction 167
        Dietary Fat Effects on Cell Function 168
        PUFA Effects on Hepatic Lipid Synthesis and Metabolism 168
    PUFA Effects on Hepatic Gene Expression 169
        PUFA-Mediated Suppression of L-Pyruvate Kinase Gene Transcription 169
        PUFA-Mediated Suppression of S14 Gene Transcription 171
        PUFA Regulation of Peroxisome Proliferator Activated Receptor (PPAR) .. 172
    Summary and Conclusions 173
    References 173

Chapter 15
Session V: Future Directions and Implications of Research on Dietary Fat and
Genetics 177
Diane F. Birt

    Discussion 177
    References 179

Abstracts 181

Index 245

FAT AND CANCER

The Epidemiologic Evidence in Perspective

Laurence N. Kolonel

Cancer Research Center of Hawaii
University of Hawaii
Honolulu, Hawaii 96813

CLASSIFICATION OF FAT IN EPIDEMIOLOGIC RESEARCH

Fat has been the subject of much epidemiologic research on the role of nutritional factors in the etiology of cancer.[1,2] Some studies reported only on fat as a single entity, whereas others distinguished among different types of fat in an effort to refine the analyses. Fat (i.e., fatty acids) can be categorized in many ways, e.g., by their chemical structure, by sources in the diet, or by human nutritional requirements. Chemically, fatty acids may be classified as unsaturated or saturated, depending on whether or not they contain any double bonds between adjacent carbon atoms. Unsaturated fats may be monounsaturated (only a single double bond, as in oleic acid) or polyunsaturated (multiple double bonds, as in linoleic acid). Polyunsaturated fatty acids can be further broken down into the omega-6 or omega-3 series, determined by the location of the first double bond from the methyl end of the molecule (e.g., arachidonic vs. docosapentaenoic acid, respectively) (Figure 1). Unsaturated fatty acids can occur as *cis* or *trans* isomers, which are different geometric configurations of the same molecule (e.g., oleic vs. elaidic acid, respectively) (Figure 2). Most naturally-occurring fatty acids are in the *cis* form. *Trans* fatty acids are largely a result of the industrial hydrogenation of liquid oils, as occurs in the production of margarine. Because the biological effects of dietary constituents depend on their chemical structures, these different forms of fat may have independent relationships to disease. Most of the epidemiologic studies that have examined different chemical forms of fat have distinguished between saturated and polyunsaturated fats, probably reflecting the earlier experience in cardiovascular disease research. More recently, studies have also been reporting on monounsaturated fats. There are few epidemiologic data on *cis* and *trans* isomers of fatty acids in relation to cancer risk.

Another way in which dietary fats have been classified is by their food sources. The primary distinction that has been made in the epidemiologic literature is between animal (including dairy) and vegetable fats. Most saturated fats come from animal products, with

Dietary Fat and Cancer, edited by AICR
Plenum Press, New York, 1997

- Saturated

    Example: $CH_3-(CH_2)_{14}-COOH$
    Palmitic acid

- Monounsaturated

    Example: $CH_3(CH_2)_7CH=CH(CH_2)_7COOH$
    Oleic acid

- Polyunsaturated – Omega-6

    Example: $CH_3(CH_2)_4CH=R=R=R=CH(CH_2)_3COOH$
    where $R = CHCH_2CH$
    Arachidonic acid

- Polyunsaturated – Omega-3

    Example: $CH_3CH_2CH=R'=R=R'=R'=CH(CH_2)_2COOH$
    where $R'=CH(CH_2)_2CH$
    Docosapentaenoic acid

Figure 1. Classification and chemical structures of fatty acids.

the notable exception of coconut and palm oils. Unsaturated fats are prominent in both animal and vegetable foods. Thus, separation of animal from vegetable fats primarily distinguishes between high and low saturated fat intakes. Another food-based distinction is between fish and meat sources of fat. Fish, especially the fattier cold water fishes, such as salmon and mackerel, contain high levels of omega-3 fatty acids, particularly eicosapentaenoic acid and decosahexaenoic acid. An omega-3 fatty acid found in red meat, butter and vegetable oils, but not fish, is alpha-linolenic acid.

For the most part, humans do not require fat sources in their diets because the body can synthesize most of the fatty acids it needs from other constituents, including carbohydrate and protein. Thus, fatty acids may also be categorized as essential or non-essential,

- Example: *cis* Isomer of $C_{18}H_{34}O_2$

    $H-C-(CH_2)_7-CH_3$
    $\parallel$
    $H-C-CH_2)_7COOH$
    Oleic acid

- Example: *trans* Isomer of $C_{18}H_{34}O_2$

    $CH_3-(CH_2)_7-C-H$
    $\parallel$
    $H-C-(CH-_2)_7-COOH$
    Elaidic acid

Figure 2. Isomers of unsaturated fatty acids.

although this distinction has seldom been used in epidemiologic studies of fat and cancer. Linoleic acid is an essential fatty acid, since it cannot be synthesized endogenously. Arachidonic acid, which can be synthesized endogenously from linoleic acid, is also considered essential. Although alpha-linolenic acid cannot be synthesized endogenously, a specific requirement for this fatty acid to maintain human health has not been established.

PLAUSIBILITY OF THE FAT–CANCER HYPOTHESIS

The notion that dietary fat may be carcinogenic in humans is entirely plausible. Many experiments in laboratory animals have demonstrated effects of fat on carcinogenesis. Although many early studies did not control for caloric intake when increasing or decreasing fat in the diets of experimental animals (thus raising the question whether the results reflected a caloric rather than a specific fat effect), more recent studies confirm that there is an independent effect of fat on carcinogenesis in experimental models.[3,4] In several studies using rodent models, a reduction of fat in the diet was shown to lower the incidence of carcinogen-induced tumors of the breast and colon; conversely, an increase in dietary fat was shown to have a promotional effect on cancer.[1] However, because of species variations, one cannot extrapolate directly to humans the findings from animal experiments.

A second basis for the plausibility of the fat hypothesis is that reasonable mechanisms can be proposed.[1,5] For example, the lipid composition of cell membranes is determined in part by the distribution of fatty acids in the diet. Alterations in cell membrane composition affect permeability, which controls movement of fats and other chemicals into and out of the tissues. Other effects of fat that may contribute to carcinogenesis include its influence on the immune system and on the synthesis of prostaglandins which, for example, may affect tumor growth. Fat may also act indirectly on carcinogenesis through its contribution to obesity, itself an established risk factor for certain cancers, notably those of the endometrium, kidney, breast (postmenopausal), and possibly colon.[2]

SOME ISSUES RELATED TO EPIDEMIOLOGIC STUDIES OF FAT AND CANCER

Epidemiologic studies on the role of dietary fat in cancer are complicated by difficulties in accurately assessing exposure levels. Because of the long latency period for most cancers, determining the appropriate time of exposure can be problematical. Even if a risk factor, such as dietary fat, acts to promote rather than initiate tumors, the relevant period of the exposure is likely to be at least several years prior to clinical diagnosis. Since adequate biochemical measures of long-term dietary fat exposure have not yet been established, epidemiologists have relied on diet history methods. Diet histories have proved to be extremely valuable in epidemiologic research, and have been validated in many instances.[6] Nevertheless, they are imprecise measures of dietary fat intake. Recall is subject to measurement error, which makes relationships to cancer more difficult to demonstrate. In addition, because of the extreme variability of the diet in most developed countries, individuals must recollect their consumption of many different foods, most of which are not eaten every day. In case-control studies, another concern is whether the cases may over- or under-estimate their intakes relative to controls, due to a perception that diet played a role in the development of their illness. Finally, because of its caloric

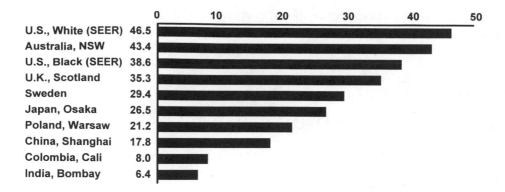

	0	10	20	30	40	50
U.S., White (SEER)	46.5					
Australia, NSW	43.4					
U.S., Black (SEER)	38.6					
U.K., Scotland	35.3					
Sweden	29.4					
Japan, Osaka	26.5					
Poland, Warsaw	21.2					
China, Shanghai	17.8					
Colombia, Cali	8.0					
India, Bombay	6.4					

*Age-adjusted, World Standard Population

Figure 3. Male colorectal cancer incidence* in selected populations, 1983–1987.

density, dietary fat is very highly correlated with energy intake, especially in western developed countries. Thus, separating these two factors in epidemiologic studies of diet and cancer can be extremely difficult, and sometimes impossible.

EPIDEMIOLOGIC ASSOCIATIONS OF DIETARY FAT AND CANCER

Colorectal Cancer

Internationally, colorectal cancer incidence rates show very wide variations.[7] As shown in Figure 3, there is more than a 7-fold difference between the rates for men in Bombay and the U.S. The incidence also varies substantially (more than 4-fold) among different ethnic groups in the U.S.,[8] where the rate is lowest in the American Indian popu-

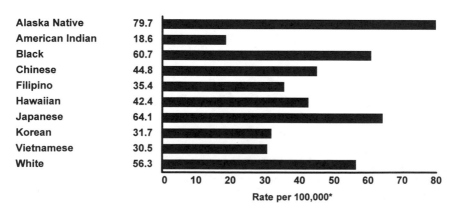

		0	10	20	30	40	50	60	70	80
Alaska Native	79.7									
American Indian	18.6									
Black	60.7									
Chinese	44.8									
Filipino	35.4									
Hawaiian	42.4									
Japanese	64.1									
Korean	31.7									
Vietnamese	30.5									
White	56.3									

Rate per 100,000*

*Age-adjusted, 1970 U.S. Standard Population

Figure 4. Male colorectal cancer incidence* by ethnicity, SEER, 1988–1992.

lation of New Mexico and highest in Alaskan natives (Figure 4). The rates for recent Asian immigrant groups, notably Vietnamese and Koreans, are noticeably lower than for the more established Asian populations, such as the Chinese and Japanese.

Such differences among and within countries could certainly be a result of dietary differences. Thus, it is reasonable to examine the correlations between colon cancer rates and corresponding dietary fat intakes. As shown in Figure 5, there is a positive correlation between colon cancer incidence and the mean percent of calories as fat in both men and women.[9] Such ecologic analyses have certain limitations. First, these dietary estimates are based on per capita intakes, which are not direct measures in individuals, but rather crude estimates based on the total availability of fat and calories in a country, without regard for wastage, food fed to animals, etc. Second, the estimates do not distinguish between age groups or between sexes (fat calorie values are the same for both men and women in the figure). Third, such correlations do not control for potential confounding factors. Nevertheless, these positive correlations are suggestive of a possible relationship and deserving of further study.

Epidemiologic study designs based on data collected from individuals, most notably case-control and cohort studies, overcome the major limitations of ecologic analyses. A summary of the literature on dietary fat and colorectal cancer is given in Table 1, which separates those studies that estimated actual fat intake from those that only reported on the intake of high-fat foods. Among 16 case-control studies that assessed dietary fat,[10–25] ten reported a positive association with colorectal cancer.[10,12–14,16–18,20,22,23] Of the 25 case-control studies that examined high-fat food intake,[10–13,15,18,19,21,23,25,26–40] 19 reported a positive association.[10,12,13,15,19,21,23,25,27–30,32,34–39] The findings from the cohort studies are less consistent. Although four[42,43,46,47] of the seven studies based on high-fat food consumption[41–47]

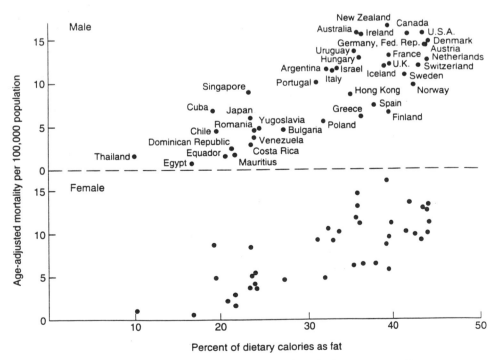

Figure 5. Correlation between % calories as fat and colon cancer mortality.

Table 1. Summary of epidemiologic studies on fat and colorectal cancer

Exposure	Study Type	Total No. of Reports	No. of Positive Reports	% Positive Reports
Dietary Fat	Case-Control	16	10	62.5
	Cohort	5	1	20.0
High-Fat Foods	Case-Control	25	19	76.0
	Cohort	7	4	57.1

show a positive relationship, only one[43] of the five studies that assessed dietary fat intake[43,45-47,48] was positive. Many case-control and cohort studies, especially the earlier ones, did not adjust for energy intake. Recent studies have included caloric adjustments and most have found no association with fat.[15,21,24,25,45-47]

The findings from cohort studies are generally more compelling than those from case-control studies, in that the dietary information in cohort studies is collected prior to the occurrence of illness and thus is not subject to the potential recall bias in case-control studies. Thus, although the data overall are suggestive of a positive relationship of dietary fat intake to colorectal cancer, the less consistent support from the cohort studies, especially those that assessed fat itself, and the weak support from studies that controlled for energy intake temper this conclusion. No particular type of fat has been clearly implicated in the positive reports, although the association with red meat intake in some studies[21,25,47] suggests the possibility of a role of saturated fats.

To illustrate the results on dietary fat and colorectal cancer, the findings from one positive study are presented in Table 2. These data show a clear dose-response relationship between animal fat intake and colon cancer in a cohort of female nurses.[43]

Breast Cancer

Like colon cancer, female breast cancer shows substantial variation in incidence among countries (Figure 6), with a 4-fold range between women in China and the U.S.[7] A similar variation can be seen among different ethnic groups within the U.S.,[8] where the lowest incidence is found in Koreans (a group comprised largely of recent immigrants) and the highest in whites (Figure 7). An effect of migration is illustrated by data on Japanese women in Hawaii, who have a much higher incidence than women in Japan, but also a much higher incidence in the second compared with the first generation in Hawaii[49]

Table 2. Association of animal fat with colon cancer in a prospective cohort of women*

Quintile of Intake	Relative Risk[+]	95% C.I.
1	1.0	–
2	1.2	(0.7 - 2.1)
3	1.3	(0.7 - 2.2)
4	1.6	(0.9 - 2.6)
5	1.9	(1.1 - 3.2)
P for trend	0.01	

*Adapted from reference 43
[+]Adjusted for total energy intake

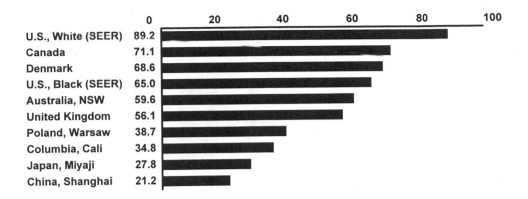

Figure 6. Female breast cancer incidence* in selected populations, 1983–1987.

(Figure 8). These data are persuasive in suggesting the importance of exogenous factors, such as diet, in the etiology of breast cancer.

The correlation between per capita fat intake as a percent of calories and breast cancer mortality in several countries[9] is shown in figure 9. This correlation is even stronger than that for colon cancer (Figure 5).

Table 3 summarizes the results from case-control and cohort studies of dietary fat and breast cancer. Of the 22 case-control analyses[50–70] that assessed dietary fat *per se* (one study[54] included separate analyses for two ethnic groups), nine studies[50,54,58,60,62–64,68,70] (41%) showed a positive association, though in nearly all instances the association was not strong. (In the Hawaii study,[54] the positive association was seen only in the Japanese women). Of 19 case-control studies that reported on high-fat foods (chiefly meat and dairy products),[50,54,55,57,60,63,67–70,71–79] a somewhat higher proportion (63%) showed a positive association.[50,54,55,60,63,68,70,71,73,74,77,79] The data from cohort studies are not dissimilar, three[81,83,87] out of eight[80–87] showing an association with fat, and four[87,88,90,91] out of six[85,87,88–91] showing an association with high-fat foods. However, comparable to the case-

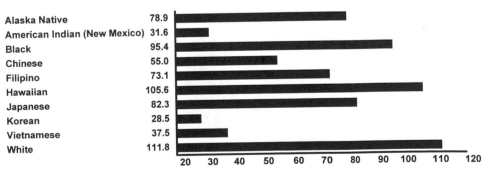

Figure 7. Female breast cancer incidence* by ethnicity, SEER, 1988–1992.

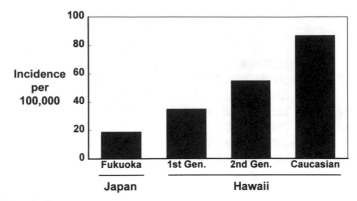

Figure 8. Breast cancer incidence in Japanese migrants and comparison populations.

control studies, these associations were not strong in most instances, and in a recent pooled analysis of seven of the cohort studies,[92] no relationship was found for either total fat (Table 4) or animal fat (Table 5). Whether a particular type of fat is responsible for the association is also not clear, though some studies have implicated saturated fat.[54,58,60,63] It is of interest that three case-control studies[69,73,79] have reported an inverse association between breast cancer and intake of monounsaturated fat (largely olive oil).

Thus, at the present time, it is not at all clear whether or not an important relationship exists between dietary fat and breast cancer. The migrant data (Figure 8) suggest the

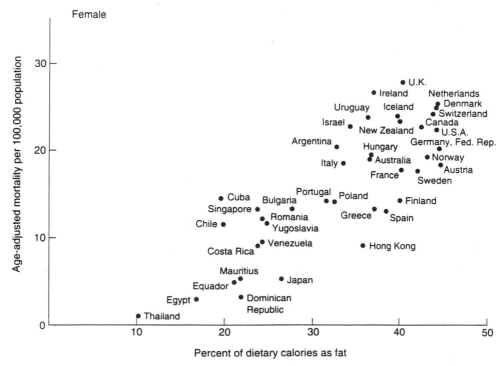

Figure 9. Correlation between % calories as fat and breast cancer mortality.

Table 3. Summary of epidemiologic studies on fat and breast cancer

Exposure	Study Type	Total No. of Reports	No. of Positive Reports	% Positive Reports
Dietary Fat	Case-Control	22	9	40.9
	Cohort	8	3	37.5
High-Fat Foods	Case-Control	19	12	63.2
	Cohort	6	4	66.7

Table 4. Association of total fat with breast cancer in a pooled analysis of seven cohort studies*

Intake Quintile	Relative Risk[+]	95% C.I.
1	1.00	—
2	1.01	(0.89-1.14)
3	1.12	(1.01-1.25)
4	1.07	(0.96-1.19)
5	1.05	(0.94-1.16)
P for trend	0.21	

*Adapted from reference 92.
[+]Adjusted for total energy intake.

possibility that early life exposures influence the ultimate risk of breast cancer, since the rate in the first generation (born in Japan) is much lower than that in the second generation (born in Hawaii). A high fat intake in pre-adolescence could contribute to adult risk for breast cancer by advancing the age at onset of menarche, an established risk factor for breast cancer.[93] Such an effect of dietary fat would not be detected by the dietary assessment methods of current case-control or cohort studies.

Prostate Cancer

International variation in prostate cancer incidence exceeds that of any other major cancer site, with the exception of nonmelanotic skin cancers, for which comparable data are not available.[7] As shown in Figure 10, the rate in U.S. black men is 60 times higher than that in men living in Shanghai. Within the U.S., the rates are lowest among Korean

Table 5. Association of animal fat with breast cancer in a pooled analysis of seven cohort studies*

Intake Quintile	Relative Risk[+]	95% C.I.
1	1.0	—
2	0.96	(0.85-1.09)
3	0.96	(0.81-1.13)
4	0.92	(0.78-1.09)
5	0.99	(0.87-1.13)
P for trend	0.70	

*Adapted from reference 92.
[+]Adjusted for total energy intake.

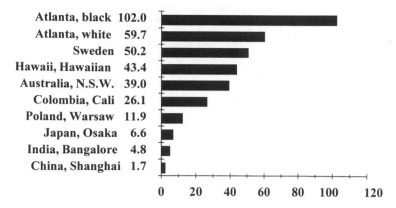

* Age-adjusted, World Standard Population

Figure 10. Prostate cancer incidence* in selected populations, 1983–1987.

men and highest among African American men, representing a 4-fold range in incidence[8] (Figure 11).

High correlations between dietary fat intake and prostate cancer have been reported in ecologic analyses. These include international comparisons based on per capita estimates of intake,[94,95] as well as direct dietary assessments in representative samples of men from different ethnic groups.[96] In one analysis, the association was found for animal or saturated fat, rather than for vegetable fat.[97] Prostate cancer incidence also differs greatly among migrants, as suggested by incidence data[7] on Chinese men in different locations (Figure 12).

Compared with colorectal and breast cancer, analytic epidemiologic studies of dietary fat and prostate cancer have been quite consistent in their findings. As summarized in Table 6, most studies show a positive relationship. Of nine case-control studies[98–106] that assessed dietary fat intake *per se*, eight[98–100,102–106] showed a positive association, and all five studies that reported on high fat foods[104,107–110] were positive. Similarly, both cohort

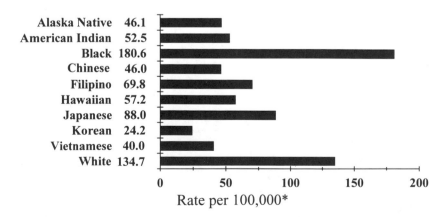

* Age-adjusted, 1970 U.S. Standard Population

Figure 11. Prostate cancer incidence* by ethnicity, SEER, 1988–1992.

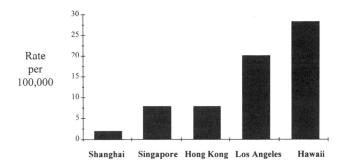

* Age-adjusted, World Standard Population

Figure 12. Prostate cancer incidence* among Chinese men in different geographic locations, 1983–1987.

studies that assessed dietary fat[111,112] found a positive association, as did six[111,112,114–116,118] of eight studies[111–118] that examined high-fat foods. The nature of the fat relationship is not clear, however. Most studies that examined types of fat implicate saturated or animal fat in particular,[98,99,102,105] although a few suggest that other types of fat may be important as well.[103,111,112]

Although the findings support the fat hypothesis, especially the significance of saturated fat, they could also implicate some other constituent(s) of these foods. One such possibility is the heterocyclic amines produced when meats are cooked at very high temperatures, as occurs during broiling, barbecuing or pan frying.[119] Although heterocyclic amines have not been documented as human carcinogens, they are carcinogenic in animals.[120]

The results from a recent case-control study of dietary fat and prostate cancer[105] are illustrative and are shown in Table 7. In this study, the risk of prostate cancer increased monotonically with saturated fat intake. Furthermore, the association was seen for each of four ethnic groups separately, though the association was strongest among the Japanese men (Table 8). In an attributable risk computation, the investigators found that saturated fat could account for 20% of the incidence in whites, 20% in blacks, and 6% in Asians.

Lung Cancer

Dietary fat and cholesterol have also been positively associated with lung cancer. Although cigarette smoking is the primary etiologic factor for lung cancer in most populations, other factors, such as diet, may also contribute to lung cancer risk. Ecologic associations between dietary fat or cholesterol and lung cancer have been reported.[96,121,122] These

Table 6. Summary of epidemiologic studies on fat and prostate cancer

Exposure	Study Type	Total No. of Reports	No. of Positive Reports	% Positive Reports
Dietary Fat	Case-Control	9	8	88.9
	Cohort	2	2	100.0
High-Fat Foods	Case-Control	5	5	100.0
	Cohort	8	6	75.0

Table 7. Association of saturated fat with prostate cancer
in a multiethnic case-control study*

Intake Quintile	Odds Ratio[+o]	95% C.I
1	1.0	—
2	1.1	(0.7-1.6)
3	1.1	(0.7-1.6)
4	1.8	(1.1-2.9)
5	2.8	(1.5-5.2)
P for trend	<.01	

*Adapted from reference 105.
[+]Based on advanced cases, and controls with normal PSA values.
[o]Adjusted for age, residence, education and monounsaturated fat.

associations have been confirmed in six[123–128] out of seven case-control studies,[123–129] and three[130–132] out of six cohort studies,[130–135] as summarized in Table 9. These studies include analyses based on assessments of fat and cholesterol intake, as well as the consumption of high-fat or high-cholesterol foods. The data are not entirely consistent, however. For example, some studies found an effect in men only,[124,125] whereas others reported a positive association in women.[127,128] Similarly, the association was limited to current smokers in one report,[125] but was found among non-smoking women in another.[128] The findings of one of these studies are shown in Table 10. In this investigation,[128] a dose-response gradient was seen for both total and saturated fat, but not cholesterol.

Other Cancers

Other cancer sites for which there is some evidence that dietary fat may have an effect include the pancreas,[136–137] endometrium,[138] ovary,[139] kidney,[140] bladder,[141] and gall-bladder.[142] However, for none of these sites is the number of reports substantial, and for no site is the evidence for a fat effect compelling at this time.

Table 8. Ethnic-specific association of saturated fat with prostate
cancer in a multiethnic case-control study*

Intake Quintile	Odds Ratio[+o]			
	Black	White	Chinese	Japanese
1	1.0	1.0	1.0	1.0
2	0.5	0.9	1.1	1.7
3	0.6	0.9	1.1	1.7
4	0.8	2.0	2.6	2.9
5	1.4	2.4	3.0	8.7
P for trend	0.08	0.17	0.33	0.02

*Adapted from reference 105.
[+]Based on advance cases and controls with normal PSA values.
[o]Adjusted for age, residence, education and monounsaturated fat.

Table 9. Summary of epidemiologic studies on fat/cholesterol and lung cancer

Study Type	Total No. of Reports	No. of Positive Reports	% Positive Reports
Case-Control	7	6	85.7
Cohort	6	3	50.0

*Includes studies that assessed fat/cholesterol intake, as well as consumption of high fat/high cholesterol foods.

CONCLUSIONS AND IMPLICATIONS FOR PUBLIC HEALTH

It seems likely that fat is a major component of the complex relationship between diet and cancer, although the evidence at this time is not definitive. A fuller understanding of the role of dietary fat in human cancer no doubt will come from continued research in this area, including additional analytic epidemiologic studies, as well as animal and in vitro studies on carcinogenesis. A few intervention trials based on substantial dietary fat reduction currently are underway. The results of these studies should be illuminating.

The current dietary recommendation regarding fat from the U.S. National Cancer Institute, the National Academy of Sciences and other prestigious bodies is to maintain fat intake at or below 30% of caloric intake.[2] As the evidence is not yet definitive, but is continuing to accumulate at a rapid pace, this recommendation will need to be reconsidered, and possibly revised, from time to time. Currently, the American Institute for Cancer Research (AICR), through its international arm, the World Cancer Research Fund (WCRF), is sponsoring an updated review of the evidence on diet and cancer, including the role of dietary fat. This review will take a global perspective rather than the necessarily narrower view of governmental agencies. The report is expected to be published in the spring of 1997.

Because the sites for which dietary fat has been implicated are major contributors to cancer morbidity and mortality, and because of the known relationship of dietary fat to other chronic conditions, including coronary heart disease and obesity, ongoing efforts to reduce the level of fat in the diets of populations in which it is high can have a substantial impact on public health.

Table 10. Association of dietary fat with lung cancer among non-smoking women in a case-control study*

Intake Quintile	Odds Ratio[+]		
	Total Fat	Saturated Fat	Cholesterol
1	1.0	1.0	1.0
2	1.4	1.5	0.6
3	1.4	1.6	0.7
4	2.2	2.3	1.1
5	2.8	4.9	1.1
P for trend	0.02	<.01	0.22

*Adapted from reference *128*.
[+]Adjusted for age, smoking history, prior lung disease, and total calories.

REFERENCES

1. Committee on Diet, Nutrition and Cancer: Diet, Nutrition and Cancer. National Research Council, Washington, D.C.: National Academy Press (1982).
2. Committee on Diet and Health: Diet and Health—Implications for Reducing Chronic Disease Risk. National Research Council, Washington, D.C.: National Academy Press (1989).
3. L.S. Freedman, C. Clifford, and M. Messina. Analysis of dietary fat, calories, body weight, and the development of mammary tumors in rats and mice: a review. *Cancer Res.* 50:5710–5719 (1990).
4. D.F. Birt, H.J. Pinch, T. Barnett, A. Phan, and K. Dimitroff. Inhibition of skin tumor promotion by restriction of fat and carbohydrate calories in SENCAR mice. *Cancer Res.* 53:27–31 (1993).
5. C. Ip, D.F. Birt, A.E. Rogers, and C. Mettlin (eds). Dietary Fat and Cancer. Alan R. Liss, New York, pp. 531–728. (1986).
6. J.H. Hankin. Dietary intake methodology. *in:* "Research: Successful Approaches" E.R. Monsen, ed., Am. Dietet. Assoc., Chicago (1991).
7. D.M. Parkin, C.S. Muir, S.L. Whelan, Y.T. Gao, J. Ferlay, and J. Powell (eds). Cancer Incidence in Five Continents, vol. VI, IARC Sci Publ No. 120, Lyon, IARC (1992).
8. B.A. Miller, L.N.Kolonel, L. Bernstein, J.L. Young, Jr., G.M. Swanson, D. West, C.R. Key, J.M. Liff, C.S. Glover, G.A. Alexander, et al., eds. Racial/Ethnic Patterns of Cancer in the United States 1988–1992, National Cancer Institute. NIH Pub. No. 96–4104. Bethesda, MD (1996).
9. K.K. Carroll. Nutrition and Cancer: Fat. *in:* "Nutrition, Toxicity, and Cancer" I.R. Rowland, ed., CRC Press, Boca Raton, FL, pp. 439–453 (1991).
10. M. Jain, G.M. Cook, F.G. Davis, M.G. Grace, G.R. Howe, and A.B. Miller. A case-control study of diet and colorectal cancer. *Int. J. Cancer* 26:757–768 (1980); A.B. Miller, G.R. Howe, M. Jain, K.J.P. Craib, and L. Harrison. Food items and food groups as risk factors in a case-control study of diet and colorectal cancer. *Int. J. Cancer* 32:155–161 (1983).
11. G. Macquart-Moulin, E. Riboli, J. Cornee, B. Charnay, P. Berthezene, and N. Day. Case-control study on colorectal cancer and diet in Marseilles. *Int. J. Cancer* 38:183–191 (1986).
12. J.D. Potter, and A.J. McMichael. Diet and cancer of the colon and rectum: A case-control study. *J. Natl. Cancer Inst.* 76:557–569 (1986); K.A. Steinmetz, and J.D. Potter. Food group consumption and colon cancer in the Adelaide case-control Study. II. Meat, poultry, seafood, dairy foods, and eggs. *Int. J. Cancer* 53:720–727 (1993).
13. S. Kune, G.A. Kune, and L.F. Watson. Case-control study of alcoholic beverages as etiologic factors: the Melbourne colorectal cancer study. *Nutr. Cancer* 9:43–56 (1987).
14. J.L. Lyon, A.W. Mahoney, D.W. West, J.W. Gardner, K.R. Smith, A.W. Sorenson, and W. Stanish. Energy intake: its relationship to colon cancer risk. *J. Natl. Cancer Inst.* 78:853–861 (1987).
15. A.J. Tuyns, R. Kaaks, and M. Haelterman. Colorectal cancer and the consumption of foods: a case-control study in Belgium. *Nutr. Cancer* 11:189–204 (1988); A.J. Tuyns, M. Haelterman, and R. Kaaks. Colorectal cancer and the intake of nutrients: oligosaccharides are a risk factor, fats are not. A case-control study in Belgium. *Nutr. Cancer* 10:181–196 (1987).
16. S. Graham, J. Marshall, B. Haughey, A. Mittelman, M.Y. A. Swanson, M. Zielezny, T. Byers, G. Wilkinson, and D. West. Dietary epidemiology of cancer of the colon in western New York. *Am. J. Epidemiol.* 128:490–503 (1988).
17. J.L. Freudenheim, S. Graham, J.R. Marshall, B.P. Haughey, and G. Wilkinson. A case-control study of diet and rectal cancer in western New York. *Am. J. Epidemiol.* 131:612–624 (1990).
18. M.L. Slattery, M.C. Schumacher, K.R. Smith, D.W. West, and N. Abd-Elghany. Physical activity, diet, and risk of colon cancer in Utah. *Am. J. Epidemiol.* 128:989–999 (1988).
19. H.P. Lee, L. Gourley, S.W. Duffy, J. Esteve, J. Lee, and N.E. Day. Colorectal cancer and diet in an Asian population—a case control study among Singapore Chinese. *Int. J. Cancer* 43:1007–1016 (1989).
20. M. Gerhardsson de Verdier, U. Hagman, G. Steineck, A. Rieger, and S.F. Norell. Diet, body mass and colorectal cancer: a case-referent study. *Int. J. Cancer* 1990;46:832–838 (1990).
21. E. Benito, A. Stiggelbout, F.X. Bosch, A. Obrador, J. Kaldor, M. Mulet, and N. Munoz. Nutritional factors in colorectal cancer risk: A case-control study in Majorca. *Int. J. Cancer* 49:161–67 (1991).
22. A.S. Whittemore, A.H. Wu-Williams, M. Lee, Z. Shu, R.P. Gallagher, J. Deng-ao, Z. Lun, W. Xianghui, C. Kun, D. Jung, C-Z. Teh, L. Chengde, X.J. Yao, R.S. Paffenberger, Jr., and B. Henderson. Diet, physical activity, and colorectal cancer among Chinese in North American and China. *J. Natl. Cancer Inst.* 82:915–26 (1990).
23. R.K. Peters, M.C. Pike, D. Garabrant, and T.M. Mack. Diet and colon cancer in Los Angeles County, California. *Cancer Causes Control* 1992;3:457–473 (1992).

24. F. Meyer, and E. White. Alcohol and nutrients in relation to colon cancer in middle-aged adults. *Am. J. Epidemiol.* 138:225–236 (1993).

25. E. Kampman, D. Verhoeven, L. Sloots, and P. van't Veer P. Vegetable and animal products as determinants of colon cancer risk in Dutch men and women. *Cancer Causes Control* 6:225–234 (1996).

26. P. Stocks. Cancer incidence in North Wales and Liverpool region in relation to habits and environment. Brit. Emp. Cancer Campaign. 35th Annual Report, Suppl. to Part 2, pp. 1–127 (1957).

27. E.L. Wynder, T. Kajitani, S. Ishikana, H. Dodo, and A. Takano. Environmental factors of cancer of the colon and rectum II. Japanese epidemiological data. *Cancer* 23:1210-1220 (1969).

28. W. Haenszel, J.W. Berg, M. Segi, M. Kurihara, and F.B. Locke. Large bowel cancer in Hawaiian Japanese. *J. Natl. Cancer Inst.* 51:1765-1779 (1973).

29. E. Bjelke. Epidemiologic studies of cancer of the stomach, colon, and rectum; with special emphasis on the role of diet. *Scand. J. Gastroenterol.* 9:124–229 (1974).

30. R. Phillips. Role of lifestyle and dietary habits in risk of cancer among Seventh-Day Adventists. *Cancer Res.* 35:3513–22 (1975).

31. B. Modan, V. Barell, F. Lubin, M. Modan, R.A. Greenberg, and S. Graham. Low-fiber intake as an etiologic factor in cancer of the colon. *J. Natl. Cancer Inst.* 55:15–18 (1975).

32. L.G. Dales, G.D. Friedman, H.K. Ury, S. Grossman, and S.R. Williams. A case-control study of relationships of diet and other traits to colorectal cancer in American Blacks. *Am. J. Epidemiol.* 109:132–144 (1970).

33. W. Haenszel, F.B. Locke, and M. Segi. A case-control study of large bowel cancer in Japan. *J. Natl. Cancer Inst.* 64:17–22 (1980).

34. O. Manousos, N.E. Day, D. Trichopoulos, F. Gerovassilis, A. Tzonou, and A. Polychronopoulou. Diet and colorectal cancer: A case-control study in Greece. *Int. J. Cancer* 32:1–5 (1983).

35. C. LaVecchia, E. Negri, A. Decarli, B. D'Avanzo, L. Gallotti, A. Gentile, and S. Franceschi. A case-control study of diet and colo-rectal cancer in northern Italy. *Int. J. Cancer* 41:492–498 (1988).

36. T.B. Young, and D.A. Wolf. Case-control study of proximal and distal colon cancer and diet in Wisconsin. *Int. J. Cancer* 42:167–175 (1988).

37. R.K. Peters, D.H. Garabrant, M.C. Yu, and T.M. Mack. A case-control study of occupational and dietary factors in colorectal cancer in young men by subsite. *Cancer Res.* 49:5459–5468 (1989).

38. J. Hu, Y. Liu, Y. Yu, T. Zhao, S. Liu, and Q. Wang. Diet and cancer of the colon and rectum: a case-control study in China. *Int. J. Epidemiol.* 20:312–367 (1991).

39. E. Bidoli, S. Franceschi, R. Talamini, S. Barra, and C. La Vecchia. Food consumption and cancer of the colon and rectum in north-eastern Italy. *Int. J. Cancer* 50:223–229 (1992).

40. D. Zaridze, V. Filipchenko, V. Kustov, V. Serdyuk, and S. Duffy. Diet and colorectal cancer: results of two case-control studies in Russia. *Eur. J. Cancer* 29A:112–115 (1993).

41. T. Hirayama. A large-scale cohort study on the relationship between diet and selected cancers of digestive organs. WR Bruce, P Correa, M Lipkin, et al., eds. Gastrointestinal Cancer: Endogenous Factors; Banbury report 7. Cold Spring Harbor Laboratory, NY 409–26 (1981).

42. R.L. Phillips, and D.A. Snowdon. Dietary relationship with fatal colorectal cancer among Seventh-day Adventists. *J. Natl. Cancer Inst.* 74:307–317 (1985).

43. W.C. Willett, M.J. Stampfer, G.A. Colditz, B.A. Rosner, and F.E. Speizer. Relation of meat fat and fiber intake to the risk of colon cancer in a prospective study among women. *N. Engl. J. Med.* 323:1664–1672 (1990).

44. M.J. Thun, E.E. Calle, M.M. Namboodiri, W.D. Flanders, R.J. Coates, T. Byers, P. Boffetta, L. Garfinkel, and C.W. Heath, Jr. Risk factors for fatal colon cancer in a large prospective study. *J. Natl. Cancer Inst.* 84:1491-1500 (1992).

45. R.M. Bostick, J.D. Potter, L.H. Kushi, T.A. Sellers, K.A. Steinmetz, D.R. McKenzie, S.M. Gapstur, and A.R. Folsom. Sugar, meat, and fat intake, and non-dietary risk factors for colon cancer incidence in Iowa women (United States). *Cancer Causes Control* 5:38–52 (1994).

46. R.A. Goldbohm, P.A. van den Brandt, P. Van 't Veer, H.A.M. Brants, E. Dorant, F. Sturmans, and R.J.J. Hermus. A prospective cohort study on the relation between meat consumption and the risk of colon cancer. *Cancer Res.* 54:718–723 (1994).

47. E. Giovannucci, E.B. Rimm, M.J. Stampfer, G.A. Colditz, A. Ascherio, and W.C. Willett. Intake of fat, meat, and fiber in relation to risk of colon cancer in men. *Cancer Res.* 54:2930–2997 (1994).

48. G.N. Stemmermann, A.M.Y. Nomura, and L.K. Heilbrun. Dietary fat and the risk of colorectal cancer. *Cancer Res.* 44:4633–4637 (1984).

49. L.N. Kolonel, M.W. Hinds, and J.H. Hankin. Cancer Patterns among Migrant and Native-born Japanese in Hawaii in Relation to Smoking, Drinking, and Dietary Habits. *in:* "Genetic and Environmental Factors in Experimental and Human Cancer", H.V. Gelboin, et al., eds., Japan Sci. Soc. Press, Tokyo, pp. 327–340 (1980).

50. M. Ewertz, and C. Gill. Dietary factors and breast cancer risk in Denmark. *Int. J. Cancer* 46:779–784 (1990).

51. S. Graham, J. Marshall, C. Mettlin, T. Rzepka, T. Nemoto, and T. Byers. Diet in the epidemiology of breast cancer. *Am. J. Epidemiol.* 116:68–75 (1982).

52. S. Graham, R. Hellmann, J. Marshall, J. Freudenheim, J. Vena, M. Swanson, M. Zielezny, T. Nemoto, N. Stubbe, and T. Raimondo. Nuritional epidemiology of post-menopausal breast cancer in western New York. *Am. J. Epidemiol.* 134:552–556 (1991).

53. T. Hirohata, T. Shigematsu, A.M. Nomura, Y. Nomura, A. Horie, and T. Hirohata. Occurrence of breast cancer in relation to diet and reproductive history: a case-control study in Fukuoka, Japan. *Natl. Cancer Inst. Monogr.* 69:187-190 (1985).

54. T. Hirohata, A.M. Nomura, J.H. Hankin, L.N. Kolonel, and J. Lee. An epidemiologic study on the association between diet and breast cancer. *J. Natl. Cancer Inst.* 78:595–600 (1987).

55. D.M. Ingram, E. Nottage, and T. Robert. The role of diet in the development of breast cancer: a case-control study of patients with breast cancer, benign epithelial hyperplasia and fibrocystic disease of the breast. *Br. J. Cancer* 64:187-191 (1991).

56. K. Katsuoyanni, W. Willett, D. Trichopoulos, P. Boyle, A. Trichopoulou, S. Vasilaros, J. Papadiamantis, and B. MacMahon. Risk of breast cancer among Greek women in relation to nutrient intake. *Cancer* 61:181-185 (1988).

57. H.P. Lee, L. Gourley, S.W. Duffy, J. Esteve, J. Lee, and N.E. Day. Dietary effects on breast cancer risk in Singapore. *Lancet* 337:1197–1200 (1991).

58. A.B. Miller, A. Kelly, N.W. Choi, V. Matthews, R.W. Morgan, L. Munan, J.D. Burch, J. Feather, G.R. Howe, and M. Jain. A study of diet and breast cancer. *Am. J. Epidemiol.* 107:499–509 (1978).

59. M. Pryor, M.L. Slattery, L.M. Robison, and M. Egger. Adolescent diet and breast cancer in Utah. *Cancer Res.* 49:2161-2167 (1989).

60. S. Richardson, M. Gerber, and S. Cenee. The role of fat, animal protein and some vitamin consumption in breast cancer: a case-control study in southern France. *Int. J. Cancer* 48:1-9 (1991).

61. T.E. Rohan, A.J. McMichael, and P.A. Baghurst. A population-based case-control study of diet and breast cancer in Australia. *Am. J. Epidemiol.* 128:478–489 (1988).

62. Y. Shun-Zhang, L. Rui-Fang, X. Da-Dao, and G.R. Howe. A case-control study of dietary and nondietary risk factors for breast cancer in Shanghai. *Cancer Res.* 50:5017–5021 (1990).

63. P. Toniolo, E. Riboli, F. Protta, M. Charrel, and A.P.M. Cappa. Calorie-providing nutrients and risk of breast cancer. *J. Natl. Cancer Inst.* 81:278–286 (1989).

64. P. Van't Veer, F.J. Kok, H.A.M. Brants, T. Ockhuizen, F. Sturmans, and R.J.J. Hermus. Dietary fat and the risk of breast cancer. *Int. J. Epidemiol.* 19:12-18 (1990); P. Van't Veer, E.M. van Leer, A. Rietdijk,F.J. Kok, E.G. Schouten, R.J. Hermus, and F. Sturmans. Combination of dietary factors in relation to breast cancer occurrence. *Int. J. Cancer* 47:649–653 (1991).

65. D. Zaridze, Y. Lifanova, D. Maximovitch, N.E. Day, and S.W. Duffy. Diet, alcohol consumption and productive factors in a case-control study of breast cancer in Moscow. *Int. J. Cancer* 48:493–501 (1991).

66. K. Katsouyanni, A. Trichopoulou, S. Stuver, Y. Garas, A. Kritselis, G. Kyriakou, M. Stoikidou, P. Boyle, and D. Trichopoulos. The association of fat and other macronutrients with breast cancer: a case-control study from Greece. *Br. J. Cancer* 1994;70:537–541 (1994).

67. L. Holmberg, E.M. Ohlander, T. Byers, M. Zack, A. Wolk, R. Bergstrom, L. Bergkvist, E. Thurfjell, A. Bruce, and H-O. Adami. Diet and breast cancer risk—results from a population-based case-control study in Sweden. *Arch. Intern. Med.* 154:1805-1811 (1994).

68. J.M. Iscovich, R.B. Iscovich, G. Howe, S. Shiboski, and J.M. Kaldor. A case-control study of diet and breast cancer in Argentina. *Int. J. Cancer* 44:770–776 (1989).

69. J.M. Martin-Moreno, W.C. Willett, L. Gorgojo, J.R. Banegas, F. Rodriquez-Artalejo, J.C. Fernandez-Rodriquez, P. Maisonneuve, and P. Boyle. Dietary fat, olive oil intake and breast cancer risk. *Int. J. Cancer* 58:774–780 (1994).

70. X-Y. Qi, A-Y. Zhang, G-L. Wu, and W-Z Pang. The association between breast cancer and diet and other factors. *Asia-Pacific J. Pub. Hlth.* 7:98-104 (1994).

71. T.G. Hislop, A.J. Coldman, J.M. Elwood, G. Brauer, and L. Kan. Childhood and recent eating patterns and risk of breast cancer. *Cancer Detect. Prev.* 9:47–58 (1986).

72. I. Kato, S. Miura, F. Kasumi, T. Iwase, H. Tashiro, Y. Fujita, H. Koyama, T. Ikeda, K. Fujiwara, K. Saotome, K. Asaishi, R. Abe, M. Nihei, T. Ishida, T. Yokoe, H. Yamamoto, and M. Murata. A case-control study of breast cancer among Japanese women: with special reference to family history and reproductive and dietary factors. *Breast Cancer Res. Treat.* 24:51-59 (1992).

73. C. La Vecchia, A. Decarli, S. Franceschi, A. Gentile, E. Negri, and F. Parazzini. Dietary factors and the risk of breast cancer. *Nutr. Cancer* 10:205–214 (1987).

74. M.G. Le, L.H. Moulton, C. Hill, and A. Kramar. Consumption of dairy produce and alcohol in a case-control study of breast cancer. *J. Natl. Cancer Inst.* 77:633–636 (1986).
75. J.H. Lubin, P.E. Burns, W.J. Blot, R.G. Ziegler, A.W. Lees, and J.F. Fraumeni, Jr. Dietary factors and breast cancer risk. *Int. J. Cancer* 28:685–689(1981).
76. E.L. Matos, D.B. Thomas, N. Sobel, and D. Vuota. Breast cancer in Argentina: case-control study with special reference to meat eating habits. *Neoplasma* 38:357–366 (1991).
77. R. Talamini, C. La Vecchia, A. Decarli, S. Franceschi, E. Grattoni, E. Grigoletto, A. Liberati, and G. Tognoni. Social factors, diet and breast cancer in a northern Italian population. *Br. J. Cancer* 49:723–729 (1984).
78. P. Van't Veer, J.M. Dekker, J.W.J. Lamers, F.J. Kok, E.G. Schouten, H.A.M. Brants, F. Sturmans, and R.J.J. Hermus. Consumption of fermented milk products and breast cancer: a case-control study in the Netherlands. *Cancer Res* 49:4020–4023 (1989).
79. A. Trichopoulou, K. Katsouyanni, S. Stuver, L. Tzala, C. Gnardellis, E. Rimm, and D. Trichopoulos. Consumption of olive oil and specific food groups in relation to breast cancer risk in Greece. *J. Natl. Cancer Inst.* 87:110–116 (1995).
80. S. Graham, M. Zielezny, J. Marshall, R. Priore, J. Freudenheim, J. Brasure, B. Haughey, P. Nasca, and M. Zdeb. Diet in the epidemiology of postmenopausal breast cancer in the New York State cohort. *Am. J. Epidemiol.* 136:1327–1337 (1992).
81. G.R. Howe, C.M. Friedenreich, M. Jain, and A.B. Miller. A cohort study of fat intake and risk of breast cancer. *J. Natl. Cancer Inst.* 83:336–340 (1991).
82. D.Y. Jones, S. Schatzkin, S.B. Green, G. Block, L.A. Brinton, R.G. Ziegler, R. Hoover, and P.R. Taylor. Dietary fat and breast cancer in the National Health and Nutrition/Examination Survey I Epidemiologic Follow-up Study. *J. Natl. Cancer Inst.* 79:465–471(1987).
83. P. Knekt, D. Albanes, R. Seppanen, A. Aromaa, R. Jarvinen, L. Hyvonen, L. Teppo, and E. Pukkala. Dietary fat and risk of breast cancer. *Am. J. Clin. Nutr.* 52:903–908 (1990).
84. L.H. Kushi, T.A. Sellers, J.D. Potter, C.L. Nelson, R.G. Munger, S.A. Kaye, and A.R. Folsom. Dietary fat and postmenopausal breast cancer. *J. Natl. Cancer Inst.* 84:1092–1099 (1992).
85. P.A. van den Brandt, P. Vant Veer, R.A. Goldbohm, E. Dorant, A. Volovics, R.J.J. Hermus, and F. Sturmans. A prospective cohort study on dietary fat and the risk of postmenopausal breast cancer. *Cancer Res.* 53:75–82 (1993).
86. W.C. Willett, D.J. Hunter, M.J. Stampfer, G. Colditz, J.E. Manson, D. Spiegelman, B. Rosner, C.H. Hennekens, and F.E. Speizer. Dietary fat and fiber in relation to risk of breast cancer. An 8-year follow-up. *JAMA* 268:2037–2044 (1992).
87. P. Toniolo, E. Riboli, R.E. Shore, and B.S. Pasternack. Consumption of meat, animal products, protein, and fat and risk of breast cancer: a prospective cohort study in New York. *Epidemiol.* 5:391–397 (1994).
88. T. Hirayama. Epidemiology of breast cancer with special reference to the role of diet. *Prev. Med.* 7:173–195 (1978).
89. L.J. Kinlen. Meat and fat consumption and cancer mortality: a study of strict religious orders in Britain. *Lancet* 1:946–949 (1982).
90. P.K. Mills, J.F. Annegers, and R.L. Phillips. Animal product consumption and subsequent fatal breast cancer risk among Seventh-Day Adventists. *Am. J. Epidemiol.* 127:440–453 (1988).
91. L.J. Vatten, K. Solvoll, and E.B. Lolken. Frequency of meat and fish intake and risk of breast cancer in a prospective study of 14,500 Norwegian women. *Int. J. Cancer* 46:12–15 (1990).
92. D.J. Hunter, D. Spiegelman, H-O. Adami, L. Beeson, P.A. van den Brandt, A.R. Folsom, G.E. Fraser, A. Goldbohm, S. Graham, G.R. Howe, L.H. Kushi, J.R. Marshall, A. McDermott, A.B. Miller, F.E. Speizer, A. Wolk, S-S Yaun, and W. Willett. Cohort studies of fat intake and the risk of breast cancer—a pooled analysis. *New Engl. J. Med.* 334:356–361 (1996).
93. F. De Waard, and D. Trichopoulos. A unifying concept of the aetiology of breast cancer. *Int. J. Cancer* 41:666–669 (1988).
94. M.A. Howell. Factor analysis of international cancer mortality data and per capita food consumption. *Br. J. Cancer* 29:328–336 (1974).
95. B. Armstrong, and R. Doll. Environmental factors and cancer incidence and mortality in different countries, with special reference to dietary practices. *Int. J. Cancer* 15:617–631 (1975).
96. L.N. Kolonel, J.H. Hankin, J. Lee, S. Chu, A. Nomura, and M.W. Hinds. Nutrient intakes in relation to cancer incidence in Hawaii. *Br. J. Cancer* 44:332–339 (1981).
97. D. P. Rose, A.P. Boyar, and E.L. Wynder. International comparisons of mortality rates for cancer of the breast, ovary, prostate, and colon, and per capita food consumption. *Cancer* 58:2363–2371 (1986).
98. S. Graham, B. Haughey, J. Marshall, R. Priore, T. Byers, T. Rzepka, C. Mettlin, and J.E. Pontes. Diet in the epidemiology of carcinoma of the prostate gland. *J. Natl. Cancer Inst.* 70:687–692 (1983).

 99. M. Y. Heshmat, L. Kaul, J. Kovi, M.A. Jackson, A.G. Jackson, G.W. Jones, M. Edson, J.P. Enterline, R.G. Worrell, and S.L. Perry. Nutrition and prostate cancer: a case-control study. *Prostate* 6:7-17 (1985).
100. R.K. Ross, H. Shimizu, A. Paganini-Hill, G. Honda, and B.E. Henderson. Case-control studies of prostate cancer in blacks and whites in Southern California. *J. Natl. Cancer Inst.* 78:869–874 (1987).
101. Y. Ohno, O. Yoshida, K. Oishi, K. Okada, H. Yamabe, and F.H. Schroeder. Dietary beta-carotene and cancer of the prostate: a case-control study in Kyoto, Japan. *Cancer Res.* 48:1331-1136 (1988).
102. L.N. Kolonel, C.N. Yoshizawa, and J.H. Hankin. Diet and prostatic cancer: a case-control study in Hawaii. *Am. J. Epidemiol.* 327:999-1012 (1988).
103. D.W. West, M.L. Slattery, L.M. Robison, T.K. French, and A.W. Mahoney. Adult dietary intake and prostate cancer risk in Utah: a case-control study with special reference to aggressive tumors. *Cancer Causes Control* 2:85–94 (1991).
104. A.R.P. Walker, B.F. Walker, N.G. Tsotetsi, C. Sebitso, D. Siwedi, and A.J. Walker. Case-control study of prostate cancer in black patients in Soweto, South Africa. *Br. J. Cancer* 65:438–441 (1992).
105. A.S. Whittemore, L.N. Kolonel, A.H. Wu, E.M. John, R.P. Gallagher, G.W. Howe, J.D. Burch, J. Hankin, D.M. Dreon, D.W. West, C-Z. Teh, and R.S. Paffenbarger, Jr. Prostate cancer in relation to diet, physical activity and body size in blacks, whites and Asians in the U.S. and Canada. *J. Natl. Cancer Inst.* 87:652–661 (1995).
106. T.E. Rohan, G.R. Howe, J.D. Burch, and M. Jain. Dietary factors and risk of prostate cancer: a case-control study in Ontario, Canada. *Cancer Causes Control* 6:145-154 (1995).
107. L.M. Schuman, J.S. Mandel, A. Radke, U. Seal, and F. Halberg. Some selected features of the epidemiology of prostatic cancer: Minneapolis-St. Paul, Minnesota case-control study, 1976-1979. *in*: "Trends in Cancer Incidence: Causes and Implications" K. Magnus, ed.. Hemisphere Publ. Corp. Washington, pp. 345–354 (1982).
108. R. Talamini, C. La Vecchia, A. Decarli, E. Negri, and S. Franceschi. Nutrition, social factors and prostatic cancer in a Northern Italian population. *Br. J. Cancer* 53:817–821 (1986).
109. C. Mettlin, S. Selenskas, N. Natarajan, and R. Huben. Beta-carotene and animal fats and their relationship to prostate cancer risk. *Cancer* 64:605–612 (1989).
110. R. Talamini, S. Franceschi, C. La Vecchia, D. Serraino, S. Barra, and E. Negri. Diet and prostatic cancer: a case-control study in Northern Italy. *Nutr. Cancer* 18:277–286 (1992).
111. E. Giovannucci, E.B. Rimm, G.A. Colditz, M.J. Stampfer, A. Ascherio, C.C. Chute, and W.C. Willett. A prospective study of dietary fat and risk of prostate cancer. *J. Natl. Cancer Inst.* 85:1571-1579 (1993).
112. P.H. Gann, C.H. Hennekens, F.M. Sacks, F. Grodstein, E.L. Giovannucci, and M.J. Stampfer. Prospective study of plasma fatty acids and risk of prostate cancer. *J. Natl. Cancer Inst.* 86:281-286 (1994).
113. T. Hirayama. Epidemiology of prostate cancer with special reference to the role of diet. *Natl. Cancer Inst. Monogr.* 53:149-155 (1979).
114. A.A. Snowdon, R.L. Phillips, and W. Choi. Diet, obesity, and risk of fatal prostate cancer. *Am. J. Epidemiol.* 120–244-250 (1984).
115. R.K. Severson, A.M.Y. Nomura, J.S. Groves, and G.N. Stemmermann. A prospective study of demographics, diet and prostate cancer among men of Japanese ancestry in Hawaii. *Cancer Res.* 49:1857-1860 (1989).
116. P.K. Mills, W.L. Beeson, R.L. Phillips, and G.E. Fraser. Cohort study of diet, lifestyle, and prostate cancer in Adventist men. *Cancer* 64:598–604 (1989).
117. A.W. Hsing, J.K. McLaughlin, L.M. Schuman, E. Bjelke, G. Gridley, S. Wacholder, H.T. Co-Chien, and W.J. Blot. Diet, tobacco use, and fatal prostate cancer: results from the Lutheran brotherhood cohort study. *Cancer Res.* 50:6836–6840 (1990).
118. L. Le Marchand, L.N. Kolonel, L.R. Wilkens, B.C. Myers, and T. Hirohata. Animal fat consumption and prostate cancer: a prospective study in Hawaii. *Epidemiol.* 5:276–282 (1994).
119. T. Sugimura. Carcinogenicity of mutagenic heterocyclic amines formed during the cooking process. *Mutation Res.* 150:33–41 (1985).
120. IARC: Some Naturally Occurring Substances: Food Items and Constituents, Heterocyclic Aromatic Amines and Mycotoxins. IARC Monographs on the Evaluation of Carcinogenic Risks to Humans. Vol. 56 IARC, Lyon, France, 599 pp. (1993).
121. K.K. Carroll, and H.T. Khor. Dietary fat in relation to tumorigenesis. *Prog. Biochem. Pharmacol.* 10:308–353 (1975).
122. J. Xie, E. Lesaffre, and H. Kesteloot. The relationship between animal fat intake, cigarette smoking and lung cancer. *Cancer Causes Control* 2;79–83 (1991).
123. M.W. Hinds, L.N. Kolonel, J.H. Hankin, and J. Lee. Dietary cholesterol and lung cancer risk in a multiethnic population in Hawaii. *Int. J. Cancer* 32:727–732 (1983).
124. T.E. Byers, S. Graham, B.P. Haughey, J.R. Marshall, and M.K. Swanson. Diet and lung cancer risk: findings from the Western New York Diet Study. *Am. J. Epidemiol.* 125:351–363 (1987).

125. M.T. Goodman, L.N. Kolonel, C.N. Yoshizawa, and J.H. Hankin. The effect of dietary cholesterol and fat on the risk of lung cancer in Hawaii. *Am. J. Epidemiol.* 128:1241-1255 (1988).

126. C. Mettlin. Milk drinking, other beverage habits and lung cancer risk. *Int. J. Cancer* 43:608–612 (1989).

127. M. Jain, J.D. Burch, G.R. Howe, H.A. Risch, and A.B. Miller. Dietary factors and risk of lung cancer: results from a case-control study, Toronto, 1981-1985. *Int. J. Cancer* 45:287–293 (1990).

128. M.C. Alavanja, C.C. Brown, C. Swanson, and R.C. Brownson. Saturated fat intake and lung cancer risk among nonsmoking women in Missouri. *J. Natl. Cancer Inst.* 85:1906-1916 (1993).

129. A. Kalandidi, K. Katsouyanni, N. Voropoulou, G. Bastas, R. Saracci, and D. Trichopoulos. Passive smoking and diet in the etiology of lung cancer among non-smokers. *Cancer Causes Control* 1:15–21 (1990).

130. G.E. Fraser, W.L. Beeson, and R.L. Phillips. Diet and lung cancer in California Seventh-Day Adventists. *Am. J. Epidemiol.* 133:683–693 (1991).

131. P. Knekt, R. Seppanen, R. Jarvinen, J. Virtamo, L. Hyvonen, E. Pukkala, and L. Teppo. Dietary cholesterol, fatty acids, and the risk of lung cancer among men. *Nutr. Cancer* 16:267–275 (1991).

132. R.B. Shekelle, A.H. Rossof, and J. Stamler. Dietary cholesterol and incidence of lung cancer: the Western Electric study. *Am. J. Epidemiol.* 134:480–484 (1991).

133. G. Kvale, E. Bjelke, and J.J. Gart. Dietary habits and lung cancer risk. *Int. J. Cancer* 31:397–405 (1983).

134. L.K. Heilbrun, A.M. Nomura, and G.N. Stemmermann. Dietary cholesterol and lung cancer risk among Japanese men in Hawaii. *Am. J. Clin. Nutr.* 39:375–379 (1984).

135. Y. Wu, W. Zheng, T.A. Sellers, L.H. Kushi, R.M. Bostick, and J.D. Potter. Dietary cholesterol, fat, and lung cancer incidence among older women: the Iowa Women's Health Study (United States). *Cancer Causes Control* 5:395–400 (1994).

136. J.P. Durbec, G. Chevillotte, J.M. Bidart, P. Berthezene, and H. Sarles. Diet, alcohol, tobacco and risk of cancer of the pancreas: a case-control study. *Br. J. Cancer* 47:463–470 (1983).

137. P. Ghadirian, A. Simard, J. Baillargeon, P. Maisonneuve, and P. Boyle. Nutritional factors and pancreatic cancer in the Francophone community in Montreal, Canada. *Int. J. Cancer* 47:1-6 (1991).

138. N. Potischman, C.A. Swanson, L.A. Brinton, M. McAdams, R.J. Barrett, M.L. Berman, R. Mortel, L.B. Twiggs, G.D. Wilbanks, and R.N. Hoover. Dietary associations in a case-control study of endometrial cancer. *Cancer Causes Control* 4:239–250 (1993).

139. X.O. Shu, Y.T. Gao, J.M. Yuan, R.G. Ziegler, and L.A. Brinton. Dietary factors and epithelial ovarian cancer. *Br. J. Cancer* 59:92–96 (1989).

140. M. Maclure, and W. Willett. A case-control study of diet and risk of renal adenocarcinoma. *Epidemiol.* 1:430–440 (1990).

141. G. Steineck, U. Hagman, M. Gerhardsson, and S.E. Norell. Vitamin A supplements, fried foods, fat and urothelial cancer. A case-referent study in Stockholm in 1985-1987. *Int. J. Cancer* 45:1006-1011 (1990).

142. K. Kato, S. Akai, S. Tominaga, and I. Kato. A case-control study of bilary tract cancer in Niigata Prefecture, Japan. *Japan J. Cancer Res.* (Gann) 80:932–938 (1989).

DIETARY LIPIDS AND THE CANCER CASCADE

Steven K. Clinton[*]

Dana-Farber Cancer Institute
and
Harvard Medical School
Department of Medical Oncology
44 Binney Street
Boston, Massachusetts 02115

INTRODUCTION

Investigators have established a large database from epidemiologic studies and experiments in animal models which strongly implicate diet and nutrition in the development and progression of many human malignancies. Among the dietary components frequently associated with the cancers common to affluent nations are the concentration and source of dietary lipid. These data are of sufficient strength to have lead many expert committees and organizations to recommend modifications in dietary fat intake as part of population based cancer prevention programs. Whether a meaningful reduction in cancer incidence among Americans can be achieved through modification of dietary lipid is the subject of intense debate. The testing of dietary lipid and cancer hypotheses in human intervention studies, considered the definitive approach, has progressed slowly for many reasons. In particular, the expenses associated with large studies having sufficient power to detect an effect of diet on cancer incidence prohibit their implementation. In contrast to interventions with pharmaceutical or chemopreventive agents where intake of the drug is precisely controlled and compliance determined by an objective test (blood or urine drug concentration), assessment of adherence with nutritional interventions presents major obstacles. Instituting a change in dietary fat among individuals, objectively quantitating the change, controlling for changes in associated foods and nutrients over time are complex problems and the tools available to address these issues remain very imprecise. Many of the chapters in this volume focus upon laboratory efforts which will significantly impact upon our future ability to implement efficient and definitive clinical trials addressing the role of dietary lipids in cancer etiology and prevention. The elucidation of the molecular, biochemical, and cellular processes modulated by lipids will translate into novel assays to assess the intake of dietary fat concentration and fatty acid pattern. In parallel, these

---

[*] Tel: 617–632–2935; Fax: 617–632–2933; steven_clinton@macmailgw.dfci.harvard.edu.

Dietary Fat and Cancer, edited by AICR
Plenum Press, New York, 1997

efforts will provide insight into mechanisms which link dietary lipids with the events associated with cancer initiation and progression. This chapter will briefly review the recent progress in understanding the cancer cascade and identify areas where dietary lipids may interact with these processes.

The fundamental components of the cancer cascade appear to be very similar in diverse tissues despite the unique features that each of the different types of cancer may exhibit in the clinic (Figure 1). Defects in the precisely controlled balance between proliferation, differentiation, and cell death leading to an accumulation of abnormal cells are a central theme of cancer progression. A second characteristic, which has been examined only rarely in the nutrition literature, is the ability of cancer cells to migrate from their location of origin and establish growth and invasion at a distant metastatic site. Indeed, the insidious progression of metastases and the multi-organ failure associated with this process is the mechanism of death for most patients when therapeutic interventions have failed. The cancer process, which begins with a mutation in a single cell and culminates in a life-threatening malignancy requires many years and even decades to evolve, providing ample opportunity for long standing dietary patterns to influence the rate of progression and the biological behavior of the malignancy.

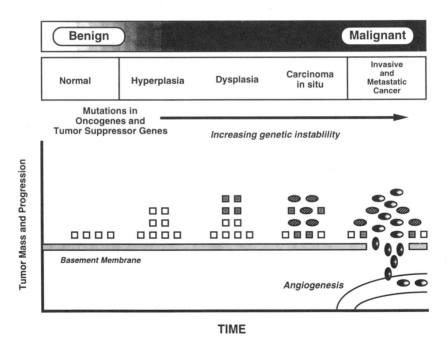

Figure 1. The cancer cascade begins when one cell within the normal population experiences a genetic mutation that increases its ability to proliferate. The resulting accumulation of cells, each having a normal cytological appearance, is termed hyperplasia. In time, some cells may experience an additional mutation that further disrupts growth control and differentiation. A dysplastic lesion is composed of cells that continue to proliferate excessively and begin to exhibit abnormalities in shape, orientation, and the appearance of cellular organelles. As additional mutations accumulate over time a subpopulation of cells becomes more abnormal in cytological appearance. A lesion composed of cells that have the cytological features of cancer but have not broken through the basement membrane is termed carcinoma in situ. These lesions may persist for many years or decades before additional mutations allow the development of the complete malignant phenotype characterized by disruption of the basement membrane and local invasion, angiogenesis, motility of tumor cells, and metastatic spread through the lymphatics and vascular system.

THE GENETIC BASIS OF CANCER PREDISPOSITION AND SUSCEPTIBILITY

Cancer is viewed by many as a genetic disease, and indeed an understanding of the central role of genetics in cancer initiation and progression is fundamental to our further elucidation of the roles nutrition plays in the cancer cascade. It is helpful to initially categorize the genetic contribution into two distinct but interacting components: (a) familial or germ line genes which exhibit Mendelian inheritance patterns and influence cancer risk, and (b) the acquired defects in critical genes of cells composing the evolving tumor. The identification of novel inherited genes which predispose to cancer is rapidly advancing (Table 1). Among those recently discovered are BRCA1 and BRCA2, which are dominant tumor suppressor genes that increase risk of developing breast cancer at an early age (1). For example, a women who inherits a mutation in BRCA1 will have a 200-fold greater risk of breast cancer than the general population by age 40. The excess risk decreases somewhat with age, but remains elevated at approximately 15-fold by age 70. Genes such as these are typically identified in familial cancer syndromes characterized by family members developing a specific pattern of malignancies at a much younger age than is observed in the general population. The time to cancer formation is compressed since each individual affected by the mutant inherited gene has a head start in the race to collect the pattern of mutations required for the development of the malignancy. The mutant genes commonly encode proteins which play a central role in cell cycle regulation or in the ability to repair DNA damage and maintain genome integrity. These genes typically exhibit a high penetrance, meaning that the probability of developing a cancer with the inherited gene is high. For example, those with BRCA1/BRCA2 have a lifetime penetrance of 85%. Fortunately, evolutionary pressure in the form of early mortality, assures that these types of genes remain rare and may account for only 5% of human cancers.

Perhaps far more important are common genetic polymorphisms involving genes encoding proteins which modulate processes linked to predisposition or susceptibility. A number

Table 1. A representative sample of inherited mutations in genes which predispose affected individuals to a very high risk of cancer at an early age

Major Cancer	Gene	Gene Category	Syndrome or Cancer Pattern
Retinoblastoma	RB	Tumor suppressor	Retinoblastoma
			Sarcoma
			Various
Breast Cancer	BRCA1	Tumor suppressor	Hereditary breast and ovary
	BRCA2	Tumor suppressor	Hereditary breast and ovary
	p53	Tumor suppressor	Li-Fraumeni Syndrome
Colon Cancer	APC	Tumor suppressor	Adenomatous polyposis
	MSH2	Mismatch repair	HNPCC, endometrium, other HNPCC
	MLH1	Mismatch repair	HNPCC
	PMS1,2	Mismatch repair	
Renal Cancer	WT1	Tumor suppressor	Wilms' tumor
	VHL	Tumor suppressor	Von Hipple-Lindau
Melanoma	MTS1	Tumor suppressor	Familial melanoma
Neuroendocrine	NF-1	Tumor suppressor	Neurofibromatosis (brain)
	NF-2	Tumor suppressor	Neurofibromatosis (brain)
	RET	Oncogene	Multiple endocrine neoplasia (thyroid and others)

HNPCC = Hereditary non-polyposis cancer of the colon.

of examples can be cited. Inheritance of a fair complexion predisposes to skin cancer if the individual is exposed to the key environmental carcinogen, sunlight. Similarly, genes which determine the capacity of an individual to activate or detoxify chemical carcinogens, may have a profound influence on the frequency and pattern of acquired mutations in critical oncogenes and tumor suppressor genes. For example, about 50% of humans inherit two deleted copies of the GSTM1 gene which encodes the enzyme glutathione-S-transferase M1 that participates in the inactivation of specific chemical carcinogens (2). The enzyme mediates a detoxification step involving the carcinogens found in cigarette smoke, which is the primary cause of bladder cancer in America. Those with the homozygous deletion of GSTM1 experience a significant increase in the risk of bladder cancer if they are smokers (Figure 2). In contrast, this polymorphism poses little risk in those not exposed to the carcinogen. Polymorphisms of genes which control growth factor expression, anti-tumor immune responses, hormone secretion and receptors, and nutrient metabolism will number in the thousands. Many of these common polymorphisms will influence risk of cancer and potentially interact with dietary factors to define individual risk.

The identification of inherited genes involved in predisposition or susceptibility will progress rapidly over the next decade and favorably impact upon the future direction of nutritional investigations. Initially, the advances in genetics will lead to the development of simple and low-cost genetic tests which will allow the categorization of individuals into groups of defined risk. The challenge to implementing genetic testing for cancer susceptibility is not scientific or technological. The more difficult issues involve establishing effective psychosocial support systems as well as providing a guarantee that results of genetic testing remain confidential, do not impact unfairly upon employment opportunities, or limit the availability of insurance and access to health care. An additional concern for those considering genetic testing is the availability of prevention strategies. For example, women with the BRCA1/BRCA2 genes may be offered bilateral prophylactic mastectomy following puberty, an effective option for breast cancer prevention, but an approach that has limited appeal for many. It is becoming more evident that the laboratory advances in cancer genetic testing will not be utilized or impact upon the cancer burden unless additional support is directed towards early detection of cancers at a curable stage, chemoprevention, and prevention strategies involving diet and nutrition.

The identification of high-risk individuals defines a highly motivated and compliant cohort for the more rapid and efficient testing of diet and cancer prevention hypotheses,

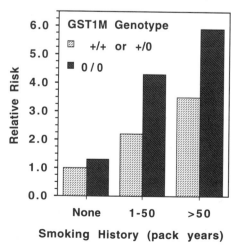

Figure 2. The risk of bladder cancer in smokers and non-smokers based on a common polymorphisms of the carcinogen metabolizing enzyme glutathione S-transferase M1 (GSTM1) is an excellent example of a critical interaction between an environmental exposure and a genetic polymorphism of a cancer susceptibility gene (2). Smoking is the major cause of bladder cancer in Americans. However, those having the homozygous deleted genotype have a much higher risk of bladder cancer if they are smokers. In contrast the genotype has no bearing on risk in a non-smoker.

including studies involving lipids. One hypothetical example may involve those with BRCA1/BRCA2. It is reasonable to postulate that diet may alter the age at which the breast tumors develop or the overall penetrance observed during a lifetime. Delaying the onset of tumor formation by one or two decades by diet or chemoprevention would have significant impact upon how we advise the affected individuals. If we extrapolate from studies in strains of mice having a genetically defined increased risk of breast cancer, it would appear that diet has the capacity to achieve this modest goal if initiated early (3, 4).

The advances in genetics will also impact upon the laboratory based nutritional scientist investigating diet-cancer hypothesis in animal models and cell culture. Murine homologues of the critical genes will be identified and manipulated using transgenic and "knock-out" strategies to develop new animal models. These approaches will provide critical tools for precisely defining mechanisms whereby diet modulates carcinogenesis and provide a system for evaluating efficacy of dietary intervention strategies prior to proposing clinical studies. The recent development of the p53 "knock-out" mouse quickly progressed to studies showing that survival could be significantly prolonged by moderate energy restriction (5).

CELLULAR PROLIFERATION, APOPTOSIS, AND IMMORTALITY

The early phase of tumor formation involves the accumulation of a small mass composed of a cell population that has become insensitive to the normal constraints on proliferation. The cells follow a reproductive pathway which has been reprogrammed through the accumulation of mutations in critical genes. The vast majority of these mutations involve genes referred to as proto-oncogenes or tumor suppressor genes. Proto-oncogenes code for proteins which typically play a central role in mediating the signals which enhance cell proliferation. When mutated at critical sites, the proto-oncogenes are referred to as oncogenes. These genes have become consitutively activated and provide constant growth promoting internal signals to the cell. Many of these genes encode receptors for growth factors, enzymes mediating cytoplasmic signal transduction, nuclear transcription factors, or proteins activating the cell cycles. In contrast to oncogenes, the tumor suppressor genes code for proteins which normally inhibit cell replication and they contribute to the cancer cascade when a mutation results in their inactivation, thereby allowing replication to proceed unchecked. Typically, human cancer cells will exhibit a pattern of several activated oncogenes and mutant tumor suppressor genes by the time the malignancy is detected (Table 2).

The origins of the mutations in most human cancers remain poorly characterized. In some cases, mutations are presumed to result from specific exposures to environmental agents, including: sunlight, tobacco smoke, radon, natural carcinogens such as aflatoxins, man-made chemicals such as benzene or formaldehyde, and even some of the drugs used in cancer chemotherapy. The impact of nutrition upon the frequency and pattern of mutations provides an intriguing area for future investigations. There are many potential mechanisms whereby diet and nutrition may impact upon the accumulation of mutational events. Diet may be one important vehicle for exposure to mutagenic agents, either as a carrier of carcinogens naturally found in foods, those incorporated as contaminating substances during production and processing of food, or through cooking methods that directly contribute to the formation of carcinogenic agents. For example, the cooking of meats at high temperatures for prolonged periods leads to the formation of carcinogenic heterocyclic amines (6–8). Smoked or charcoal broiled foods contain high concentrations

Table 2. Proto-oncogenes encode proteins that play a central role in regulating cell proliferation. Alteration in the structure or expression of the proto-oncogenes allows them to function as oncogenes capable of tranforming the susceptible cell into one exhibiting specific characteristics of the malignant phenotype. Tumor suppressor genes code for proteins that are involved in the inhibition of cell proliferation, which when inactivated result in a loss of the normal constraints on cell proliferation. This table illustrates the various classes of oncogenes and tumor suppressor genes and provides representative examples

Class	Functional Group	Gene Name	Mechanism (cancer types)
Oncogenes	Growth Factors and Receptors	sis	Platelet Derived Growth Factor
		c-fms	MCSF receptor (leukemia)
		erb-B	EGF receptor (brain, breast)
		RET	? receptor (thyroid)
		neu(erb-B2)	? receptor (breast, salivary)
	Cytoplasmic Signaling	Ras family	Membrane GTPase (many cancers)
		src	tyrosine kinase
		raf	serine/threonine kinase
	Nuclear Proteins	myc family	DNA binding protein
		fos/jun	Transcription factors.
		myc family	DNA binding proteins
		erb A	T3 receptor/DNA binding protien
	Other	Bcl-1	Cyclin D1
		Bcl-2	Blocks apoptosis
Tumor Suppressor Genes	Cytoplasmic	APC	Unknown function (Gastrointestinal cancers)
	Nuclear	p53	Cell cycle and apoptosis regulator (many cancers)
			Nuclear cell cycle regulator (many cancers)
		RB	Transcription regulator (renal)
		WT1	
	Unknown	BRCA1/2	Unknown functions (breast, other)
		VHL	Probable transcription regulator (kidney, other)
		NF1/NF2	Probable cytoplasmic RAS binding protein (neurofibromatosis)

of carcinogenic polycyclic aromatic hydrocarbons (9). We also suspect that reactive oxygen compounds may also be produced *in vivo* as a result of normal metabolism or during specific physiologic processes such as the immune response (10). It is reasonable to postulate that the formation of endogenous mutagenic substances may also be directly influenced by the nutritional status of the host.

The evolution of complex organisms required the development of precisely regulated and carefully orchestrated controls on proliferation and differentiation in order to insure that muticellular tissues and organs can successfully develop and perform their functions within the highly integrated internal environment. Apoptosis, also known as programmed cell death, is one critical internal quality control system which guards the host against disruptive individual cell behavior and uncontrolled proliferation. Indeed, apoptosis is a highly conserved cellular function and may be a common phenomena in cells affected by damage in critical genes. It is probable that a mutant cell escaping DNA repair or cellular suicide to form a cancer is a rare event (Figure 3). Cancer biologists are now

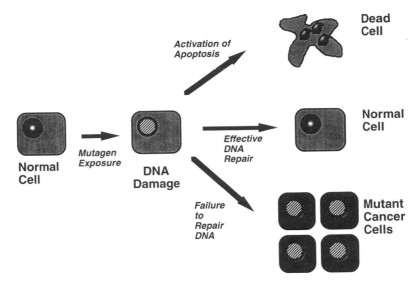

Figure 3. Exposure of a cell to mutagenic stress may result in three outcomes. Cells have many types of DNA re-pair processes which may survey and correct the mutations which allows the cell to function normally. If the cell determines that the damage to the genome is extensive and unrepairable, the "cellular suicide" or apoptosis path-way can be activated. In some cases the mutations in critical genes escape these processes and the cell proceeds along the malignant cascade.

beginning to identify the genes that participate in the activation of the apoptotic program when damage has occurred to cellular DNA. For example, following exposure to mu-tagenic stress in the form of UV-irradiation, x-ray irradiation, chemicals or other agents the induction or stabilization of the p53 protein is necessary to check the advance of a cell through mitosis. When the DNA damage is minimal and can be repaired, the p53 protein can arrest the cell in the G1 phase of the cell cycle, allowing the DNA repair machinery to correct the damage (11). If DNA damage is significant or unrepairable, p53 induces cells to activate the apoptosis program. For this reason, p53 acts as a "sensor" for damaged DNA and is referred to as the "guardian of the genome". It is therefore not surprising that mutations or deletions in p53 are common in cancer cells. In contrast to p53, inappropriate overexpression of the Bcl-2 protein blocks the suicide pathway. Therefore mutations which result in the consitutive activation of Bcl-2 prevent a cell with DNA damage from undergoing apoptosis and increases the chance of proceeding along the cancer cascade. The emerging data reveal that cell cycle regulation and apoptosis are intimately interre-lated and controlled by a number of key proteins which participate in carcinogenesis. Very little effort has thus far been directed towards understanding the influence of diet and nu-trition on the pathways involved in regulating the delicate balance between cell cycle ar-rest, DNA repair, and apoptosis following mutagenic stress.

In addition to apoptosis, mammalian cells have another safeguard against uncon-trolled proliferation. We now appreciate that normal cells, when dispersed and grown in culture, exhibit a very predictable number of cell doublings (50–60) after which prolifera-tion stops. This process has been termed cellular senescence. Mutations in key genes, such as p53 and RB, may allow cells to maintain proliferation through the crisis phase and achieve immortality, a feature characteristic of the many cancer cell lines used in the labo-ratory. Some of the most exciting research in recent years has focused upon the molecular process whereby the cell determines or counts the number of doublings. The answer appears to be found in the telomere, which is the distal end of the chromosome (12). Every time a cell divides a small piece of the telomere is lost and the overall length decreases with each cell cycle until a threshold has been exceeded which signals the cell to undergo senescence, since continued shortening leads to complete genetic instability and cell death. One mechanism whereby a cancer cell can escape this additional cellular obstruction in the cancer cascade seems to be mediated by an enzyme called telomerase(12). This enzyme is present in almost all cancer cells but absent in normal cells. Its function is to replace the telomeric DNA fragments that have been lost with each cycle of replication, which allows the cell to maintain the integrity of the telomere allowing continued proliferation. The activity of telomerase is a potential target of new cancer therapies but also should be considered as one factor which may be regulated by diet to influence tumor growth and progression, other degenerative diseases, or perhaps the aging process.

ANGIOGENESIS AND METASTASIS

A cell which has lost the normal controls on proliferation and apoptosis, as well as achieved immortality, still cannot develop into a significant tumor mass without a steady supply of oxygen, energy, substrates for the synthesis of macromolecules, and circulating growth factors and hormones. In addition, the removal of metabolic waste products is necessary for continued survival of the replicating and metabolically active tumor cells. As a nascent tumor mass expands, each cell becomes farther from the nearest capillary. Indeed, experimental evidence suggests that a tumor cannot exceed a few millimeters in size without an expansion of the capillary network and can remain as a small *in situ* cancer for years (13). In some cases, these tumors will abruptly develop a vasculature and begin to expand and invade local tissues. This process has been termed tumor angiogenesis or neovascularization (14). Indeed, perhaps 20% of a tumor mass is composed of host endothelial cells and most cancer cells are within 2–3 cell diameters from a tumor capillary. The onset of the angiogenic phase in a subpopulation of the *in situ* tumors appears to be a critical step in facilitating the further growth, invasion, and metastatic spread of a cancer. The development of tumor vasculature is a highly regulated process which depends upon a shift in the balance between the secretion and activation of angiogenic growth factors and the inhibition of anti-angiogenic factors within the microenvironment of the tumor. These key regulatory factors play important roles in healthy tissues by maintaining endothelial cells in the quiescent state and by activating neovascularization during physiologic processes such as wound healing or during the menstrual cycle. In the last few years scientists have identified approximately 15 proteins that have angiogenic or anti-angiogenic properties (13). The tumor cells can promote neovascularization by secreting angiogenic factors, mobilizing precursor proteins deposited in the tissue matrix, or inducing the secretion of these proteins by other cell types such as macrophages or fibroblasts. Recent evidence also suggests that decreased expression of anti-angiogenic factors may be equally as important as the enhanced expression of angiogenic factors (15). The proliferating endothelial cells forming the tumor capillary bed not only provide the infrastructure needed for continued oxygen and nutrient supply as well as waste removal but also secrete proteins which can directly modify the proliferation of tumor cells, their propensity to invade local tissue, or migrate into the blood stream. Additional efforts to determine how diet and nutrition may interact with tumor cells and the vascular cells to modify the balance between angiogenesis and dormancy should provide valuable insight into the cancer process. Preliminary data from our laboratory suggests that the activity of anti-angiogenic substances such as platelet factor 4 or angiostatin can be enhanced by dietary modulation.

The invasion of tumor cells into critical organs and the spread of tumor cells to sites distant from the primary cancer is the biological property which contributes most to morbidity and mortality in cancer patients. The metastatic process depends critically on the interactions between the tumor cell and the many cellular and matrix elements that contribute to the tumor microecosystem. Metastasis is inefficient in laboratory models where it can be quantitated, and presumably in humans, because may obstacles are in place to inhibit dissemination (16, 17). In order for a tumor to successfully metastasize, the cells must develop mobility and detach from their primary tumor, traverse the vascular or lymphatic endothelium, survive in the vascular environment until the cell adheres to a new capillary bed, cross the endothelium and basement membrane, and finally invade the local tissue and reestablish proliferation and angiogenesis. It is likely that only one in tenthousand tumor cells entering the circulation are successful in establishing metastatic growth (16, 17). Some investigators have begun to investigate how diet influences the specific steps involved in the metastatic process. For example, mice fed diets rich in linoleic acid show increased lung metastases from breast cancer cells injected into the circulation (18, 19). These observations have been extended and further characterized using a variety of approaches. The capacity of human breast cancer cell lines to invade through a reconstituted basement membrane *in vitro* is enhanced by linoleic acid (20). This effect is mediated by 12-HETE and the induction of the 92 kDa isoform of type IV collagenase (21). Furthermore, long-chain omega-3 fatty acids appear to counteract the effects of linoleic acid and suppress metastatic properties of human breast cancer xenografts in immune deficient mice (22, 23). Much more effort is needed before we will fully understand how the concentration and type of dietary lipids modify the multiple steps of tumor metastases. Areas of future investigation include the role of lipids on tumor cell-basement membrane interactions, cell-cell adhesion, motility factors and receptors, tumor cell interactions with tissue matrix, tumor cell-interactions with endothelium, and the secretion and activation of enzymes involved in degradation of extracellular matrix.

THE RELATIONSHIP BETWEEN DIETARY LIPIDS, ENERGY INTAKE, AND CANCER IN EXPERIMENTAL MODELS

Studies completed over the later half of this century have clearly demonstrated that dietary factors, including the concentration and source of lipids, have profound effects on the initiation and promotion of a variety of cancers in laboratory animal models (24–32). The role of lipids has been investigated most extensively in models of colon, breast, and prostate cancer. With regards to colon cancer, the emerging data from epidemiologic studies (33–35) are generally supportive of the studies in animal models showing that dietary fat concentration enhances risk of malignancy (36). Furthermore, specific fatty acids and metabolic pathways controlling the synthesis and degradation of prostaglandins and leukotrienes are critically involved in modulating steps of the cancer cascade (37).

The role of diet in prostate cancer has not been extensively evaluated although reports are beginning to accumulate. The geographic epidemiology of prostate cancer is very similar to that of colon or breast cancer and high rates are typically found in affluent nations of North American and Western Europe with lower rates in Japan and many developing nations (38). Populations exhibiting higher rates of prostate cancer are those consuming "affluent" dietary patterns rich in lipids (35, 39–42). Cohort studies also suggest an association between prostate cancer risk and diets rich in lipids, meats, and dairy products (42–45). Furthermore, dietary fat is more closely related to advanced or lethal prostate cancer than overall incidence (43). The investigation of dietary hypotheses in the genesis of prostate cancer has been hampered by the relative lack of animal models which are well characterized, easy to use, and consistently reproducible in many laboratories over time. Many studies employ rat prostate cancer cell lines grown as transplantable tumors in syngeneic hosts or the human prostate cancer cell lines grown as xenografts in immunodeficient mice. Carcinogen induced models are also under development (46–48). Prostate tumors, like most cancers, require a supply of essential fatty acids (27, 28, 49). The growth rate of the human LNCaP prostate cell line as a xenograft in immune deficient mice was recently shown to be greater in those fed a diet containing 40% of energy from fat compared to those fed less than 20% energy from lipid (50). In contrast, the well differentiated Dunning R3327 shows no change in growth with dietary corn oil over the range of 10 to 40 % of energy (28, 51). Prostate cancer is now the most common visceral malignancy in American males (38) and the evidence generated suggests that diet is one of the major etiologic factors warranting continued and vigorous investigation.

Breast cancer presents an intriguing and complex problem for those concerned with the relationship between dietary lipids and cancer risk. The ecological data derived from nations shows a significant association between rate of per capita fat consumption and breast cancer incidence even after adjustment for gross national product and average age of menarche (52). A pooled analysis of the case-control studies suggested a modest but significant positive association between total fat and saturated fat intake in post-menopausal breast cancer (53). However, we are faced with accumulating data from prospectively evaluated cohort studies which have failed to detect any relationship between dietary lipids and breast cancer (54).

In contrast to the cohort studies, the relationship between dietary fat and breast cancer in experimental models has provided a large body of evidence supporting a role for lipids as modulators of breast carcinogenesis (3, 32). Fifty years ago studies by two groups emerged using well defined diets and nutritional protocols providing strong evidence that dietary lipids can profoundly modify risk of breast carcinogenesis (24, 25, 55). Some investigators have proposed that the tumor promoting effects of a high-fat diet are primarily mediated by increased energy intake (56, 57). A dietary pattern that is rich in lipids is often associated with increased energy intake in humans. Furthermore, increasing the lipid concentration in rodent diets may enhance palatability of the food leading to increased energy intake. For these reasons precise measurement of food intake and body weight is critical in laboratory studies. Despite the potential for misinterpretation of results if energy intake is not carefully documented in laboratory investigations, many studies have provided definitive data showing that fat concentration and energy intake have independent effects on breast carcinogenesis. For example, Tannenbaum and coworkers fed diets containing lipid at high- or low-concentrations while controlling energy intake by instituting multiple levels of enforced restriction in mice at risk of spontaneous mammary carcinoma (25). Their studies clearly show that high fat diets enhance breast tumorigenesis over the range of imposed restrictions in energy intake (Fig. 4). Further evidence of the independent effects of fat and energy on breast cancer were derived from our studies of 7,12-dimethylbenz(α)anthracene-induced breast cancer in rats (29, 58). Rats were fed 12, 24, and 48% of energy as corn oil through out the study and the energy intake and body weights were unaffected by dietary fat (Fig. 5). In contrast, palpable tumor incidence, pathologically confirmed breast cancer, and risk of adenocarcinoma were all increased in proportion to fat intake. This large study involved 360 rats and food intake was not con-

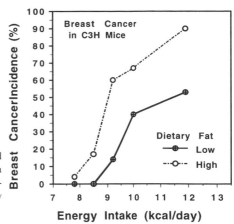

Figure 4. The effects of low- and high-fat diets at controlled levels of energy intake on spontaneous mammary carcinoma in C3H mice (25). This study illustrates that dietary fat concentration enhances mammary carcinogenesis independently with energy intake.

trolled but was carefully quantitated for each individual rat. Fig. 6 shows a frequency distribution of average daily intake (kcal/day) for the rats on this study. A regression analysis generated the curve shown on the right, indicating the very potent ability of freely selected energy intake to enhance mammary carcinogenesis in this model (which has also been observed for colon cancer (59)). These examples have been selected from a literature which now includes over 100 studies of dietary fat and breast cancer in diverse models (60, 61). Overall, these studies support the conclusion that both fat and energy have direct effects on mammary carcinogenesis. Indeed, fat has been shown to enhance breast cancer in studies where tumors are caused by the mouse mammary tumor virus and in a diverse array of models where tumors are induced by direct and indirect acting chemical carcinogens, irradiation, or hormones (60, 61).

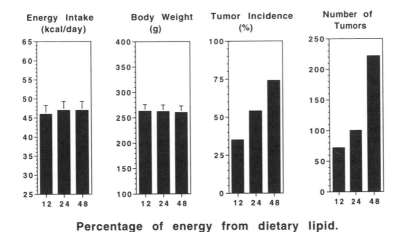

Percentage of energy from dietary lipid.

Figure 5. The effects of dietary lipid at 12, 24, and 48% of energy on energy intake, body weight, and tumorigenesis in rats given the breast carcinogen, 7,12-dimethylbenz(α)anthracene (66). Dietary lipid concentration had no significant influence on energy intake and final body weight. In contrast, palpable mammary tumor incidence and frequency, and pathologically confirmed adenocarcinoma incidence were significantly increased. This study is representative of many in the literature reporting an increase in mammary carcinogenesis by dietary fat with no significant effect on growth and food intake.

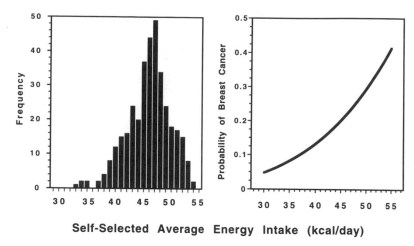

Figure 6. The relationship between self-selected energy intake and breast cancer risk in 351 rats treated with the breast carcinogen, 7,12-dimethylbenz(α)anthracene (66). The frequency distribution on the left shows the mean daily energy intake for each rat. Regression analysis generated the curve on the right which illustrates the odds of developing a tumor over the range of self-selected energy intake observed in this study. The odds of having tumor of any histologic type is increased by approximately 10% for each 1-kcal increase in self-selected energy intake.

The role of lipids in carcinogenesis involves several interacting and partially overlapping areas of investigation. The proportion of fat in the total diet was briefly addressed above. The composition of the dietary lipids, particularly relative to the fatty acid profile, has also proven to be a critical variable modulating experimental carcinogenesis (62). The role of essential, monounsaturated, saturated, omega-3, and trans fatty acids are the subject of many ongoing investigations. In general, we can conclude that tumor cells, like all mammalian cells require essential fatty acids for their maximal growth. However, many studies suggest that omega-6 fatty acids, such as linoleic acid, have a tumor promoting activity even when essential fatty acid requirements have been satisfied (31). The effects of omega-3 polyunsaturated fatty acids on carcinogenesis remain to be resolved. Overall, it appears that the balance between the omega-3 and omega-6 fatty acids influences tumorigenesis, perhaps through regulation of the prostaglandin and leukotriene pathways. Although, frequently discussed, there is no strong evidence linking the consumption of trans fatty acids with cancer risk. An area of increasing interest concerns the role of conjugated linoleic acid, derived from meat and dairy products, as an "anti-cancer" fatty acid (63–65).

FUTURE DIRECTIONS: DIETARY LIPIDS AND THE CANCER CASCADE

The role of dietary lipid concentration, specific fatty acids, fatty acid profiles, and lipid sources in carcinogenesis remains a complex and exciting area of research. In the past, those studying these relationships with population based epidemiologic approaches and those utilizing animal models or cellular, biochemical, and molecular techniques have generally worked independently. Increasingly, the advances in the nutrition and cancer arena will be derived from collaborations between the investigators using different approaches and from many other disciplines, such as genetics. There are major limitations in the ability of current diet assessment tools to precisely estimate the intakes of lipids or fatty acids in humans. The interactions between the laboratory scientists and the epidemiologists will provide new approaches that will allow the prospective cohort studies to maximize the knowledge that can be obtained. Studies employing blood or tissue samples for the evaluation of endpoints indicative of fatty acid intake or metabolism will be critical. The development and validation of laboratory assays to assess nutritional status will reduce our dependence upon relatively imprecise questionnaires to estimate intake. Blood and tissue samples from cohort studies also allow for investigation of novel endpoints correlated with different stages of tumorigenesis. Cancer researchers and geneticists are providing novel tools for the detection of mutant oncogenes and tumor suppressor genes as well as the expression of many genes contributing to the carcinogenesis process. These tools, employing techniques such as immunohistochemistry, *in situ* hybridization, and PCR based technology are becoming more consistent, specific, and quantitative while requiring increasingly smaller tissue samples. Coupling dietary intake data with nutritional biomarkers, molecular and histopathologic endpoints associated with premalignancy, or markers of aggressive tumor biology will provide the answers we need to define and evaluate preventive strategies based upon nutrition. It is also abundantly clear that the rapid progress in the characterization of the human genome will provide many new opportunities for nutritional intervention for cancer prevention. Defining inherited mutant genes exhibiting high penetrance and predisposing to cancer will allow investigators to assess the ability of diet to modulate risk in specific subpopulations. Furthermore, genetic testing will provide subgroups that are highly motivated for participation in diet and nutritional protocols designed to reduce or eliminate risk. Genotyping of individuals in cohort studies based upon the more common polymorphisms in cancer susceptibility genes will directly allow assessment of many critical diet and genetic interactions in risk. Genetic advances will also have an enormous impact upon our ability to evaluate dietary hypotheses using precisely controlled laboratory investigations. For example, the ability to precisely manipulate murine genes and develop transgenic and knock-out mice allows investigators to rapidly evaluate interactions between diet and genetics which may influence the cancer cascade (5).

The investigations summarized in this volume provide ample support for our optimism that cancer prevention though dietary modification is readily obtainable. Furthermore, interactions between nutritional scientists, cancer biologists, geneticists, and epidemiologists will provide the data necessary for the further refinement of population based dietary guidelines, but also for the more precise definition of cancer prevention and health maintenance strategies based upon nutrition for individuals. Prostate cancer is a disease which provides a powerful illustration of how a prevention program focusing upon diet and nutrition could potentially impact upon our society. Prostate cancer is a disease of aging, with the relative risk increasing by approximately 100-fold between the ages of 40 and 70. A life-style intervention that modestly delays the current incidence curve will translate into a significant impact on morbidity and mortality. For example, Fig. 7 illustrates the age related rates of prostate cancer (the pre-PSA era) and a parallel curve that could hypothetically result from a prevention program. Indeed, a 30 to 50% reduction in the growth of prostate tumors can be achieved in animal models by dietary interventions, such as modest energy restriction. In our hypothetical illustration, the risk of prostate cancer at age 70 would be reduced by over 90% in response to a dietary intervention that modestly impacts the rate prostate cancer progression. These goals should be readily achievable based upon the scientific data that is rapidly accumulating. The success of our

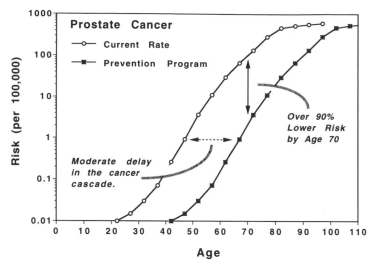

Figure 7. These data are derived from the era prior to extensive use of PSA screening and illustrate the risk of prostate cancer with age. The risk is very low prior to age 40 but rapidly increases over the next 4 decades. A dietary or nutritional program that reduces the rate of progression of prostate cancer by 30 to 50% may have an enormous impact upon public health. Even a modest delay in tumor progression could reduce the incidence of prostate cancer in the 70 year old cohort by 90%.

efforts in the area of cancer prevention will depend upon supportive and cooperative interactions between academia, the medical profession, government, industry, and charitable organizations.

REFERENCES

1. Narod, S., Ford, D., Devilee, P., Barkardottir, R., Eyfjord, J., Lenoir, G., Serova, O., Easton, D., and Goldgar, D. Genetic heterogeneity of breast-ovarian cancer revisited. Breast Cancer Linkage Consortium. Am. J. Human Genetics, *57:* 957–958, 1995.
2. Bell, D. A., Taylor, J. A., Paulson, D. F., Robertson, C. N., Mohler, J. L., and Lucier, G. W. Genetic risk and carcinogen exposure: a common inherited defect of the carcinogen-metabolism gene glutathione S-transferase M1 (GSTM1) that increases susceptibility to bladder cancer. J. Natl. Cancer Inst., *85:* 1159–1164, 1993.
3. Tannenbaum, A. Nutrition and Cancer. *In:* F. Homburger (ed.), The Physiopathology of Cancer. New York: Hoeber-Harper, 1959.
4. Clinton, S. K. Nutrition in the etiology and prevention of cancer. *In:* J. F. Holland, E. Frei, B. C. Bast, D. W. Kufe, D. L. Morton, and R. R. Weichselbaum (eds.), Cancer Medicine, pp. 370–395. Philadelphia: Lea and Febiger, 1993.
5. Hursting, S. D., Perkins, S. N., and Phang, J. M. Calorie restriction delays spontaneous tumorigenesis in p53-knockout transgenic mice. Proc. Natl. Acad. Sci. USA, *91:* 7036–7040, 1994.
6. Sinha, R., Rothman, N., Brown, E. D., Mark, S. D., Hoover, R. N., Caporaso, N. E., Levander, O. A., Knize, M. G., Lang, N. P., and Kadlubar, R. F. Pan-fried meat containing high levels of heterocyclic aromatic amines but low levels of polycyclic aromatic hydrocarbons induces cytochrome P4501A2 activity in humans. Cancer Res., *54:* 6154–6159, 1994.
7. Ohagaki, H., Takayama, S., and Sugimura, T. Carcinogenicies of heterocyclic amines in cooked food. Mutation Res., *259:* 399–410, 1991.
8. Hatch, R. T., Knize, M. G., Moore, D. H., and Felton, J. S. Quantitative correlation of mutagenic and carcinogenic potencies for heterocyclic amines from cooked foods and additional aromatic amines. Mutation Res., *271:* 269–287, 1992.

9. Lijinsky, W., and Shubik, P. Benzo(a)pyrene and other polynuclear hydrocarbons in charcoal-broiled meat. Science, *145:* 53, 1964.

10. Ames, B. N., Shigenaga, M. K., and Hagen, T. M. Oxidants, antioxidants, and the degenerative diseases of aging. Proc. Natl. Acad. Sci. USA, *90:* 7915–7922, 1993.

11. Prokocimer, M., and Rotter, V. Structure and function of p53 in normal cells and their aberrations in cancer cells: projection on the hematologic cell lineages. Blood, *84:* 2391–2421, 1994.

12. Greider, C. W., and Blackburn, E. H. Telomeres, telomerase, and cancer. Sci. Am., *275:* 92–97, 1996.

13. Folkman, J. Angiogenesis in cancer, vascular, rheumatoid and other disease. Nature Medicine, *1:* 27–31, 1995.

14. Folkman, J. What is the evidence that tumors are angiogenesis dependent? [editorial]. Journal of the National Cancer Institute, *82:* 4–6, 1990.

15. O'Reilly, M. S., Holmgren, L., Chen, C., and Folkman, J. Angiostatin induces and sustains dormancy of human primary tumors in mice. Nature Medicine, *2:* 689–692, 1996.

16. Fidler, I. J. Critical factors in the biology of human cancer metastasis: twenty-eigth G.H.A. Clowes Memorial Award Lecture. Cancer Res., *50:* 6130, 1990.

17. Mareel, M. M., Van Roy, F. M., and Bracke, M. E. How and when do tumor cells metastasize? Critical Reviews in Oncogenesis, *4:* 559–594, 1993.

18. Buckman, D. K., Chapkin, R. S., and Krickson, K. L. Modulation of mouse mammary tumor growth and linoleate enhanced metastasis by oleate. J. Nutr., *120:* 148–157, 1990.

19. Hubbard, N. E., and Erickson, K. L. Enhancement of metastasis from a transplantable mouse mammary tumor by dietary linoleic acid. Cancer Res., *47:* 6171–6175, 1987.

20. Connolly, J. M., and Rose, D. P. Effects of fatty acids on invasion through reconstituted basement membrane (Matrigel) by a human breast cancer cell line. Cancer Letters, *75:* 137–142, 1993.

21. Liu, X.-H., Connolly, J. M., and Rose, D. P. Eicosanoids as mediators of linoleic acid-stimulated invasion and type IV collagenase production by a metastatic human breast cancer cell line. Clin. Exp. Metastasis, *14:* 145–152, 1996.

22. Rose, D. P., Connolly, J. M., Rayburn, J., and Coleman, M. Influence of diets containing different levels of eicosapentaenoic or docosahexaenoic acid on the growth and metastasis of human breast cancer cells in nude mice. J. Natl. Cancer Inst., *87:* 587–592, 1995.

23. Rose, D. P., and Connolly, J. M. Effects of dietary omega-3 fatty acids on human breast cancer growth and metastasis in nude mice. J. Natl. Cancer Inst., *85:* 1743–1747, 1993.

24. Tannenbaum, A. The genesis and growth of tumors III. Effects of a high-fat diet. Cancer Res., *2:* 468–475, 1942.

25. Tannenbaum, A. The dependence of tumor formation on the composition of the calorie-restricted diet as well as on the degree of restriction. Can. Res., *5:* 616–625, 1945.

26. Carroll, K. K., and Kohr, H. T. Effects of level and type of dietary fat on incidence of mammary tumors induced in female Sprague-Dawley rats by 7,12-dimethylbenz(a)anthracene. Lipids, *6:* 415–420, 1971.

27. Rose, D. P., and Connolly, J. M. Effects of fatty acids and eicosanoid synthesis inhibitors on the growth of two human prostate cancer cell lines. The Prostate, *18:* 243–254, 1991.

28. Clinton, S. K., Palmer, S. S., Spriggs, C. E., and Visek, W. J. The growth of Dunning transplantable prostate adenocarcinomas in rats fed diets varying in fat content. J. Nutr., *118:* 1577–1585, 1988.

29. Clinton, S. K., Imrey, P. B., Alster, J. M., Simon, J., Truex, C. R., and Visek, W. J. The combined effects of dietary protein and fat on 7,12-dimethylbenz(a)anthracene-induced breast cancer in rats. J. Nutr., *114:* 1213–1223, 1984.

30. Chan, P. C., and Cohen, L. A. Effect of dietary fat, antiestrogen, and antiprolactin on the development of mammary tumors in rats. J. Natl. Cancer Inst., *52:* 25–30, 1974.

31. Ip, C., Carter, C. A., and Ip, M. M. Requirement of essential fatty acid for mammary tumorigenesis in the rat. Cancer Res., *45:* 1997–2001, 1985.

32. Welsch, C. W. Relationship between dietary fat and experimental mammary tumorigenesis: A review and critique. Cancer Res., *52:* 2040s-2048s, 1992.

33. Willett, W. The search for the causes of breast and colon cancer. Nature, *338:* 389–393, 1989.

34. Giovannucci, E., Rimm, E. B., Stampfer, J. J., Colditz, G. A., Ascherio, A., and Willett, W. C. Intake of fat, meat, and fiber in relation to risk of colon cancer in men. Cancer Res. *54:* 2390–2397, 1994.

35. Rose, D. P., Boyer, A. P., and Wynder, E. L. International comparisons of mortality rates for cancer of the breast, ovary, prostate, and colon, and per capita food consumption. Cancer, *58:* 2363–2371, 1986.

36. Reddy, B. S. Diet and colon cancer; evidence from human and animal model studies. *In:* B. S. Reddy and L. A. Cohen (eds.), Diet, Nutrition and Cancer; A Critical Evaluation. Vol. 1, pp. 47–66. Boca Raton: CRC Press, 1986.

37. Reddy, B. S. Inhibitors of the arachidonic acid cascade and their chemoprevention of colon carcinogenesis. *In:* L. Wattenberg, M. Lipkin, C. W. Boone, and G. J. Kelloff (eds.), Cancer Chemoprevention, pp. 153–164. Boca Raton: CRC Press, 1992.

38. Wingo, P. A., Tong, T., and Bolden, S. Cancer statistics. 1995 [published erratum appears in CA Cancer J. Clin. 1995; 45:127–128]. CA Cancer J. Clin., *45:* 8–30, 1995.

39. Blair, A., and Fraumeni, J. F. Geographic patterns of prostate cancer in the United States. J. Natl. Cancer Inst., *61:* 1379–1384, 1978.

40. Armstrong, B., and Doll, R. Environmental factors and cancer incidence and mortality in different countries, with special reference to dietary practices. Int. J. Cancer, *15:* 617–631, 1975.

41. Carroll, K. K., and Kohr, H. T. Dietary fat in relation to tumorigenesis. Prog. Biochem. Pharmacol., *10:* 308–353, 1975.

42. Giovannucci, E. Epidemiological characteristics of prostate cancer. Cancer, *75:* 176–177, 1995.

43. Giovannucci, E., Rimm, E. B., Colditz, G. A., Stampfer, M. J., Ascherio, A., Chute, C. C., and Willett, W. C. A prospective study of dietary fat and risk of prostate cancer. J. Natl. Cancer Inst., *85:* 1571–1579, 1993.

44. Whittemore, A. S., Kolonel, L. N., Wu, A. H., John, E. M., Gallagher, R. P., Howe, G. R., Burch, D., Hankin, J., Dreon, D. M., West, D. W., Teh, C., and Paffenbarger Jr., R. S. Prostate cancer in relation to diet, physical activity, and body size in blacks, whites, and asians in the United States and Canada. J. Natl. Cancer Inst., *87:* 652–661, 1995.

45. Whittemore, A. S., Wu, A. H., Kolonel, L. N., et al. Family history and prostate cancer risk in black, white, and Asian men in the United States and Canada. Am. J. Epidemiol., *141:* 732–740, 1995.

46. Bosland, M. C., and Prinsen, M. K. Induction of dorsolateral prostate adenocarcinomas and other accessory sex gland lesions in male Wistar rats by a single administration of N-methyl-N-nitrososurea, 7,12-dimethylbenz(a)anthracene, and 3,2'-dimethyl-4-aminobiophenyl after sequential treatment with cyproterone acetate and testosterone propionate. Cancer Res., *50:* 691–699, 1990.

47. Pollard, M., and Luckert, P. H. Promotional effects of testosterone and dietary fat on prostate carcinogenesis in genetically susceptible rats. Prostate, *6:* 1–5, 1985.

48. Pollard, M., Luckert, P. H., and Sporn, M. B. Prevention of primary prostate cancer in Lobund-Wistar rats by N-(4-hydroxyphenyl)retinamide. Cancer Res., *51:* 3610–3611, 1991.

49. Rose, D. P., and Cohen, L. A. Effects of dietary menhaden oil and retinyl acetate on the growth of DU 145 human prostatic adenocarcinoma cells transplanted into athymic nude mice. Carcinogenesis, *9:* 603–605, 1988.

50. Wang, Y., Corr, J. G., Thaler, H. T., Tao, Y., Fair, W. R., and Heston, W. D. W. Decreased growth of established human prostate LNCaP tumors in nude mice fed a low-fat diet. J. Natl. Cancer Inst., *87:* 1456–1462, 1995.

51. Clinton, S. K., Mulloy, A. L., Li, S. P., Mangian, H. J., and Visek, W. J. Dietary fat and protein intake differ in modulation of prostate tumor growth, prolactin secretion and metabolism, and prostate gland prolatin binding capacity in rats. J. Nutr. (in press), 1997.

52. Prentice, R. L., and Sheppard, L. Consistency of the epidemiology data, and disease prevention that may follow from a practical reduction in fat consumption. Cancer Causes and Control, *1:* 81–97, 1990.

53. Howe, G. R., Hirohata, T., Hislop, G., Iscovich, J., Yuan, J., Katsouyanni, K., Lubin, F., Marubini, E., Modan, B., Rohan, T., Toniolo, P., and Shunzhang, Y. Dietary factors and risk of breast cancer: Combined analysis of 12 case-control studies. J. Natl. Cancer Inst., *82:* 561–569, 1990.

54. Hunter, D., Spiegelman, D., Adami, H., Beeson, L., van den Brandt, P., Folsom, A., Goldbohm, A., Graham, S., Howe, G., Kushi, L., Marshall, J., McDermott, A., Miller, A., Speizer, F., Wolk, A., Yaun, S., and Willett, W. Cohort studies of fat intake and risk of breast cancer: a pooled analysis. New England J. Med., *334:* 356–361, 1996.

55. Boutwell, R. K., Brush, M. K., and Rusch, H. P. The stimulating effect of dietary fat on carcinogenesis. Cancer Res., *9:* 741, 1949.

56. Kritchevsky, D., and Klurfeld, D. M. Caloric effects in experimental mammary tumorigenesis. Am.J. Clin. Nutr., *45:* 236–242, 1987.

57. Boissoneault, G. A., Elson, C. E., and Pariza, M. W. Net energy effects of dietary fat on chemically-induced mammary carcinogenesis in F344 rats. J. Natl. Cancer Inst., *76:* 335–338, 1986.

58. Clinton, S. K., Alster, J. M., Imrey, P. B., Nandkumar, S., Truex, C. R., and Visek, W. J. The combined effects of dietary protein, fat, and energy intake during an initiation phase study of 7,12-dimethyl-benzanthracene-induced breast cancer in rats. J.Nutr., *116:* 2290–2302, 1986.

59. Clinton, S. K., Imrey, P. B., Mangian, H. J., Nandkuymar, S., and Visek, W. J. The combined effects of dietary fat, protein, and energy intake on azoxymethane-induced intestinal and renal carcinogenesis. Cancer Res., *52:* 857–865, 1992.

60. Albanes, D. Total calories, body weight, and tumor incidence in mice. Cancer Res., *47:* 1987–1992, 1987.

61. Freedman, L. S., Clifford, C., and Messina, M. Analysis of dietary fat, calories, body weight and the development of mammary tumors in rats and mice: a review. Cancer Res., *50:* 5710–5719, 1990.

62. Carroll, K. K., and Hopkins, G. J. Dietary polyunsaturated fat versus saturated fat in relation to mammary carcinogenesis. Lipids, *14:* 155–158, 1979.

63. Bonorden, W., Storkson, J., Liu, W., Albright, K., and Pariza, M. Fatty acid inhibition of 12-O-tetradecanoylphorbol-13-acetate (TPA)-induced phospholipase C activity. FASEB J., *7:* A618, 1993.

64. Chin, S. F., Liu, W., Storkson, J. M., Ha, Y. L., and Pariza, M. W. Dietary sources of conjugated dienoic isomers of linoleic acid, a newly recognized class of anticarcinogens. J. Food Comp. Anal., *5:* 185–197, 1992.

65. Ip, C., Chin, S. F., Scimeca, J. A., and Pariza, M. W. Mammary cancer prevention by conjugated dienoic derivative of linoleic acid. Cancer Res., *51:* 6118–6124, 1991.

66. Clinton, S. K., Imrey, P. B., Alster, J. M., Simon, J., Truex, C. R., and Visek, W. J. The combined effects of dietary protein and fat on 7,12-dimethyl-benzanthracene-induced breast cancer in rats. J. Nutr., *114:* 1213–1223, 1984.

MOLECULAR STUDIES ON THE ROLE OF DIETARY FAT AND CHOLESTEROL IN BREAST CANCER INDUCTION

Michael C. Archer, Ahmed El-Sohemy, Laurie L. Stephen, and
Alaa F. Badawi

Department of Nutritional Sciences
Faculty of Medicine
University of Toronto
150 College Street
Toronto, Canada M5S 3E2

INTRODUCTION

It has been known for many years that rates of breast cancer vary five to ten-fold between countries[1]. Furthermore, migrants from low to high-risk countries adopt the rates of their new country[2]. This type of evidence, albeit indirect, strongly suggests that environmental factors play a role in breast cancer development. There is also substantial evidence that genetic and hormonal factors contribute significantly.[3,4] Although there are many hypotheses to account for the role of the environment, diet is undoubtedly one of the most important.[5] A number of years ago, Doll and Peto estimated that as much as 50% of breast cancer in North America might be prevented by changes in diet.[6]

Specific dietary factors responsible for breast cancer have not yet been definitively identified from epidemiological investigations. For the past twenty years or more, the major dietary factor thought to increase breast cancer risk has been fat.[7] The international correlations are striking,[8] but many other factors are correlated with fat intake.[9] Recent case-control and cohort studies have either been negative or the associations have been weak.[10,11] It has been argued, however, that there are insufficient differences between the diets in the populations studied to allow detection of associations,[12] the measurement errors associated with dietary questionnaires are large,[13] and there are many confounding dietary variables.

Although cholesterol has long been implicated in the development of cardiovascular disease, descriptive and analytic epidemiological studies of its role in cancer development have not been consistent.[14] No relationship between dietary cholesterol and breast cancer risk has been established.[10] However, the strong correlation between the intake of cholesterol and animal fat and protein make it difficult to determine the independent effect of cholesterol. Serum levels have been used as a surrogate measure of dietary intake and as an independent risk factor for cancer risk. Some studies have shown that elevated serum cholesterol is a risk factor for breast cancer,[15] while others report an inverse association.[16] There are a number of difficulties, however, in interpreting associations between serum cholesterol and human cancer. Placed on the same hypercholesterolemic diet, some individuals show a marked increase in serum cholesterol (responders), while others show no such response (non-responders), or even a decrease in serum levels.[17] The well established effects of other dietary components such as fat and fiber on serum cholesterol may be further confounding. Moreover, the presence of a tumor has been shown to lower markedly serum cholesterol levels and can obscure observations that are made in case-control studies.[18]

In order to assess the biological plausibility of epidemiological associations and to study the mechanisms of action of dietary factors that may affect cancer development, a great deal of emphasis is placed on experiments using rodents. The most widely used and useful model of human breast cancer development is the rat, using either 7,12-dimethyl-benz(a)anthracene (DMBA) or N-methylnitrosourea (MNU) as carcinogens.[19] The MNU model is simpler mechanistically than the DMBA model because MNU is a direct acting carcinogen that does not require the enzymes of drug metabolism for its activation. Furthermore, MNU has a short half-life (1h) in the animal following injection, so that cancer initiation is rapid. Single doses of either of these agents administered at about 50 days of age, are able to produce essentially 100% breast adenocarcinoma incidence with a short latency in susceptible strains.[20] At this age, the animals are maturing sexually, and readily develop cancer due to active organogenesis and high proliferation rates in the glandular epithelium.[21] The Ha-ras oncogene, activated by a G to A transition mutation is found in ~85% of MNU-induced rat mammary tumors.[22] In contrast, DMBA-induced tumors contain few, if any, ras mutations.[23] Human breast cancers do not commonly harbor ras mutations, but 60–70% of tumors overexpress the ras protein product p21 (reviewed in 24). In spite of these differences at the genetic level, MNU-induced malignant rat mammary tumors share a very similar pathogenesis and have many features in common with the development of intraductal and infiltrating ductal carcinomas in humans.[25] Furthermore, the effects of hormonal and dietary factors on mammary tumorigenesis appear to be similar in rats and humans.

DIETARY FAT AND RAT MAMMARY TUMORIGENESIS

Although, as mentioned above, the effect of different levels of dietary fat on breast cancer development in humans is unclear, a high fat diet has been demonstrated in many studies to enhance the formation of spontaneous, transplantable and carcinogen-induced tumors in rodents.[26] There is evidence that dietary fat exerts its effect during the promotion phase of breast cancer development.[26] It is not clear, however, at what stage it acts during the 6 months or so of promotion. In order to address this problem, we have examined the ability of dietary fat to accelerate the growth rate of preneoplastic cells during rat mammary adenocarcinogenesis.

First, we developed a highly sensitive technique to quantify the frequency of mutations of the Ha-ras gene in cells isolated from individual mammary glands a short time after administration of a carcinogenic dose of MNU.[27] Briefly, since a G to A transition at codon 12 of the Ha-ras gene eliminates the GAGG sequence recognized by the restriction endonuclease Mnl1, we first enrich our mammary gland DNA samples in the mutant allele by an Mnl1 digestion. This is followed by PCR amplification using a 3' primer containing two mismatched bases, that together with the codon 12 mutation, create an Xmn1 restriction site in the mutant allele. The normal allele generates a 71 bp fragment, while the mutant allele generates a 53 bp fragment. Using liquid hybridization and gel retardation techniques, we have been able to detect Ha-ras mutations in individual mammary glands. Furthermore, the method allowed us to estimate the mutant cell fraction in a semi-quantitative manner. We were able to show that Ha-ras mutations are present in the mammary glands of rats 30 days following MNU treatment, with mutant cell fractions generally in the 10^{-6} to 10^{-5} range. We also showed that the fraction of cells containing a mutated Ha-ras allele increases by a factor of 10–100 between 30 and 60 days, undoubtedly due to clonal expansion of the initiated cell population.

We have used this sensitive method for measuring Ha-ras mutations to determine when during tumor development dietary fat exerts its promotional effect.[28] We chose to use a protocol already developed by Cohen and Chan[29] in which F344 rats administered 50 mg/kg MNU at 50 days of age followed by a diet containing 20% lard had a tumor incidence of 79% 26 weeks later, whereas rats treated in the same way, but fed a diet containing 4% lard had a tumor incidence of only 29%. Other components of these diets were also identical to those described by Cohen and Chan.[29] At 30 and 75 days following MNU treatment, four rats per group were sacrificed and up to eight macroscopically normal mammary glands were excised per rat. Figure 1 illustrates the analysis of cells harboring Ha-ras mutations by the assay described above.

While there appeared to be differences in the mutant cell portion of the animals on the different diets even at 30 days, no definitive conclusions could be drawn by simple visual comparison of these gels. Therefore, we quantified the mutant cell fraction in each gland using a PhosphorImager by comparison of band intensities with those of calibration standards as we previously described.[27] We then determined the proportion of glands

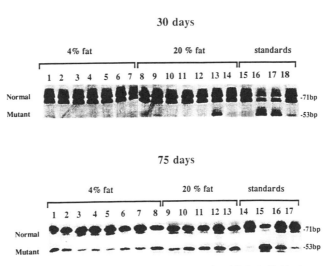

Figure 1. Examples of Ha-ras mutations (G to A transitions at codon 12) in individual mammary glands of F344 rats fed low- or high-fat diets 30 and 75 days following a single MNU dose of 50 mg/kg. Lanes for 30 day analysis: 1–7, animal on 4% fat diet; 8–14, animal on 20% fat diet; 15, untreated control; 16–18, standards of tumor cell line DNA diluted into DNA isolated from the mammary glands of untreated rat at ratios of 10^{-1}, 10^{-2}, and 10^{-3} respectively. Lanes for 75 day analysis: 1–8, animal on 4% fat diet; 9–13, animal on 20% fat diet; 14, untreated control; 15–17, standards as above. Reprinted with permission of Oxford University Press from ref. 28.

within each animal in which the mutant cell fraction exceeded the median value for all of the glands analyzed at each time point. The high-fat and low-fat groups were compared based on these proportions, allowing for extra-binomial variation due to correlations between glands of the same animal. We were able to show that the glands from the animals on the high-fat diet had a significantly higher mutant cell fraction than those on the low-fat diet both at 30 days (31% of glands greater than median on low-fat diet compared to 69% on high-fat diet, $p<0.04$) and 75 days (19% of glands greater than median on low-fat diet compared to 80% on high-fat diet, $p<0.04$).

These results show for the first time that dietary fat acts early in carcinogenesis to stimulate the growth of cells harboring Ha-ras mutations. The acceleration in growth of these cells at 30 days, suggests that some fat component or metabolite is acting directly, or indirectly via a growth factor, acts on the ras-transformed cells to stimulate their growth relative to normal mammary epithelial cells. Another possibility is that other genetic changes occur early during carcinogenesis in these cells that make them responsive to growth stimulation by the promoter, although such additional genetic changes have not yet been identified in this tumor model.

In a related area, we have recently been investigating the effects of different levels and types of dietary fat on the expression of genes that may play a role in cancer development. We have begun by examining the genes for cyclooxygenase (Cox) 1 and 2. The corresponding gene products play an important role in regulating prostaglandin (PG) biosynthesis. Elevated levels of PGs have been implicated as risk factors in breast cancer development in both humans[30] and rodents.[31]

In preliminary studies, we have compared the effects of diets containing different levels of ω-3 polyunsaturated fatty acids (PUFA) with diets containing the same levels of ω-6 PUFA. The former appear to inhibit the promoting effects of the latter on mammary carcinogenesis. Female Sprague-Dawley rats were fed AIN 93G diets containing either menhaden oil (rich in ω-3) or safflower oil (rich in ω-6) at low (7%) or high (21%) levels

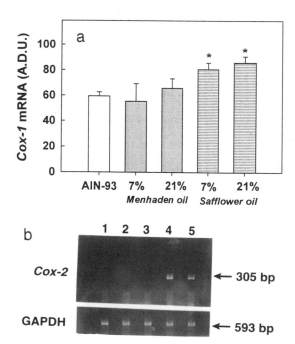

Figure 2. The expression of Cox-1 and Cox-2 mRNA in rat mammary tissue in response to different levels and types of dietary fatty acids. a. Cox-1 expression levels measured by Northern blot analysis and quantified by measuring band intensities relative to β-actin in arbitrary density units (A.D.U.). Data are means ± SD for 5 animals; * significantly different from controls (p< 0.001) by Student's t test. b. Cox-2 expression levels in the same mammary glands measured by RT-PCR. The PCR products were analysed using 2% agarose gels; Cox-2 and the internal control, glyceraldehyde phosphate dehydrogenase (GAPDH), gave 305 and 593 bp fragments respectively. Lane 1 AIN 93G; lanes 2 and 3, 7% and 21% menhaden oil; lanes 3 and 4, 7% and 21% safflower oil.

for 3 weeks. Controls were fed the standard AIN 93G diet that contains 7% soybean oil. This contains a somewhat lower proportion of ω-6 PUFA than safflower oil, but is richer in mono-unsaturated and saturated fatty acids. Mammary gland mRNA was examined for Cox-1 levels by Northern blot analysis and for Cox-2 levels by RT-PCR.

Figure 2a illustrates that Cox-1 expression levels were approximately the same in animals fed the control diet or diets containig high or low levels of menhaden oil. Although Cox-1 is known to be constitutively expressed,[32] both high and low levels of safflower oil yielded small, but significantly higher Cox-1 levels. The effect of safflower oil on Cox-2 expression, however, was much more marked. Cox-2 transcripts were not detectable in animals fed the control diet or the diets containing menhaden oil, but were clearly visible in animals fed safflower oil (Figure 2b). While these results are preliminary, they do illustrate that dietary factors that are known to play a role in mammary carcinogenesis, may exert their effects, at least in part, by altering the expression of genes that affect cancer development.

DIETARY CHOLESTEROL AND RAT MAMMARY TUMORIGENESIS

In a different, but related area, we have recently been investigating the effects of cholesterol and cholesterol oxidation products on rat mammary adenocarcinogenesis. Cholesterol is known to oxidize readily in air to yield a variety of products that have been detected in foods such as processed meats and dairy products.[33] Some of the oxidation products are known to be cytotoxic, mutagenic, immunosupressive and carcinogenic.[34] Furthermore, cholesterol α and β-epoxides have been detected in breast fluid aspirates of women with benign breast disease.[35] In spite of these observations, the role of cholesterol oxidation in experimental mammary carcinogenesis had not been explored. Thus, we treated female Sprague-Dawley rats with N-methylnitrosourea (MNU) to initiate carcinogenesis, then fed groups either a control diet that contained no cholesterol (AIN-76), or the control diet supplemented with 0.3% cholesterol or oxidized cholesterol prepared by heating cholesterol at 110°C for 48h. The cumulative tumor incidences for the three groups are shown in Figure 3.

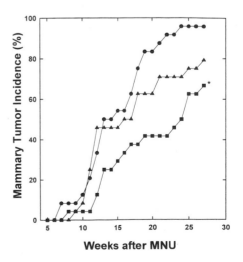

Figure 3. Mammary adenocarcinoma incidence in Sprague-Dawley rats initiated with MNU then fed a control diet (●), 0.3% oxidized cholesterol (▲), or 0.3% cholesterol (■). *Significant difference in the final tumor incidence from the control group ($p < 0.05$). Reprinted with permission of Oxford University Press from ref 36.

Tumor incidence after 26 weeks in the cholesterol group (67%) was significantly lower than in the control group (96%, p<0.05), but the oxidized cholesterol group (79%) was not significantly different from the control or cholesterol groups. The average number of tumors per animal was also significantly lower in the cholesterol group (1.5 ± 0.22) than in the control (2.8 ± 0.27) or oxidized cholesterol groups (2.3 ± 0.37), $p \leq 0.005$. The oxidized cholesterol produced an effect that was intermediate between the control and cholesterol-fed groups. We analyzed the oxidized cholesterol by gas chromatography and discovered that 2% of the cholesterol had been oxidized by the method employed. This level of oxidation seemed to have reduced the inhibitory effects of cholesterol somewhat. We speculated that the oxidation products may have acted as weak tumor promoters, but that any significant effects would have been masked by the remaining unoxidized cholesterol. A very recent report showed that high levels of pure cholesterol oxidation products did not initiate mammary tumorigenesis, although effects on tumor promotion were not studied.[37] Our most interesting result, however, was that the animals fed a diet containing cholesterol had a significantly lower tumor incidence than the controls that received no cholesterol.

In a previous study by Cohen and Chan,[29] cholesterol at a level of 2% in the diet of Fischer rats had no effect on MNU-induced mammary tumor development, regardless of the amount of fat in the diet. The diets used in their study, however, had higher levels of saturated fat and lower levels of unsaturated fat than in our experiments. Animals fed saturated fats have a lower tumor incidence than those fed unsaturated fat,[26] so that an inhibitory effect of cholesterol may have been masked in the experiment of Cohen and Chan. Two other factors relating to the rat strain used may have played a more important role in determining the different outcome of our experiment. First, Fischer rats are of intermediate sensitivity to mammary carcinogenesis.[38] We used the highly sensitive Sprague-Dawley strain that may have enabled us to detect inhibitory effects more easily. Second, Fischer rats do not respond to a hypercholesterolemic diet by increasing their serum cholesterol level.[39] In contrast, in our Sprague-Dawley rats, serum low-density lipoprotein (LDL) cholesterol levels were significantly greater in the cholesterol group (185 ± 38 mg/dl) than in the control (55 ± 4 mg/dl). This result led us to hypothesize that the elevated serum LDL cholesterol in the animals fed cholesterol is associated with the inhibition of tumorigenesis that is observed. Serum LDL can enter cells via the LDL receptor and the internalized cholesterol down regulates the levels of 3-hydroxy-3- methylglutaryl-CoA (HMG-CoA) reductase, the rate limiting enzyme in the cholesterol biosynthetic pathway.[40] This enzyme converts HMG-CoA to mevalonate that is required for DNA synthesis and cell proliferation.[41] Thus, our hypothesis suggests that exogenous cholesterol may inhibit de novo cholesterol synthesis in preneoplastic mammary epithelial cells, thereby inhibiting their proliferation and subsequent development into neoplasms.

It is of interest that inhibition of HMG-CoA reductase inhibits isoprenoid synthesis, resulting in a reduction of the post-translational isoprenylation of ras and other G proteins involved in signal transduction pathways.[42] Isoprenylation of the ras proto-oncogene product p21 has attracted much attention because of its involvement in cell growth and differentiation. Dietary cholesterol inhibits the development of mammary tumors initiated by MNU, most of which would normally harbor the mutated Ha-ras gene.[22] Since DMBA-induced rat mammary tumors that appear not be inhibited by dietary cholesterol[43] have a low level of Ha-ras mutations,[23] it is possible that feedback inhibition of HMG-CoA reductase by dietary cholesterol preferentially inhibits the isoprenylation of the mutated p21 protein compared to the normal p21 protein. A further possibility is that DMBA-induced mammary tumors lack or have low levels of LDL receptors thereby not allowing serum LDL to enter the cell. We are addressing these possibilities in our current research.

ACKNOWLEDGMENTS

These investigations were supported by grants from the Medical Research Council of Canada, the Canadian Breast Cancer Research Initiative and the Natural Sciences and Engineering Research Council of Canada. MCA is the recipient of a Natural Sciences and Engineering Research Council of Canada Industrial Research Chair, and acknowledges support from the member companies of the Program in Food Safety, University of Toronto. LLS is the recipient of a National Institute of Nutrition Fellowship.

REFERENCES

1. International Agency for Research on Cancer. Cancer Incidences on Five Continents. IARC Sci. Publ., 5: 882–883, 1987.
2. Haenszel, W. Migrant studies. In: D. Schottenfeld and J.F. Fraumeni (Eds) Cancer Epidemiology and Prevention, W.B. Saunders, Philadelphia, PA, pp. 194–207, 1982.
3. Eby, N., Chang-Claude, J. and Bishop, D.T. Familial risk and genetic susceptibility for breast cancer. Cancer Causes Control, 5: 458–470, 1994.
4. Henderson, B.E., Ross, R.K., Pike, M.C. and Casagrande, J.T. Endogenous hormones as a major factor in human cancer. Cancer Res., 42: 3232–3239, 1982.
5. Miller, A.B., Berrino, F., Hill, M., Pietinen, P., Riboli, E. and Wahrendorf, J. Diet in the aetiology of cancer: a review. Eur. J. Cancer, 30A: 207–220, 1994.
6. Doll, R. and Peto, R. The causes of cancer. J. Natl. Cancer Inst., 66: 1191–1308, 1981.
7. Willet, W. The search for the causes of breast and colon cancer. Nature, 338: 389–394, 1989.
8. Carroll, K.K., Braden, L.M., Bell, J.A. and Kalamegham, R. Fat and cancer. Cancer, 58: 1818–1825, 1986.
9. Mendola, P., Marshall, J., Graham, S., Laughlin, R.H. and Freudenheim, J.L. Dietary correlates of fat. Nutr. Cancer, 23: 161–169, 1995
10. Hunter, D.J., Spieglman, D., Adami, H-O., Beeson, L., van den Brandt, P.A., Folsom, A.R., Fraser, G.E., Goldbohm, A., Graham, S., Howe, G.R., Kushi, L.H., Marshall, J.R., McDermott, A., Miller, A.B., Speizer, F.E., Wolk, A., Yaun, S-S. and Willet, W. Cohort studies of fat intake and the risk of breast cancer- a pooled analysis. N. Engl. J. Med. 334: 356–61, 1996.
11. Willett, W.C., Hunter, D.J., Stampfer, M.J., Colditz, G., Manson, J.E., Spiegelman, D., Rosner, B., Hennekens, C.H. and Speizer, F.E. Dietary fat and fiber in relation to risk of breast cancer: an 8-year follow up. JAMA, 268: 2037–2044, 1992.
12. Goodwin, P.J. and Boyd, N.F. Critical appraisal of the evidence that dietary fat intake is related to breast cancer in humans. J. Natl. Cancer Inst., 79: 473–485, 1987.
13. Bingham, S.A., Gill, C., Welch, A., Day, K., Cassidy, A., Khaw, K.T., Sneyd, M.J., Key, T.J.A., Roe, L. and Day, N.E. Comparison of dietary assessment methods in nutritional epidemiology: weighed records v. 24h recalls, food-frequency questionnaires and estimated-diet records. Br. J. Nutr., 72: 619–643, 1994.
14. McMichael, A.J., Jensen, O.M., Parkin, D.M. and Zaridze, D.G. Dietary and endogenous cholesterol and human cancer. Epidemiol. Rev., 6: 192–216, 1984.
15. Cowan, L.D., O'Connell, D.L., Criqui, M.H., Barrett-Connor, E., Bush, T.L., et al. Cancer mortality and lipid and lipoprotein levels. The Lipid Research Clinic Program mortality follow-up study. Am. J. Epidemiol., 131: 468–482, 1990.
16. Vatten, L.J. and Foss, O.P. Total serum cholesterol and triglycerides and risk of breast cancer: a prospective study of 24,329 Norwegian women. Cancer Res., 50: 2341–2346, 1990.
17. Beynen, A.C., Katan, M.B. and Van Zutphen, L.F.M. Hypo- and hyperresponders: individual differences in the response of serum cholesterol concentration to changes in the diet. Ad. Lipid Res., 22: 115–171, 1987.
18. Vitols, S., Bjorkholm, M., Gahrton, G. and Peterson, C. Hypocholesterolaemia in malignancy due to elevated low-density-lipoprotein-receptor activity in tumour cells: evidence from studies in patients with leukemia. Lancet, 2: 1150–1154, 1985.
19. Rogers, A.E. and Lee, S.Y. Chemically-induced mammary gland tumors in rats: modulation by dietary fat. In: Ip, C. et al.. Dietary Fat and Cancer. Alan R. Liss, New York, NY, pp. 255–269, 1988.
20. Thompson, H.J., Adlakha, H. and Singh, M. Effect of carcinogen dose and age at administration on induction of mammary carcinogenesis by 1-methyl-1-nitrosourea. Carcinogenesis, 13: 1535–1539, 1992.

21. Russo, J. and Russo, I.H. Towards a physiological approach to breast cancer prevention. Cancer Epi. Bio. Prev., 3: 353–364, 1994.

22. Sukumar, S., Notario, V., Martin-Zanca, D. and Barbacid, M. Induction of mammary carcinomas in rats by nitroso-methylurea involves malignant activation of H-ras-1-locus by single point mutations. Nature, 306: 658–661, 1983.

23. Waldmann, V., Suchy, B. and Rabes, H.M. Cell proliferation and prevalence of ras gene mutations in 7,12-dimethylbenz(a)anthracene (DMBA)- induced rat mammary tumors. Res. Exp. Med., 193: 143–151, 1993.

24. Zhang, P-L., Calaf, G. and Russo, J. Allele loss and point mutation in codons 12 and 61 of the c-Ha-ras oncogene in carcinogen-transformed human breast epithelial cells. Mol. Carcinogenesis., 9: 46–56, 1994.

25. Russo, J., Gusterson, B.A., Rogers, A.E., Russo, I.H., Wellings, S.R. and Zwieten, M.J. Comparative study of human and rat mammary tumorigenesis. Lab. Invest., 62: 244–278, 1990.

26. Welsch, C.W. Relationship between dietary fat and experimental mammary tumorigenesis: a review and critique. Cancer Res., 52: 2040s-2048s, 1992.

27. Lu, S-J. and Archer, M.C. Ha-ras activation in mammary glands of N-methyl-N-nitrosourea-treated rats genetically resistant to mammary adenocarcinogenesis. Proc. Natl. Acad. Sci. USA, 89:1001–1005, 1992.

28. Hu, Z., Chaulk, J.E., Lu, S-L., Xu, Z. and Archer, M.C. Effect of dietary fat or tamoxifen on the expansion of cells harboring Ha-ras oncogenes in mammary glands from methylnitrosourea-treated rats. Carcinogenesis, 16: 2281–2284, 1995.

29. Cohen, L.A. and Chan, P-C. Dietary cholesterol and experimental mammary cancer development. Nutr. Cancer, 4: 99–106, 1982.

30. Karmali, R.A., Welt, S., Thaler, H.T., and Lefevre F. Prostaglandins in breast cancer. Relationship to disease stage and hormone status. Br. J. Cancer., 48: 689–696, 1983.

31. Tan, W.C., Privett, O.S., and Goldyne, M.E. Studies of prostaglandins in rat mammary tumors induced by 7,12-dimethylbenz(a)anthracene. Cancer Res., 34: 3229–3231, 1974.

32. O'Neill, G.P., and Ford-Hutchinson, A. Expression of mRNA for cyclooxygenase-1 and cyclooxygenase-2 in human tissues. FEBS Lett., 2: 156–160, 1993.

33. Sander, B.D., Addis, P.B., Park, S.W. and Smith, E.D. Quantification of cholesterol oxidation products in a variety of foods. J. Food Protein, 52: 109–114, 1989.

34. Morin, R.J., Hu, B., Peng, S-K. and Sevanian, A. Cholesterol oxides and carcinogenesis. J. Clin. Lab. Anal., 5: 219–225, 1991.

35. Petrakis, N.L., Gruenke, L.D. and Craig, J.C. Cholesterol and cholesterol epoxides in nipple aspirates of human breast fluid. Cancer Res., 41: 2563–2565, 1981.

36. El-Sohemy, A., Bruce, W.R. and Archer, M.C. Inhibition of rat mammary tumorigenesis by dietary cholesterol. Carcinogenesis, 17: 159–162, 1996.

37. El-Bayoumy, K., Ji, B-Y., Upadhyaya, P., Chae, Y-H., Kurtzke, C., Rivenson, A., Reddy, B.S., Amin, S. and Hecht, S.S. Lack of tumorigenicity of cholesterol epoxides and esterone-3,4-quinone in the rat mammary gland. Cancer Res., 56: 1970–1973, 1996.

38. Chan, P-C., Head, J.F., Cohen, L.A. and Wynder, E.L. Influence of dietary fat on the induction of mammary tumors by N-nitrosourea: associated hormone changes and differences between Sprague-Dawley and F344 rats. J. Natl. Cancer Inst., 59: 1279–1283, 1977.

39. Klurfeld, D.M. and Kritchevsky, D. Serum cholesterol and 7,12-dimethylbenz[a]anthracene-induced mammary carcinogenesis. Cancer Lett., 14: 273–278, 1981.

40. Brown, M.S. and Goldstein, J.L. A receptor-mediated pathway for cholesterol homeostasis. Science, 232: 34–47, 1986.

41. Siperstein, M.D. Cholesterol, cholesterogenesis and cancer. Adv. Exp. Med. Biol., 369: 155–166, 1995.

42. Schmidt, R.A., Schneider, C.J. and Glomset, J.A. Evidence for post-translational incorporation of a product of mevalonic acid into Swiss 3T3 cell proteins. J. Biol. Chem., 259: 10175–10180, 1984.

43. Nakayama, M., Ju, H.R., Sugano, M., Hirose, N., Ueki, T., Doi, F. and Eynard, A.R. Effect of dietary fat and cholesterol on dimethylbenz[a]anthracene- induced mammary tumorigenesis in Sprague-Dawley rats. Anticancer Res., 13: 691–698, 1993.

FATTY ACID REGULATION OF BREAST CANCER CELL GROWTH AND INVASION

David P. Rose, Jeanne M. Connolly, and Xin-Hua Liu

Division of Nutrition and Endocrinology
American Health Foundation
Valhalla, New York

INTRODUCTION

In 1993, the American Institute for Cancer Research organized its fourth Annual Symposium around the topic "Diet and Breast Cancer". At that time, we discussed the influence of dietary fatty acids on human breast cancer cell growth, invasion and metastasis, and described the stimulatory effects of the polyunsaturated omega-6 fatty acids (n-6 FAs), and the inhibitory effects of the long-chain omega-3 fatty acids (n-3 FAs) on both cell proliferation and expression of the metastatic phenotype.[1] Since then, progress has been made in understanding the distinct, but complementary, roles of cyclooxygenase and lipoxygenase products of n-6 FA metabolism in breast cancer progression. Also, further support has been obtained for a dietary intervention trial with n-3 FA supplementation, either alone or with selective pharmacological inhibitors of eicosanoid biosynthesis, in women at high breast cancer risk and/or as a novel approach to adjuvant therapy.

A major challenge for breast cancer cell biologists is to gain a mechanistic understanding of the relationship between estrogen-independence and disease progression. In general, breast cancer patients whose tumors possess a significant level of the estrogen receptor (ER) have a better prognosis after surgical treatment than do those with ER negative tumors.[2] Estrogen receptor positive human breast cancer cells such as the MCF-7 cell line exhibit a low capacity for invasion in an *in vitro* assay,[3] and either do not metastasize, or do so poorly, when growing in athymic nude mice.[4-6] In contrast, the ER negative breast cancer cell lines are frequently highly invasive, and readily form metastases *in vivo*.[5-7] Recent experiments in our laboratory suggest that this relationship between estrogen independence and aggressive tumor behavior involves the expression of specific n-6 FA-metabolizing enzymes, and the biosynthesis of lipoxygenase and cyclooxygenase products.

Dietary Fat and Cancer, edited by AICR
Plenum Press, New York, 1997

EFFECTS OF OMEGA-6 FATTY ACIDS ON NORMAL MAMMARY EPITHELIAL AND BREAST CANCER CELL GROWTH

Stimulation *in Vitro*

The principal n-6 FA in the diet is linoleic acid (LA), which is metabolized in some host tissues and tumor cells by a system of desaturase and elongase enzymes to arachidonic acid (AA). The growth of both normal mouse mammary epithelial cells,[8,9] and nontransformed human breast epithelial cells,[10] *in vitro* is stimulated by LA, a response which requires the presence of epidermal growth factor (EGF).

Wicha et al.[11] first observed that the addition of LA to the culture medium stimulates the growth of a cell line derived from a chemically-induced rat mammary carcinoma, and subsequently we showed similar effects of the n-6 FA on the proliferation of the ER negative MDA-MB-231 and MDA-MB-435 human breast cancer cell lines *in vitro*.[1,12] Our demonstration that selective inhibitors of lipoxygenase, but not cyclooxygenase, activity could suppress the growth response of MDA-MB-231 cells to LA was consistent with a later report by Buckman and colleagues[13] that LA stimulates the *in vitro* growth of the estrogen-independent, metastatic, 4526 mouse mammary tumor cell line, and that this response requires intact lipoxygenase activity.

These growth responses in estrogen-independent cell lines are in contrast to the failure of estrogen-dependent human breast cancer cell lines to exhibit a mitogenic response when cultured in the presence of LA.[10,14] Fig. 1 illustrates the point; while growth of MCF-7 cells was stimulated by estradiol *in vitro*, there was no response to LA either in the absence or the presence of the estrogen. Grammatikos et al.[10] associated this failure of MCF-7 cells to undergo LA-mediated growth stimulation to deficient desaturase activity and AA production; a similar deficiency of Δ^6 and Δ^4 desaturases has also been described in the estrogen-dependent T47D human breast cancer cell line.[15] However, as we will discuss later in this chapter, there is emerging evidence that the primary defect may be a low level of 12-lipoxygenase activity in estrogen-dependent breast cancer cells.

Effects *in Vivo*

In a series of studies, we demonstrated the stimulatory effect of a high-fat, LA-rich, diet on the growth of MDA-MB-435 and MDA-MB-231 human breast cancer cells after

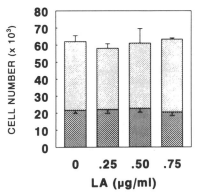

Figure 1. The growth of MCF-7 breast cancer cells for 6 days in serum-free medium without (hatched) or with (lines) the addition of 1 nM estradiol, and a range of LA concentrations. There was no growth response to LA.

their injection into the thoracic mammary fat pads of athymic nude mice.[16–19] This behavior of the estrogen-independent cell lines was in agreement with an earlier report by Gabor and Abraham[20] that feeding a LA-rich diet stimulated the growth of a transplantable mouse mammary adenocarcinoma. Both the dietary LA stimulation of MDA-MB-435 cell mammary fat pad tumor growth, and its metastasis to the lungs were suppressed by the simultaneous administration of indomethacin, at a dose which partially inhibited the production of cyclooxygenase products by the cancer cells.[21] In contrast to the results obtained from experiments *in vitro* with MDA-MB-231 cells,[12] this growth suppression was achieved without any discernible effects on lipoxygenase-mediated eicosanoid production.

SUPPRESSION OF BREAST CANCER CELL GROWTH BY n-3 FAs

There is some epidemiological support for the proposition that the long-chain n-3 FAs which are present at high concentration in some fish oils exert a protective effect against breast cancer.[22–26] It may be, however, that the critical relationship is the ratio of n-3 FAs to n-6 FAs that occurs in the diet,[26,27] rather than simply the absolute level of n-3 FA intake. In Japan, an increasing breast cancer risk has been accompanied by both an increase in dietary n-6 FA intake, and a reduction in n-3 FA consumption.[27]

We have reported that feeding docosahexaenoic acid (DHA) or eicosapentaenoic acid (EPA), the principal n-3 FAs, to nude mice inhibits the growth of MDA-MB-435 cells and their metastasis from the mammary fat pad,[28] as well as disease progression after resection of the primary tumor.[29] These results indicate that the inhibitory effects of feeding diets containing high levels of fish oil preparations on human breast cancer cell growth *in vivo*,[30,31] and on transplantable rodent mammary tumors,[32,33] are attributable to their EPA and DHA content.

The suppression of breast cancer progression and metastasis in the nude mouse model by EPA and DHA may result, at least in part, from altered eicosanoid biosynthesis by the tumor cells.[28] This could occur by effects at several points on the metabolic pathway of eicosanoid production from n-6 FAs, but a major event is most likely the incorporation of EPA and DHA into cell membrane phospholipids at the expense of AA.[28,29]

FATTY ACIDS, EICOSANOIDS, AND THE METASTATIC PHENOTYPE

Cyclooxygenases and Lipoxygenases

Free AA is the substrate for an enzyme variously referred to as prostaglandin synthase, prostaglandin-endoperoxide synthase, and cyclooyxgenase. It performs two sequential reactions: the cyclooxygenase activity of the enzyme catalyzes the conversion of AA to prostaglandin G2 (PGG2), and this in turn is converted to PGH2 by a peroxidase-mediated reaction (Fig. 2). The PGH2 is then converted to various PGs, thromboxanes, and prostacyclins. By common usage, prostaglandin synthase is regarded as equivalent to cyclooxygenase (COX), which exists as two isoenzymes. One, designated cyclooxygenase-1 (COX-1) is the constitutive form of the enzyme, and is involved in physiological "housekeeping" functions and homeostasis.[34,35] The other, COX-2, is inducible by mitogens, tetradecanoyl phorbol acetate (TPA), cytokines, and growth factors in various cell

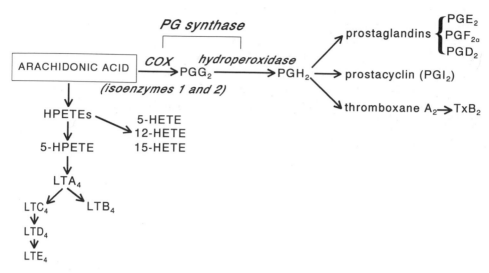

Figure 2. The biosynthesis of eicosanoids from arachidonic acid. PG: prostaglandin; COX: cyclooxygenase; HPETE: hydroperoxyeicosatetraenoic acid; HETE: hydroxyeicosatetraenoic acid.

types.[34-36] It is responsible for the production of PGs which occurs as part of the inflammatory response,[37] and has also been associated with skin,[38] and colon[39] carcinogenesis.

An alternative metabolic fate for AA is mediated by the family of lipoxygenases (LOXs) which yield the three major hydroxyeicosatetraenoic acids (HETEs), 5-, 12- and 15-HETE, respectively,[40,41] (Fig. 2). The 5-LOX catalyzes the conversion of AA to an intermediate product, 5-hydroperoxy-2-*trans*,4-*cis*-pentadiene (HPETE) which can be either reduced to 5-HETE by the action of peroxidases, or reduced by a dehydrase to form a labile epoxide, leukotriene (LT) A$_4$. This latter is the metabolic precursor of LTB$_4$, which may have a role in breast cancer cell proliferation.[12]

There are two isoforms of 12-LOX; that associated with platelets metabolizes AA exclusively to 12-HPETE and thence to 12-HETE, while the leukocyte-type 12-LOX metabolizes both AA and LA.[42] In general, it is the platelet-type 12-LOX which has been associated with tumor cells.[43-45]

Cell Growth

Several groups of investigators have employed pharmacological inhibitors of eicosanoid biosynthesis in attempts to identify the individual products of AA metabolism which are directly responsible for LA-stimulated breast cancer cell growth. Experiments with piroxicam, a selective inhibitor of COX, and esculetin, an inhibitor of 5-LOX as well as 12-LOX, led to the conclusion that enhanced MDA-MB-231 cell growth in the presence of LA was dependent on LT rather than PG production.[12] In their experiments with the 4526 metastatic mouse mammary tumor cell line, Buckman et al.[13] obtained a dose-dependent growth inhibition with the LOX inhibitors nordihydroguaiaretic acid, caffeic acid, and ethacrynic acid; also 5,6-dehydroarachidonate, a selective inhibitor of 5-LOX but not 12- or 15-LOX,[46] was particularly effective, again suggesting the involvement of an LT in breast cancer cell growth. Somewhat similar results were obtained by Earashi et al.,[47,48] although they also observed some suppression of MDA-MB-231 cell growth in the presence of high concentrations of piroxicam.

Further studies are needed to define more clearly which of the various eicosanoids are involved in breast cancer cell growth. Perhaps "rescue" experiments, in which the eicosanoid of interest is added to cells cultured in the presence of the corresponding inhibitor, would provide some clarification; as noted in the next section, this approach was effective in defining the role of 12-HETE in the LA-stimulated invasive process.

Invasion and Metastasis

The observation that a high-fat, high-LA, diet enhances the metastasis of estrogen-independent human breast cancer cell lines when these are growing as primary solid tumors in the mammary fat pads of female nude mice,[16–19] caused us to examine the effects of fatty acids on breast cancer cell invasion. We found that LA stimulates the invasion of MDA-MB-435 cells in an *in vitro* assay,[49] and that this is associated with induction of the 92 kDa isoform of type IV collagenase (matrix metalloproteinase-9; MMP-9), a key enzyme in the invasive process.[50] As might be expected, given their inhibitory effects on MDA-MB-435 cell metastasis *in vivo*,[28] the n-3 FAs diminish the invasive capacity of the cell line *in vitro*,[49] and suppress expression of MMP-9.[51]

By the use of selective inhibitors of eicosanoid biosynthesis, and the relevant pure eicosanoids, it was possible to establish the role of 12-HETE as the mediator of both LA-stimulated invasion and MMP-9 production by MDA-MB-435 cells.[52] In these experiments, esculetin, but not piroxicam, was shown to block the stimulatory effect of the n-6 FA; this action of the lipoxygenase inhibitor was reversed by 12-HETE, but not by 5-HETE or PGE$_2$. Enzyme activity analysis by zymography, and Northern blot analyses, showed that stimulation by LA and 12-HETE involve induction of MMP-9 at the level of both the enzyme and the corresponding mRNA.

While these experiments *in vitro* established a specific role for AA-derived 12-HETE in the stimulation of MDA-MB-435 cell invasion by LA, the situation *in vivo* has proven to be more complex. Dietary LA-stimulated metastasis in nude mice was reduced significantly by treatment with indomethacin, a response which was associated with low levels of several COX products, including PGE$_2$ and the stable metabolite of thromboxane A$_2$, but no decrease in tumor 12-HETE concentrations.[21] These results are consistent with those obtained by Hubbard and Erickson[53] in the 4526 mouse mammary tumor model, and with the involvement of PGE$_2$ and thromboxane A$_2$ in critical intravascular events in the metastatic cascade. Elevated PGE$_2$ production by primary breast cancers has been associated with a high risk of developing metastatic disease,[54] and was positively correlated with metastatic potential in a series of mouse mammary tumor cell lines.[55]

Although the proteolytic enzymes are usually thought of in the context of tumor invasion, they may also have a role in cell growth. It has been reported that procathepsin D, the precursor of an aspartyl proteolytic enzyme, can stimulate the growth of several human breast cancer cell lines *in vitro*.[56,57] Similarly, urokinase-type plasminogen activator (uPA) stimulates thymidine uptake and proliferation by human tumor cell lines.[58,59] It is interesting that the uPA molecule does include a growth factor domain which is homologous to EGF,[60] but its putative role as a mitogen may in fact be indirect and involve the activation of latent growth factor(s) by uPA generated plasmin.[61] We are currently studying the potential regulation of the cathepsins and the uPA-uPA receptor complex by LA-derived tumor cell eicosanoids, and in a preliminary study we have shown that LA stimulates uPA production by the MDA-MB-231 and MDA-MB-435 human breast cancer cell lines.[62]

ESTROGEN DEPENDENCE AND ARACHIDONIC ACID METABOLISM

In the Introduction to this chapter, we noted that aggressive breast cancer behavior is typically accompanied by the loss of estrogen dependence. Clarke et al.[6,63] manipulated the hormonal environment to obtain a series of variants of the MCF-7 cell line which exhibit estrogen-independent growth, and a somewhat enhanced metastatic potential. We are currently engaged in studies which take a different approach; these are based upon differences between the ER positive and the ER negative human breast cancer cell lines with respect to their mitogenic and invasive responses to n-6 FAs.

Western blot analysis showed that 12-LOX expression in MCF-7 cells is low compared with the metastatic MDA-MB-435 breast cancer cell line. When MCF-7 cells were transfected with a cDNA for human platelet 12-LOX, the resulting stable clones secreted large quantities of 12-HETE in response to AA. The cultured 12-LOX overexpressing cells grew more rapidly than the parental MCF-7 cells, they were estrogen-independent, and their growth was stimulated by LA.[64] Moreover, their capacity for invasion *in vitro* is considerably greater than that of MCF-7 cells (J.M. Connolly, unpublished data).

A preliminary comparison of COX-1 and COX-2 expression by MCF-7 and MDA-MB-231 cells and their induction by TPA has been completed (Liu and Rose, this volume, p. 221). This showed that the estrogen-independent, highly metastatic, MDA-MB-231 cell line expressed a constitutively high level of COX-2, which responded to TPA with a further, and sustained enhancement of enzyme expression. In contrast, the MCF-7 cells exhibited a relatively high level of COX-1 expression; COX-2 was barely detectable by Western blot analysis, and showed only a transient response to TPA. Again, these early results are consistent with the hypothesis that breast cancer cell endocrine status is related to the capacity for eicosanoid biosynthesis, and that enhanced synthesis of eicosanoid products of AA metabolism is a feature of the metastatic phenotype.

COMMENTARY

As research into the effects of dietary fatty acids on breast cancer cell biological behavior continues, support is accumulating for the concept that dietary intervention, most likely combined with one or more pharmacological agents, has a place in breast cancer prevention strategies, and, perhaps, as an adjuvant for the postsurgical breast cancer patient. The most promising dietary approach appears to be a reduction in total fat intake, perhaps to 25–30% of total calories and particularly targeting the n-6 FAs, together with an increase in the proportion of long-chain n-3 FAs so as to achieve an elevation in the n-3/n-6 FA ratio. Among the candidate drugs for future clinical evaluation are a specific inhibitor of COX-2, so avoiding the toxic side-effects of indomethacin, and a selective inhibitor of the lipoxygenases. Another approach might be to combine a similar dietary intervention with one of the protease inhibitors[65] or antiangiogenic drugs[66] which are currently undergoing preclinical evaluation as candidates for new approaches to chemotherapy, but which may also have a place as chemopreventive agents.

As the preliminary studies which relate tumor cell endocrine status, eicosanoid biosynthesis, and the expression of the invasive phenotype mature, it may emerge that there is a place for combination chemosuppressive or adjuvant therapy with an antiestrogen and an eicosanoid synthesis inhibitor.

ACKNOWLEDGMENTS

We thank Arlene Banow for assistance in preparing this manuscript. Our own work is supported by PHS Grant No. CA53124 from the National Cancer Institute, Grant No. CN-100 from the American Cancer Society, and Grant 94A12 from the American Institute for Cancer Research.

REFERENCES

1. Rose, D.P., J.M. Connolly, and X.-H. Liu, Dietary fatty acids and human breast cancer cell growth, invasion, and metastasis, in: Diet and Breast Cancer, E.K. Weisburger, ed., Plenum Press, New York (1994).
2. Clark, G.M., C.R. Wenger, S. Beardslee, M.A. Owens, G. Pounds, T. Oldaker, P. Vendely, M.R. Pandian, D. Harrington, and W.L. McGuire, How to integrate steroid hormone receptor, flow cytometric, and other prognostic information in regard to primary breast cancer, Cancer 71:2157–2162, (1993).
3. Thompson, E.W., R. Reich, T.B. Shima, A. Albini, J. Graf, G.R. Martin, R.B. Dickson, and M.E. Lippman, Differential regulation of growth and invasiveness of MCF-7 breast cancer cells by antiestrogens, Cancer Res. 48:6764–6768 (1988).
4. Ozzello, L. and M. Sordat, Behavior of tumors produced by transplantation of human mammary cell lines in athymic nude mice, Eur. J. Cancer 16:553–559 (1980).
5. Price, J.M., and R.D. Zang. Studies of human breast cancer metastasis using nude mice, Cancer Metastasis Rev. 8:285–297 (1989/1990).
6. Thompson, E.W., N. Brünner, J. Torri, M.D. Johnson, V. Boulay, A. Wright, M.E. Lippman, P.S. Steeg, and R. Clarke, The invasive and metastatic properties of hormone-independent but hormone-responsive variants of MCF-7 human breast cancer cells, Clin. Exp. Metastasis 11:15–26 (1993).
7. Sommers, C.L., S.W. Byers, E.W. Thompson, J.A. Torri, and E.P. Gelmann, Differentiation state and invasiveness of human breast cancer cell lines, Breast Cancer Res. Treat. 31:325–335 (1994).
8. Bandyopadhyay, G.K., W. Imagawa, D. Wallace, and S. Nandi, Linoleate metabolites enhance the in vitro proliferative response of mouse mammary epithelial cells to epidermal growth factor, J. Biol. Chem. 262:2750–2756 (1987).
9. Bandyopadhyay, G.K., W. Imagawa, D. Wallace, and S. Nandi, Proliferative effects of insulin and epidermal growth factor on mouse mammary epithelial cells in primary culture. Enhancement by hydroxyeicosatetraenoic acids and synergism with prostaglandin E_2, J. Biol. Chem. 263:7567–7573 (1988).
10. Grammatikos, S.I., P.V. Subbaiah, T.A. Victor, and W.M. Miller, n-3 and n-6 fatty acid processing and growth effects in neoplastic and non-cancerous human mammary epithelial cell lines, Br. J. Cancer 70:219–227 (1994).
11. Wicha, M., L.A. Liotta, and W.R. Kidwell, Effects of unsaturated fatty acids on the growth of normal and neoplastic rat mammary epithelial cells, Cancer Res. 39:426–435 (1978).
12. Rose, D.P., and J.M. Connolly, Effects of fatty acids and inhibitors of eicosanoid synthesis on the growth of a human breast cancer cell line in culture, Cancer Res. 50:7139–7144 (1990).
13. Buckman, D.K., N.E. Hubbard, and K.L. Erickson, Eicosanoids and linoleate-enhanced growth of mouse mammary tumor cells, Prostaglandins Leukot. Essent. Fatty Acids 44:117–184 (1991).
14. Rose, D.P., Individual dietary fatty acids and the hormone-dependent cancers: animal studies, Am. J. Clin. Nutr. in press (1997).
15. Bardon, S., M.T. Le, and J.-M. Allesandri, Metabolic conversion and growth effects of n-6 and n-3 polyunsaturated fatty acids in the T47D breast cancer cell line, Cancer Lett. 99:51–58 (1996).
16. Rose, D.P., J.M. Connolly, and C.L. Meschter, Effect of dietary fat on human breast cancer growth and lung metastasis in nude mice, J. Natl. Cancer Inst. 83:1491–1495 (1991).
17. Meschter, C.L., J.M. Connolly, and D.P. Rose, Influence of regional location of the inoculation site, and dietary fat on the pathology of MDA-MB-435 human breast cancer cell-derived tumors grown in nude mice. Clin. Exp. Metastasis 10:167–173 (1992).
18. Rose, D.P., M.A. Hatala, J.M. Connolly, and J. Rayburn, Effect of diets containing different levels of linoleic acid on human breast cancer growth and lung metastasis in nude mice, Cancer Res. 53:4686–4690 (1993).
19. Rose, D.P., J.M. Connolly, and X.-H. Liu, Effects of linoleic acid on the growth and metastasis of two human breast cancer cell lines in nude mice, and the invasive capacity of these cell lines in vitro, Cancer Res. 54:6557–6562 (1994).

20. Gabor, H., L.A. Hillyard, and S. Abraham, Effect of dietary fat on growth kinetics of transplantable mammary adenocarcinoma in BALB/c mice, J. Natl. Cancer Inst. 74:1299–1305 (1985).

21. Connolly, J.M., X.-H. Liu, and D.P. Rose, Dietary linoleic acid-stimulated human breast cancer cell growth and metastasis in nude mice and their suppression by indomethacin, a cyclooxygenase inhibitor, Nutr. Cancer 25:231–240 (1996).

22. Kaizer, L., N.F. Boyd, V. Kriukov, and D. Trichler, Fish consumption and breast cancer risk: an ecological study, Nutr. Cancer 12:61–68 (1989).

23. Kromann, N., and A. Green, Epidemiological studies in the Upernavik District, Greenland: incidence of some chronic diseases 1950–1974, Acta Med. Scand. 208:401–406 (1980).

24. Parkinson, A.J., A.L. Cruz, W.L. Heyward, L.R. Bulkow, D. Hall, L. Barstaed, and W.E. Conner, Elevated concentrations of plasma ω-3 polyunsaturated fatty acids among Alaskan Eskimos, Am. J. Clin. Nutr. 59:384–388 (1994).

25. Lanier, A.P., L.R. Bulkow, and B. Ireland, Cancer in Alaskan Indians, Eskimos and Aleuts, 1969–83: Implications for etiology and control, Public Health Rep. 104:658–665 (1989).

26. Caygill, C.P.J., A. Charlett, and M.J. Hill, Fat, fish, fish oil and cancer, Br. J. Cancer 74:159–164 (1996).

27. Kamano, K., H. Okuyama, R. Konishi, and H. Nagasawa, Effect of a high-linoleate and a high Á-linolenate diet on spontaneous mammary tumorigenesis in mice, Anticancer Res. 9:1903–1908 (1989).

28. Rose, D.P., J.M. Connolly, J. Rayburn, and M. Coleman, Influence of diets containing different levels of eicosapentaenoic or docosahexaenoic acid on the growth and metastasis of human breast cancer cells in nude mice, J. Natl. Cancer Inst. 87:587–592 (1995).

29. Rose, D.P., J.M. Connolly, and M. Coleman, Effect of omega-3 fatty acids on the progression of metastases after the surgical excision of human breast cancer cell solid tumors growing in nude mice, Clin. Cancer Res. in press.

30. Welsch, C.W., C.S. Oakley, C.-C. Chang, and M.A. Welsch, Suppression of growth by dietary fish oil of human breast carcinomas maintained in three different strains of immune-deficient mice, Nutr. Cancer 20:119–127 (1993).

31. Rose, D.P. and J.M. Connolly, Effects of dietary omega-3 fatty acids on human breast cancer growth and metastasis in nude mice, J. Natl. Cancer Inst. 85:1743–1747 (1993).

32. Karmali, R.A., J. Marsh, and C. Fuchs, Effect of omega-3 fatty acids on growth of a rat mammary tumor, J. Natl. Cancer Inst. 73:457–461 (1984).

33. Gabor, H., and S. Abraham, Effect of dietary menhaden oil on tumor cell loss and accumulation of mass of a transplantable adenocarcinoma in BALB/c mice, J. Natl. Cancer Inst. 76:1223–1229 (1986).

34. DeWitt, D.L., Prostaglandin endoperoxide synthase: Regulation of enzyme expression, Biochim. Biophys. Acta, 1083:121–134 (1991).

35. Loll, P.J., and R.M. Garavito, The isoforms of cyclooxygenase: structure and function, Expert Opin. Invest. Drugs 3:1171–1180 (1994).

36. Herschman, H.R., Regulation of prostaglandin synthase-1 and prostaglandin synthase-2, Cancer Metastasis Rev. 13:241–256 (1994).

37. Masferrer, J.L., B.S. Zweifel, P.T. Manning, S.D. Hauser, K.M. Leahy, W.G. Smith, P.C. Isakson, and K. Seibert, Selective inhibition of inducible cyclooxygenase 2 *in vivo* is antiinflammatory and nonulcerogenic, Proc. Natl. Acad. Sci. USA 91:3228–3232 (1994).

38. Müller-Decker, K., K. Scholz, F. Marks, and G. Fürstenberger, Differential expression of prostaglandin H synthase isozymes during multistage carcinogenesis in mouse epidermis, Mol. Carcinogenesis, 12:31–41 (1995).

39. Kargman, S.L., G.P. O'Neill, P.J. Vickers, J.F. Evans, J.A. Mancini, and S. Jothy, Expression of prostaglandin G/H synthase-1 and -2 protein in human colon cancer, Cancer Res. 55:2556–2559 (1995).

40. Spector, A.A., J.A. Gordon, and S.A. Moore, Hydroxyeicosatetraenoic acids (HETEs), Prog. Lipid Res. 27:271–323 (1988).

41. Yamamoto, S., Mammalian lipoxygenases: molecular and catalytic properties, Prostaglandins Leukot. Essent. Fatty Acids 35:219–229 (1989)

42. Chen, X.-S., U. Kurre, N.A. Jenkins, N.G. Copeland, and C.D. Funk, cDNA cloning, expression, mutagenesis of C-terminal isoleucine, genomic structure, and chromosomal localizations of murine 12-lipoxygenases, J. Biol. Chem. 269:13979–13987 (1994).

43. Chang, W.-C., Y.-W. Liu, C.-C. Ning, H. Suzuki, T. Yoshimoto, and S. Yamamoto, Induction of arachidonate 12-lipoxygenase mRNA by epidermal growth factor in A431 cells, J. Biol. Chem. 268:18734–18739 (1993).

44. Honn, K.V., D.G. Tang, X. Gao, I.A. Butovich, B. Liu, J. Timar, and W. Hagmann, 12-Lipoxygenases and 12(*S*)-HETE in cancer metastasis, Cancer Metastasis Rev. 13:365–396 (1994).

45. Krieg, P., A. Kinzig, M. Ress-Löschke, S. Vogel, B. Vanlandingham, M. Stephan, W.-D. Lehmann, F. Marks, and G. Fürstenberger, 12-Lipoxygenase isoenzymes in mouse skin tumor development, Mol. Carcinogenesis, 14:118–129 (1995).

46. Corey, E.J., and J.E. Munroe, Irreversible inhibition of prostaglandin and leukotriene biosynthesis from arachidonic acid by 11,12-dehydro- and 5,6-dehydroarachidonic acids, respectively, J. Am. Chem. Soc. 104:1752–1754 (1982).

47. Earashi, M., M. Noguchi, K. Kinoshita, I. Miyazaki, M. Tanaka, and T. Sasaki, Effects of eicosanoid synthesis inhibitors on the *in vitro* growth and prostaglandin E and leukotriene B secretion of a human breast cancer cell line, Oncology 52:150–155, 1995.

48. Noguchi, M., D.P. Rose, M. Earashi, and I. Miyazaki, The role of fatty acids and eicosanoid synthesis inhibitors in breast carcinoma, Oncology, 52:265–271 (1995).

49. Connolly, J.M., and D.P. Rose, Effects of fatty acids on invasion through reconstituted basement membrane ("Matrigel") by a human breast cancer cell line, Cancer Lett. 75:137–142 (1993).

50. Liu, X.-H., and D.P. Rose, Stimulation of type IV collagenase expression by linoleic acid in a metastatic human breast cancer cell line. Cancer Lett. 76:71–77 (1994).

51. Liu, X.-H., and D.P. Rose, Suppression of type IV collagenase in MDA-MB-435 human breast cancer cells by eicosapentaenoic acid *in vitro* and *in vivo*, Cancer Lett. 92:21–26 (1995).

52. Liu, X.-H., J.M. Connolly, and D.P. Rose, Eicosanoids as mediators of linoleic acid-stimulated invasion and type IV collagenase production by a metastatic human breast cancer cell line, Clin. Exp. Metastasis 14:145–152 (1996).

53. Hubbard, N.E., and K.L. Erickson, Enhancement of metastasis from a transplantable mouse mammary tumor by dietary linoleic acid, Cancer Res. 47:6171–6175 (1987).

54. Rolland, P.H., P.M. Martin, J. Jacquemier, A.M. Rolland, and M. Toga, Prostaglandin in human breast cancer: evidence suggesting that an elevated prostaglandin production is a marker of high metastatic potential for neoplastic cells, J. Natl. Cancer Inst. 64:1061–1070 (1980).

55. Fulton, A.M., and G.H. Heppner, Relationships of prostaglandin E and natural killer sensitivity to metastatic potential in murine mammary adenocarcinomas, Cancer Res. 45:4779–4784 (1985).

56. Vignon, F., F. Capony, M. Chambon, G. Freiss, M. Garcia, and H. Rochefort, Autocrine growth stimulation of the MCF-7 breast cancer cells by the estrogen-regulated 52K protein, Endocrinology 118:1537–1545 (1986).

57. Vetvick, V., J. Vektvicková, and M. Fusek, Effect of procathepsin D on proliferation of human cell lines, Cancer Lett. 79:131–135 (1994).

58. Kirchheimer, J.C., J. Wojita, G. Christ, and B.R. Binder, Proliferation of a human epidermal tumor cell line stimulated by urokinase, FASEB J. 1:125–128 (1987)

59. Kirchheimer, J.C., J. Wojita, G. Christ, and B.R. Binder, Functional inhibition of endogenously produced urokinase decreases cell proliferation in a human melanoma cell line, Proc. Natl. Acad. Sci. USA 86:5424–5428 (1989)

60. Patthy, L, Evolution of the proteases of blood coagulation and fibrinolysis by assembly of molecules, Cell 41:657–663 (1985).

61. Suto, Y., and D.B. Rifkin, Inhibition of endothelial cell movement by pericytes and smooth muscle cells. Activation of a latent transforming growth factor-beta-1-like molecule by plasmin during co-culture, J. Cell Biol. 109:309–315 (1989).

62. Long, B.J., and D.P. Rose, Modulation of plasminogen activator (PA) activity by linoleic acid in two metastatic human breast cancer cell lines, Proc. Am. Assoc. Cancer Res. 36:75 (1995).

63. Clarke, R., E.W. Thompson, F. Leonessa, J. Lippman, M. McGarvey, T.L. Frandsen, and N. Brünner, Hormone resistance, invasiveness, and metastatic potential in breast cancer, Breast Cancer Res. Treat. 24:227–239 (1993).

64. Liu, X.-H., J.M. Connolly, and D.P. Rose. The 12-lipoxygenase gene-transfected MCF-7 human breast cancer cell line exhibits estrogen-independent, but estrogen and omega-6 fatty acid-stimulated proliferation *in vitro*, and enhanced growth in athymic nude mice, Cancer Lett. - in press.

65. Low, J.A., M.D. Johnson, E.A. Bone, and R.B. Dickson, The matrix metalloproteinase inhibitor batimastat (BB-94) retards human breast cancer solid tumor growth but not ascites formation in nude mice, Clin. Cancer Res. 2:1207–1214 (1996).

66. Baillie, C.T., M.C. Winslet, and N.J. Bradley, Tumour vasculature—a potential therapeutic target, Br. J. Cancer 72:257–267 (1995).

FATTY ACIDS AND BREAST CANCER CELL PROLIFERATION

Robert W. Hardy,[1] Nalinie S. M. D. Wickramasinghe,[1] S. C. Ke,[2] and Alan Wells[1]

[1]Department of Pathology
[2]Department of Physics
University of Alabama at Birmingham
Birmingham, Alabama

INTRODUCTION

Dietary Fat and the Development of Breast Cancer

Experimental Animal and Cell Models. Experimental evidence using animal models and cultured cells has clearly linked dietary fat to breast cancer. More than 150 studies using animal models of carcinogen-induced, transplantable, "spontaneous" and metastatic breast cancer as well as *in vitro* investigations employing cultured breast cancer cells provide an impressive array of support for the conclusion that dietary fat influences breast cancer cell proliferation and tumorigenesis (1–11).

Dietary Fat Intake and Human Breast Cancer Risk. Prentice and Sheppard studied data from international comparisons, time-trends analysis, migrant study results and case control studies and concluded that a 50% reduction of dietary fat from current U.S. consumption levels would result in a 2.5-fold reduced risk of breast cancer among post-menopausal women (12). Another summary of ecological studies has found a similar strong correlation between dietary fat and breast cancer (13). However this correlation has not been found consistently in clinical trials. More than 20 large, case control and cohort studies have been performed in different countries to determine the correlation of dietary fat intake and breast cancer risk. These studies have conflicting results and attempts have been made to quantitatively summarize the combined data. Two of these composite studies have concluded that there is an association (14,15) while the other two found no association been dietary fat intake and breast cancer risk (16,17).

The failure of clinical studies to consistently uncover a significant association between dietary fat intake and breast cancer risk may be due to a number of factors:

Dietary Fat and Cancer, edited by AICR
Plenum Press, New York, 1997

- In dietary studies fat is taken as a mix of several different fatty acids, not individual fatty acids. Not all unsaturated or saturated fatty acids can be grouped into a single set of actions on tumorigenesis and tumor cell proliferation (18,19).
- Fatty acids are produced endogenously by breast cancer and other types of cells and almost certainly there is inter- and intra-individual variation in both the amount and types produced. The potential importance of this factor in the development and treatment of breast cancer has recently been recognized by Kuhajda *et al.* (20)
- In animal models, fat promotes mammary carcinogenesis under very stringent sets of conditions, which might not be duplicated in the arena of fat intake and human breast cancer risk. (21).
- A high-fat diet or increased fatty acids together with differences in hormonal milieu during childhood and/or adolescence may be important factors (critical developmental period and putative window for cancer initiation), which have not been taken into account in these studies (22).
- In dietary studies, fat is metabolized to varying degrees on an individual basis confounding the effect of the parent fat.
- It has been suggested that the modest dietary reductions in per cent fat intake may not be sufficient to see a change in tumorigenesis (23).
- In large clinical studies there is difficulty in collecting accurate dietary information (24).
- Cross-cultural and migrational studies have provided evidence for a relationship between dietary fat intake and breast cancer. However, this effect has a lengthy latency period. This is in line with the apparent long latency of breast cancer and could be a major flaw in case-control studies which do not determine long term changes from dietary fat (22).

Clearly the answer to why human dietary studies have not been as consistent as animal studies in relating breast cancer to fat consumption is a complex one. We have proceeded with the premise that determining the effects of individual fatty acids on cell proliferation will shed light on the mechanisms by which fats may modulate tumor cell behavior and thereby help in the design of future clinical trials.

Breast Cancer Cell Proliferation

Why Is Breast Cancer Cell Proliferation Important? An increased cell proliferation rate in breast tumors correlates with a shorter relapse-free survival time and worse overall patient survival (25). Five other studies have reported similar findings on the prognostic relevance of the proliferative rate of breast tumors in subjects with early or advanced disease subjected to only local-regional treatment and followed for 3.5–14 years (25–29). Furthermore, in all five studies the probability of developing local-regional or distant metastasis in patients with rapidly proliferating tumors was about twice that of patients with slowly proliferating tumors. Additionally, the 14 year relapse-free curves provide evidence for proliferative activity as an indicator of metastatic potential; not only of tumor growth rate (26). The proliferative rate retains its predictive value even in the context of other prognostic information provided by tumor size, histological grade or hormone receptors (26–30). Therefore, breast cancer cell proliferation rate is an important factor in determining relapse free survival, overall survival and the risk of developing metastasis and thus may play a causal role in breast cancer progression.

How Does Dietary Fat Affect Breast Cancer Cell Proliferation? Many studies have demonstrated that the type of dietary fat affects the cell proliferation (or promotion) stage of mammary cancer. Generally, high levels of long chain saturated fatty acids (LCSFA) suppress the development of mammary tumors compared to long chain unsaturated fatty acids in the carcinogen-induced rodent mammary tumor model (11,31–41). Supplementing saturated fatty acids with a small amount of long chain unsaturated fatty acids reverses the mammary tumor cell proliferation of such fats (36,37) but this reversal is dependent on the amount of unsaturated fat added (31). It has been suggested that this inhibition is simply due to inadequate linoleic acid content in the high LCSFA diet. However, when rats were fed high levels LCSFA they were typically fed palm oil (11% linoleic acid) or lard (9% linoleic acid) which exceed the normal recommended essential fatty acid requirement for normal growth (42). This inhibitory effect has also been demonstrated in spontaneously derived (19, 43, 44–46) and transplantable (47–56) mammary tumors in mice and rats as well as breast cancer cell culture studies (57–60). Tinsley *et al.* (18) have shown that using a spontaneous mammary tumor model in the C3H mouse a statistically significant increase in time of tumor development was associated with the ingestion of stearic acid. This increase was not observed for laurate, myristate or palmitate, while linoleate decreased the time of tumor development, suggesting a specific inhibitory effect of stearate on breast cancer cell proliferation. A similar stearate induced increase in time of tumor development was also observed by Bennet (19). The effect of dietary long chain unsaturated fat on breast cancer cell proliferation in animal models is inconsistent for oleic acid (18,32,40,44,61–66) and linoleic acid (18, 67–71). In fact, linoleic acid stimulates cell proliferation in cell culture models (57–60). Mice or rats fed relatively high dietary levels of ω-3 fatty acids (fish oil) consistently show inhibition of mammary cell proliferation in the carcinogen-induced rat mammary tumor model (31,68,72,73).

In summary, it is clear that dietary fat can alter the development of breast cancer in experimental models. It is also clear that LCSFA inhibit breast cancer cell proliferation, both in animal models and cell culture models.

What Is the Role of the Epidermal Growth Factor Receptor (EGFR) in Breast Cancer Cell Proliferation? Growth factors and hormones have been linked to the development of breast cancer (74,75). EGFR is overexpressed in 20–32% of breast cancers, and its presence is associated with a more aggressive clinical course suggesting that it has an important growth regulatory function in breast cancer (74, 76–79). Ligands to the EGFR such as EGF and TGFα initiate a cascade of intracellular events through specific interactions with cell surface receptors, which ultimately leads to DNA synthesis and cell proliferation. EGF activates its receptor by inducing receptor autophosphorylation of tyrosine residues, evoking a catalytically active receptor capable of phosphorylating endogenous substrates. The tyrosine-autophosphorylated EGFR has been shown to associate with and activate phosphatidyl inositol specific phospholipase C-γ (PLC-γ) (89), GTPase activating proteins (GAPS), Grb 2, Shc, and PI-3 kinase among others.

Exposure to hormonal estrogenic steroids appears to be critical in the tumorigenesis (or initiation) phase of human breast cancer, and in mitogenesis in the estrogen-responsive form of the disease. Estrogen stimulates the production of TGF-α in the T47D estrogen receptor positive breast cancer cell line. Although EGFR-positive breast cancer cells are usually estrogen receptor negative, it would appear that there is a subset of breast cancer cells (ER and EGFR positive, e.g. T47D) that demonstrate cooperation between the EGFR and ER stimulated cell proliferation.

Do G-Proteins Have a Role in EGF Induced Cell Proliferation? Experimental evidence suggests that at least one other type of transducer molecules, guanine nucleotide-binding proteins (G-proteins), participate in the coupling of receptor binding to activation of effector molecules involved in cell proliferation. Our recent studies (80) indicate the involvement of a pertussis toxin-sensitive G-protein in EGFR-mediated breast cancer cell proliferation. Church and Buick reported that G-proteins mediate EGF signal transduction in MDA-468 breast cancer cells (81). A pertussis toxin-sensitive G-protein, G_i, appears to be involved in coupling the EGFR to PLC-γ in rat hepatocytes (82,83), neutrophils and macrophages (84). Direct involvement of $G_{i\alpha}$ in cell division by translocation from the plasma membrane to the nucleus has been reported by Crouch in Balb/c 3T3 cells (85).

EFFECTS OF FATTY ACIDS ON EGF INDUCED BREAST CANCER CELL PROLIFERATION

How Do Individual Long Chain Fatty Acids Alter EGF Induced Breast Cancer Cell Proliferation?

We investigated the effects of long chain fatty acids on both basal and EGF induced breast cancer cell proliferation in a breast cancer cell culture system that allows the specific effects of individual fatty acids to be elucidated. The important features of this system are that all fatty acids tested were presented to the cells in association with albumin which is known to be an important physiological vehicle *in vivo* for the delivery of fatty acids to cells (86). Fatty acid-free albumin is used in these studies. This albumin has the specific fatty acid of interest added to it using the inert carrier celite (87).

We found that fatty acids affect breast cancer cell proliferation in a manner that depends on acyl chain length and whether the fatty acid is saturated. In Hs578T cells a comparison of C:14, C:16 and C:18 long chain saturated fatty acids demonstrated a progressive inhibition of cell proliferation with increasing chain length (80). Fig. 1A demonstrates that increasing the LCSFA chain length past C:18 has no further effect on cell proliferation. We confirmed that stearate inhibits breast cancer cell proliferation at physiological concentrations (15 μM), calculated from a free fatty acid concentration of 0.4 mM and that stearate is approximately 10% of the total or at least 40 μM. In contrast to long chain saturated fatty acids, long chain unsaturated fatty acids such as oleate increased both basal and EGF induced cell proliferation (Fig. 1B). Increasing the degree of unsaturation to two double bonds (linoleate) gave similar results to oleate (one double bond) (Fig. 1B). Once again these concentrations (10 μM) are well within the physiological range for these fatty acids. These results are generally consistent with the results of the animal model studies demonstrating the inhibitory effect of LCSFA on breast cancer cell proliferation.

Is Stearate Inhibition of EGF Induced Cell Proliferation a Cytotoxic or Membrane Fluidity Effect?

To determine if the observed LCSFA inhibition of breast cancer cell proliferation was a general cytotoxic or membrane fluidity effect we performed the following tests. Initially we examined the cells after stearate incubation and found no apparent differences by visual inspection. Cells were then tested for viability using trypan blue staining and

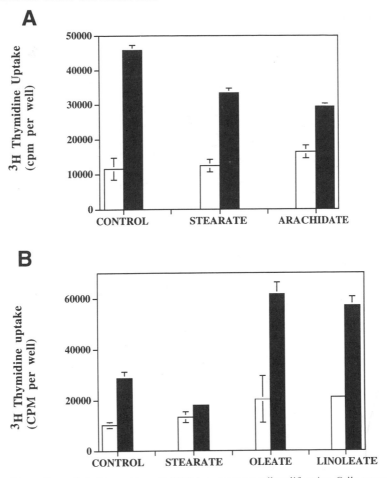

Figure 1. The effects of long chain fatty acids on Hs578T breast cancer cell proliferation. Cells were treated for 6 hours with either 15 μM stearate (C:18:0) or arachidate (C:20:0) (panel **A**) or 50 μM oleate (C:18:1) or linoleate (C:18:2) or stearate (panel **B**) or without (control). Cells were then treated with 0.3 nM EGF overnight (solid columns) or not (open columns) and ³H thymidine incorporation measured. As seen in panel **A** increasing the saturated fatty acid chain length past C:18 does not increase the inhibition of EGF induced cell proliferation. Panel **B** demonstrates that the addition of one double bond to C:18-stearate, in contrast to stearate, increases both basal and EGF-induced cell proliferation. The addition of a second double bond (linoleate) does not further increase cell proliferation.

once again no differences were found (80). If cells were being adversely affected by stearate treatment the basal levels of cell proliferation should be affected. We found that the basal proliferation rates were not altered by stearate (Fig. 1 and 80). If cell membranes were being seriously perturbed by fatty acid treatment we might expect differences in the ability of EGF to induce EGFR tyrosine phosphorylation since it has been reported to be dependent on receptor oligomerization. However we found no difference in EGFR tyrosine phosphorylation post stearate or oleate treatment (80, Fig. 2C). To test whether this is a non-specific effect, cells were preincubated with stearate under the same conditions that almost completely inhibited EGF induced cell proliferation and were treated with IGF1. Stearate pretreatment did not significantly inhibit IGF1 induced cell proliferation (80). Fi-

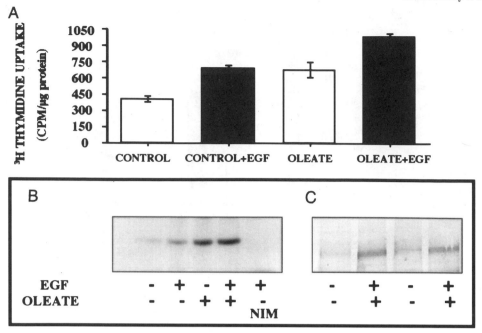

Figure 2. Oleic acid increases basal and EGF induced cell proliferation and increases the association of Gi protein with the EGFR without altering EGFR tyrosine phosphorylation. In panel **A**, Hs578T cells were treated for 6 hours with 50 μM oleic acid (oleate) or without (control). Cells were then treated with 0.3 nM EGF overnight as indicated and ^{3}H thymidine incorporation measured. In contrast to stearate, oleate increased basal and additively increased EGF induced cell proliferation. In panel **B** Hs578T cells were treated with 1 nM EGF for 1 minute, or 50 μM oleate as indicated, EGFR immunoprecipitated (the NIM lane is the nonimmune mouse IgG2a control immunoprecipitation) and ADP ribosylation was performed on the immunoprecipitates. Shown is the autoradiogram of the 41 kDa G-protein. In contrast to stearate oleate increased the amount of G-protein that associates with the EGFR in both the basal and EGF treated states. In panel **C** Hs578T cells were treated with or without 1 nM EGF for 1 minute or 50 μM oleate as indicated, lysed, run on SDS-PAGE and immunoblotted for phosphotyrosine residues. As shown in this panel the EGFR tyrosine phosphorylation is not altered by oleate pretreatment. These data are consistent with an EGFR associated G-protein signaling pathway being involved in EGF induced breast cancer cell proliferation.

nally to determine if membrane fluidity was being altered by stearate treatment we once again preincubated cells with stearate under the same conditions that almost completely inhibited EGF induced cell proliferation and determined cell membrane fluidity using electron spin resonance spectroscopy (ESR). As shown in Fig. 3 stearate pretreatment did not alter membrane fluidity.

What EGFR Induced Signaling Pathway Is Affected?

Fatty acids are known to modulate tyrosine kinase activities (88). However when cells were pretreated with stearate under the same conditions as were shown to inhibit cell proliferation there was no difference in the tyrosine phosphorylation of the EGFR (80). To test the involvement of G-proteins as potential second messengers that associate with the EGFR, we immunoprecipitated the EGFR from cells and ADP-ribosylated the immunoprecipitate in vitro using ^{32}P-NAD (83). This is a very sensitive method for de-

A

B

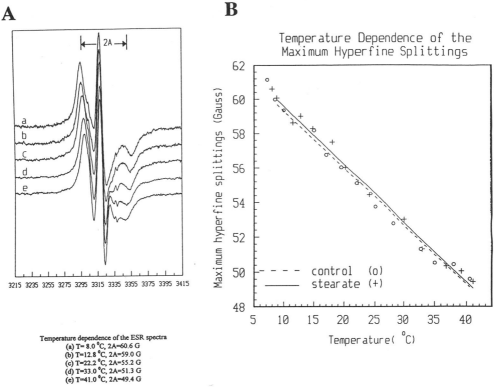

Temperature dependence of the ESR spectra
(a) T= 8.0 °C, 2A=60.6 G
(b) T=12.8 °C, 2A=59.0 G
(c) T=22.2 °C, 2A=55.2 G
(d) T=33.0 °C, 2A=51.3 G
(e) T=41.0 °C, 2A=49.4 G

Figure 3. Stearate treatment does not alter Hs578T cell membrane fluidity under conditions that it inhibits EGF induced cell proliferation. Hs578T cells were treated with 0.1 mM stearate or fatty acid free media (control) for 6 hours. Cells were harvested, treated with the spin label, 5-doxyl-stearic acid and subjected to electron spin resonance (ESR) measurements at various temperatures according to the method of Calder *et al.* (Biochem J 300:509–518, 1994). An example of the resulting spectra is shown in panel **A**. This example has only selected temperatures as space restrictions prevent plotting all temperatures. The arrows and 2A indicate the maximum hyperfine splitting or difference in magnetic field strengths, expressed on the x-axis as Gauss (G) for the two lines being measured. The hyperfine splittings are due to the nitrogen species in the spin label and 2A indicates the maximum hyperfine splitting. As the temperature increases the maximum hyperfine splitting increases. Thus when a series of these changes in maximum hyperfine splitting are collected and plotted vs temperature the resulting line is a measure of membrane fluidity. This was done for control and stearate treated cells, as shown in panel **B**. There is no significant difference in the lines, and thus no difference in the membrane fluidity of the stearate treated compared to control cells.

tecting G_i and G_o proteins. We were able to demonstrate that a 41 kDa G-protein coimmunoprecipitates with the EGFR and does so in an EGF-dependent manner. When this transient association is inhibited by either pertussis toxin or stearate, EGF-induced cell proliferation is also inhibited. The inhibition of cell proliferation by stearate paralleled the decrease in association of G-protein with the EGFR (80). In contrast, oleate, which increased cell proliferation, (Fig. 1B, Fig. 2A) also increased the association of G-protein with the EGFR and had no effect on EGFR tyrosine phosphorylation (Fig. 2B). Since oleate increased both cell proliferation and G-protein association with the EGFR independently of EGF this may indicate that oleate activates the target EGFR-initiated signaling pathway independently.

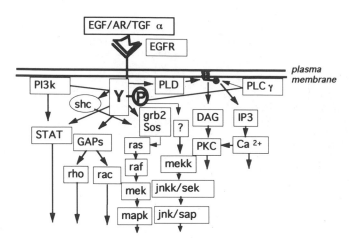

Figure 4. Epidermal growth factor receptor induced signaling cascade. Epidermal growth factor or transforming growth factor alpha (TGFα) or amphiregulin (AR) initiate complex intracellular signaling pathways via the epidermal growth factor receptor (EGFR). These pathways depend on the tyrosine phosphorylation and active kinase activity of the EGFR.

FUTURE STUDIES

Which Signaling Pathway Is Affected by Fatty Acids Altering EGFR/G-Protein Interaction?

As shown below the known signaling cascade initiated by the EGFR is complex and dependent on EGFR tyrosine phosphorylation (Fig. 4). Yang *et al.* (83) have provided evidence that phospholipase Cγ is activated by EGFR/G-protein signaling. The MAP kinase signaling pathway has also been implicated in EGFR and other cell proliferation signaling pathways. It will be important to investigate these signaling pathways. Characterization of the germane EGFR signaling pathways affected by fatty acids would be greatly facilitated by identifying the site of interaction of the G-protein and EGFR. We have begun to use mouse NR6 cells, which lack endogenous EGFR, to express EGFR constructs with specific domains removed or altered. These will be used to identify the site of association between the G-protein and EGFR which in turn will enable us to identify candidate effector molecules and signaling pathways.

Do Fatty Acids Affect Other EGFR Related Signaling Mechanisms?

EGFR (also termed c-erbB-1) shares a very high homology with, but is distinct from, c-erbB-2 (also indicated as HER-2 and *neu*) proto-oncogene. Elevated levels of c-erbB-2 are associated with tumor recurrence and shortened survival in patients with node-negative breast cancer. It is possible that fatty acids affect other members of the EGFR supergene family.

EGFR is a large transmembrane tyrosine kinase that is the receptor for the peptide hormones EGF and transforming growth factor-alpha (TGF-α). Both of these hormones can act as potent mitogens in breast tissue (75). Virtually all breast cancer cells derived from epithelial tumors produce TGF-α. It is believed to be an autocrine growth factor in breast cancer involved in the initiation or maintenance of transformation. Since acylation of TGFα is known to occur and be important in regulating its processing it will also be important to determine if fatty acids modulate EGFR cell proliferation via altered acylation of TGFα. All of these possibilities warrant in depth investigation.

Are Select Fatty Acids Safe Dietary Supplements?

Both *in vitro* (80) and *in vivo* (18,19) studies support the view that stearate specifically inhibits breast cancer cell proliferation. It is possible that stearate should be viewed in a different light than other LCSFA with respect to the possibility of its being useful as a chemopreventative agent. Most people would not consider a LCSFA as a potential chemopreventative agent because they are generally, thought of as atherogenic. This may not be justified. For example, stearate has been reported not to raise plasma total or LDL cholesterol levels while having no effect on HDL cholesterol or triglyceride levels (89–94). Furthermore, stearate may not adversely affect hemostasis (95,96). Thus, it may be possible that specific saturated fatty acids, such as stearate, could merit dietary recommendation based upon experimentally and clinically determined beneficial anticancer properties and a lack of atherogenic properties.

SUMMARY

We and others have shown that fatty acids are important regulators of breast cancer cell proliferation. In particular individual fatty acids specifically alter EGF-induced cell proliferation in very different ways. This regulation is mediated by an EGFR/G-protein signaling pathway. Understanding the molecular mechanisms of how this signaling pathway functions and how fatty acids regulate it will provide important information on the cellular and molecular basis for the association of dietary fat and cancer. Furthermore these *in vitro* studies may explain data previously obtained from *in vivo* animal studies and identify "good" as well as "bad" fatty acids with respect to the development of cancer.

ACKNOWLEDGMENTS

This research was supported by grant 95B051 from the American Institute for Cancer Research (RWH), grant DK47878 from the National Institutes of Health (RWH), a VA Merit Award (AW) and grant GM54739 from the National Institutes of Health (AW).

REFERENCES

1. Welsch CW. Relationship between dietary fat and experimental mammary tumorigenesis: A review and critique. *Cancer Res* 52(suppl):2040s-2048s, 1992.
2. Ip C, Chin SF, Scimeca JA, Pariza MW. Mammary cancer prevention by conjugated dienoic derivative of linoleic acid. *Cancer Res* 51:6118–6124, 1991.
3. Cohen LA, Kendall ME, Zang E, Meschter C, Rose DP. Modulation of N-nitrosomethylurea-induced mammary tumor promotion by dietary fiber and fat. *J Natl Cancer Inst* 83:496–501, 1991.
4. Rose DP, Connolly JM, Meschter CL. The effect of dietary fat on human breast cancer growth and lung metastasis in nude mice. *J Natl Cancer Inst* 83:1491–1495, 1991.
5. Rose DP, Connolly JM. The influence of dietary fat intake on local recurrence and progression of metastasis arising from MDA-MB-435 human breast cancer cells in nude mice after excision of the primary tumor. *Nutr Cancer* 18:113–122, 1992.

6.   Zevenbergen JL, Verschuren PM, Faalberg J, van Stratum P, Vles RO. The effect of the amount of dietary fat on the development of mammary tumors in BALB/c-MTV mice. *Nutr Cancer* 17:9–18, 1992.

7.   Mizukami Y, Nonomura A, Noguchi M, Taniya T, Thomas M, Nakamura S, Miyazaki I. The effects of high and low dietary fat and indometracin on tumor growth, hormone receptor status and growth factor expression in DMBA-induced rat breast cancer. *Int J Tissue React* 14:269–276, 1992.

8.   Meschter CL, Connolly JM, Rose DP. Influence of regional location of the inoculation site and dietary fat on the pathology of MDA-MB-435 human breast cancer cell-derived tumors grown in nude mice. *Clin Exp Metastasis* 10:167–173, 1992.

9.   Shultz TD, Chew BP, Seamon WR. Differential stimulatory and inhibitory responses of human MCF-7 breast cancer cells to linoleic acid and conjugated linoleic acid in culture. *Anticancer Res* 12:2143–2145, 1992.

10. DeWille JW, Waddell K, Steinmeyer C, Farmer JJ. Dietary fat promotes mammary tumorigenesis in MMTV/V-Ha-ras transgenic mice. *Cancer Lett* 69:59–66, 1993.

11. Mehta R, Harris S, Gunnett C, Runce O, Hartle D. The effects of patterned-calorie-restricted diets on mammary tumor incidence and plasma endothelin levels in DMBA-treated rats. *Carcinogenesis* 14:1693–1696, 1993.

12. Prentice RL, Sheppard L. Dietary fat and cancer: Consistency of the epidemiologic data, and disease prevention that may follow from a practical reduction in fat consumption. *Cancer Causes Control* 1:81–97, 1990.

13. Carroll KK. Dietary fat and breast cancer. *Lipids* 27:793–797, 1992.

14. Boyd NF, Martin LJ, Noffel M, Lockwood GA, Trichler DL. A meta-analysis of studies of dietary fat and breast cancer risk. *Br J Cancer* 68:627–636, 1993.

15. Howe GR, Hirohata T, Hislop TG, Iscovich JM, Yaun JM, Katsouyanni K, Lubin F, Marubini E, Modan B, Rohan T, Toniolo P, Shunzhang Y. Dietary factors and risk of breast cancer: Combined analysis of 12 case-control studies. *J Natl Cancer Inst* 82:561–569, 1990.

16. Byers T. Nutritional risk factors for breast cancer. *Cancer* 74:288–295, 1994.

17. Hunter et al. Cohort studies of fat intake and the risk of breast cancer-a pooled analysis *New Engl. J. Med.* 334:356–361, 1996.

18. Tinsley IJ, Schmitz JA, Pierce DA. Influence of dietary fatty acids on the incidence of mammary tumors in the C3H mouse. *Cancer Res* 41:1460–1465, 1981.

19. Bennet A.S. Effect of dietary stearic acid on the genesis of spontaneous mammary adenocarcinomas in strain A/ST mice. *Int J Cancer* 34:529–533, 1984.

20. Kuhajda FP, Jenner K, Wood FD, Randolph AH, Jacobs LB, Dick JD, Pasternack GR. Fatty acid synthesis: A potential selective target for antineoplastic therapy. *J Biol Chem* 91:6379–6383, 1994.

21. Ip C. Controversial issues of dietary fat and experimental mammary carcinogenesis. *Prev Med* 22:728–737, 1993.

22. Bradlow HL, Fishman J. Diet and cancer. *Nature* 361:390, 1993.

23. Wynder E, Cohen L, Rose D, Stellman S. Dietary fat and breast cancer. *J Clin Epidemiol* 47:217–222, 1994.

24. Cuzick J. Methodologic issues in the chemoprevention of breast cancer. *Cancer Detect Prev* 16:81–85, 1992.

25. Silvestrini R, Daidone MG, Valagussa P, Di Fronzo G, Mezzanotte G, Mariani L, Bonadonna G. ³H-thymidine-labeling index as a prognostic indicator in node-positive breast cancer. *J Clin Oncol* 8:1321–1326, 1990.

26. Tubiana M, Pejovic MH, Koscielny S, Chavaudra N, Malaise E. Growth rate, kinetics of tumor cell prolieration and long-term outcome in human breast cancer. *Int J Cancer* 44:17–22, 1989.

27. Hery M, Gioanni J, Lalanne CM, Namer M, Courdi A. The DNA labeling index: A prognostic factor in node-negative breast cancer. *Breast Cancer Res Treat* 9:207–212, 1987.

28. Silvestrini R, Daidone MG, Valagussa P, Di Fronzo G, Mezzanotte G, Bonadonna G. Cell kinetics as a prognostic indicator in node-negative breast cancer. *Eur J Cancer* 25:1165–1171, 1989.

29. Meyer JS, Province M. Proliferative index of breast carcinoma by thymidine labeling: Prognostic power independent of stage, estrogen and progesterone receptors. *Breast Cancer Res Treat* 12:191–199, 1988.

30. Paradiso A, Mangia A, Picciariello M. Fattori prognostici nel carcinoma della mammella operabile N⁻; attivita prolierative e caratteristiche clinico patologiche. *Folia Oncol* 13:1–13, 1992.

31. Branden LM, Carroll KK. Dietary polyunsaturated fat in relation to mammary carcinogenesis in rats. *Lipids* 21:285–288, 1986.

32. Cohen LA, Thompson DO, Choi K, Karmall RA, Rose DP. Dietary fat and mammary cancer. II. Modulation of serum and tumor lipid composition and tumor prostaglandins by different dietary fats: association with tumor incidence patterns. *J Natl Cancer Inst* 77:43–51, 1986.

33. Cohen LA, Thompson DO, Maeura Y, Choi K, Blank ME, Rose D. P. Dietary fat and mammary cancer. I. Promoting effects of different dietary fats on N-nitrosomethylurea-induced rat mammary tumorigenesis. *J Natl Cancer Inst* 77:33–42, 1986.

34. Dao TL, Chan PC. Hormones and dietary fat as promoters in mammary carcinogenesis. *Environ Health Perspect* 50:219–225, 1983.

35. Gabriel HF, Melhem MF, Rao KN. Enhancement of DMBA-induced mammary cancer in Wistar rats by unsaturated fat and cholestyramine. *In Vivo* 1:303–308, 1987.

36. Hopkins, GJ, Carroll KK. Relationship between amount and type of dietary fat in promotion of mammary carcinogenesis induced by 7.12-dimethylbenzanthracene. *J Natl Cancer Inst* 62:1009–1012, 1979.

37. Hopkins GJ, Kennedy TG, Carroll KK. Polyunsaturated fatty acids as promoters of mammary carcinogenesis induced in Sprague-Dawley rats by 7,12-dimethylbenzanthracene. *J Natl Cancer Inst* 66:517–522, 1981.

38. Ip C, Carter CA, Ip MM. Requirement of essential fatty acid for mammary tumorigenesis in the rat. *Cancer Res* 45:1997–2001, 1985.

39. Rogers AE, Wetsel WC. Mammary carcinogenesis in rats fed different amounts and types of fat. *Cancer Res* 41:3735–3737, 1981.

40. Sundram K, Khor HT, Ong ASH, Pathmanathan R. Effect of dietary palm oils on mammary carcinogenesis in female rats induced by 7,12-dimethylbenzanthracene. *Cancer Res* 49:1447–1451, 1989.

41. Wetsel WC, Rogers AE, Newberne PM. Dietary fat and DMBA mammary carcinogenesis in rats. *Cancer Detect Prev* 4:535–543, 1981.

42. National Research Council. Nutrient Requirements of Laboratory Animals. ed. 3, Washington D.C. National Academy of Sciences pp 9–10, 1978.

43. Rogers AE, Conner B, Boulanger C, Lee S. Mammary tumorigenesis in rats fed diets high in lard. *Lipids* 21:275–280, 1986.

44. Harman D. Free radical theory of aging: Effect of the amount and degree of unsaturation of dietary fat on mortality rate. *J Gerontol* 26:451–457, 1971.

45. Kort WJ, Zondervan PE, Hulsman LOM, Weijma IM, Hulsmann WC, Westbroek DL. Spontaneous tumor incidence in female Brown Norway rats after lifelong diets high and low in linoleic acid. *J Natl Cancer Inst* 74:529–536, 1985.

46. Tinsley IJ, Wilson G, Lowry RR. Tissue fatty acid changes and tumor incidence in C3H mice ingesting cottonseed oil. *Lipids* 17:115–117, 1982.

47. Abraham S, Faulkin LJ, Hillyard LA, Mitchell DJ. Effect of dietary fat on tumorigenesis in the mouse mammary gland. *J Natl Cancer Inst* 72:1421–1429, 1984.

48. Abraham S, Hillyard LA. Effect of dietary 18-carbon fatty acids on growth of transplantable mammary adenocarcinomas in mice. *J Natl Cancer Inst* 71:601–605, 1983.

49. Erickson KL, Thomas IK. The role of dietary fat in mammary tumorigenesis. *Food Technol* 39:69–73, 1985.

50. Gabor H, Abraham S. Effect of dietary menhaden oil on tumor cell loss and the accumulation of mass of a transplantable mammary adenocarcinoma in BALB/c mice. *J Natl Cancer Inst* 76:1223–1229, 1986.

51. Gabor H, Hillyard LA, Abraham S. Effect of dietary fat on growth kinetics of transplantable mammary adenocarcinoma in BALB/c mice. *J Natl Cancer Inst* 74:1299–1305, 1985.

52. Giovarelli M, Padula E, Ugazio G, Forni G, Cavallo G. Strain- and sex-linked effects of dietary polyunsaturated fatty acids on tumor growth and immune functions in mice. *Cancer Res* 40:3745–3749, 1980.

53. Hillyard LA, Abraham S. Effect of dietary polyunsaturated fatty acids on growth of mammary adenocarcinomas in mice and rats. *Cancer Res* 39:4430–4437, 1979.

54. Hopkins GJ, West CE. Effect of dietary polyunsaturated fat on the growth of a transplantable adenocarcinoma in C3HA^{vy}fB mice. *J Natl Cancer Inst* 58:753–756, 1977.

55. Rao GA, Abraham S. Enhanced growth rate of transplanted mammary adenocarcinoma induced in C3H mice by dietary linoleate. *J Natl Cancer Inst* 56:431–432, 1976.

56. Rao GA, Abraham S. Reduced growth rate of transplantable mammary adenocarcinoma in C3H mice fed eicosa-5,8,11,14-tetraenoic acid. *J Natl Cancer Inst* 58:445–447, 1977.

57. Rose DP, Connolly JM. Effects of fatty acids and inhibitors of eicosanoid synthesis on the growth of a human breast cancer cell line in culture. *Cancer Res* 50:7139–7144, 1990.

58. Buckman DK, Hubbard NE, Erickson KL. Eicosanoids and linoleate-enhanced growth of mouse mammary tumor cells. *Prostaglandins Leukot Essent Fatty Acids* 44:177–184, 1991.

59. Buckman DK, Erickson KL. Relationship of the uptake and β-oxidation of 18-carbon fatty acids with stimulation of murine mammary tumor cell growth. *Cancer Lett* 59:257–265, 1991.

60. Buckman DK, Chapkin RS, Erickson KL. Modulation of mouse mammary tumor growth and linoleate enhanced metastasis by oleate. *J Nutr* 120:148–157, 1990.

61. Aksoy M, Berger MR, Schmahl D. The influence of different levels of dietary fat on the incidence and growth of MNU-induced mammary carcinoma in rats. *Nutr Cancer* 9:227–235, 1987.

62. Carroll KK, Khor HT. Effects of level and type of dietary fat on incidence of mammary tumors induced in female Sprague-Dawley rats by 7,12-dimethylbenzanthracene. *Lipids* 6:415–420, 1971.

63. Dayton S, Hashimoto S, Wollman J. Effect of high-oleic and high-linoleic safflower oils on mammary tumors induced in rats by 7,12-dimethylbenzanthracene. *J Nutr* 107:1353–1360, 1977.

64. Brown RR. Effects of dietary fat on incidence of spontaneous and induced cancer in mice. *Cancer Res* 41:3741–3742, 1981.

65. Erickson KL, Schlanger DS, Adams DA, Fregeau DR, Stern JS. Influence of dietary fatty acid concentration and geometric configuration on murine mammary tumorigenesis and experimental metastases. *J Nutr* 11:1834–1842, 1984.

66. Katz EB, Boylan ES. Effect of the quality of dietary fat on tumor growth and metastasis from a rat mammary adenocarcinoma. *Nutr Cancer* 12:343–350, 1989.

67. Abou-El-Ela SH, Prasse KW, Carroll R, Bunce OR. Effects of dietary primrose oil on mammary tumorigenesis induced by 7,12-dimethylbenzanthracene. *Lipids* 22:1041–1044, 1987.

68. Abou-El-Ela SH, Prasse KW, Carroll R, Wade A, Dharwadkar S, Bunce OR. Eicosaenoic synthesis in 7,12-dimethylbenzanthracene-induced mammary carcinomas in Sprague-Dawley rats fed primrose, menhaden or corn oil diets. *Lipids* 23:948–954, 1988.

69. Cameron E, Bland J, Marcuson R. Divergent effects of ω6 and ω3 fatty acids on mammary tumor development in C3H/Heston mice treated with DMBA. *Nutr Res* 9:383–393, 1989.

70. Ghayur T, Horrobin DF. Effects of essential fatty acids in the form of evening primrose oil on the growth of the rat R3230AC transplantable mammary tumor. *IRCS Med Sci* 9:582, 1981.

71. Karmall RA, Marsh J, Fuchs C. Effects of dietary enrichment with g-linolenic acid upon growth of the R3230AC mammary adenocarcinoma. *J Nutr Growth Cancer* 2:41–51, 1985.

72. Abou-El-Ela SH, Prasse KW, Farrell BL, Carroll RW, Wade AE, Bunce OR. Effects of DL-2-difluoromethylornthine and indomethacin on mammary tumor promotion in rats fed high *n*-3 and/or *n*-6 fat diets. *Cancer Res* 49:1434–1440, 1989.

73. Carroll KK, Braden LM. Dietary fat and mammary carcinogenesis. *Nutr Cancer* 6:254–259, 1985.

74. Mansour EG, Ravdin PM, Dressler L. Prognostic factors in early breast carcinoma. *Cancer* 74:381–400, 1994.

75. Dickson BB. Stimulatory and inhibitory growth factors and breast cancer. *J Steroid Biochem Mol Biol* 37:795–811, 1990.

76. Seshadri R, Mcleay WR, Horsfall DJ, McCaul K. Prospective study of the prognostic significance of epidermal growth factor receptor in primary breast cancer. *Int J Cancer* 69:23–27, 1996.

77. Klijn JG, Berns PM, Schmitz PI, Foekens JA. The clinical significance of epidermal growth factor receptor (EGF-R) in human breast cancer: A review on 5232 patients. *Endocrinology Rev* 13:3–17,1992.

78. Sainsbury JR, Farndon JR, Needham GK, Malcolm AJ, Harris AL. Epidermal growth factor receptor status as predictor of early recurrence of and from breast cancer. *Lancet* 1:1398–1402, 1987.

79. Macias A, Azavedo E, Hagerstrom T, Klintenberg C, Perez R, Skoog L. Prognostic significance of the receptor for epidermal growth factor in human mammary carcinomas. *Anticancer Res* 7:459–464, 1987.

80. NSMD Wickramasinghe, H Jo, JM McDonald, and RW Hardy Stearate inhibition of breast cancer cell proliferation: A mechanism involving epidermal growth factor receptor and Gi-proteins.*The American Journal of Pathology* 148:987–995, 1996.

81. Church JG, Buick RN. G-protein-mediated epidermal growth factor signal transduction in a human breast cancer cell line. *J Biol Chem* 263:4242–4246, 1988.

82. Johnson RM, Garrison JC. Epidermal growth factor and Angiotensin II stimulate formation of inositol 1,4,5 -and 1,3,4-triphosphate in hepatocytes. *J Biol Chem* 262:17285–17293, 1987.

83. Yang L, Baffy G, Rhee SG, Manning D, Hansen CA, Williamson JR. Pertussis toxin-sensitive G_i protein involvement in epidermal growth factor-induced activation of phospholipase c-g in rat hepatocytes. *J Biol Chem* 33:22451–22458, 1991.

84. De Vivo M, Gershengorn MC. In ADP-ribosylating toxins and G proteins: Insights into signal transduction (Moss J, and Vaughan M. Eds) pp. 267–293, American Society for Microbiology, Washington, D.C.

85. Crouch MF. Growth factor-induced cell division is paralleled by translocation of G_ia to the nucleus. *FASEB J* 5:200–206, 1991.

86. Trigatti BL, Gerber GE. A direct role for serum albumin in the cellular uptake of long-chain fatty acids. *Biochem J* 308:155–159, 1995.

87. Spector AA, Hoak JC: An improved method for the addition of long-chain free fatty acid to protein solutions. *Anal Biochem* 32:297–302, 1969.

88. Hardy RW, Ladenson JH, Henriksen EJ, Holloszy JO, McDonald JM. Palmitate stimulates glucose transport in rat adipocytes by a mechanism involving translocation of the insulin sensitive glucose transporter (GLUT 4). *Biochem Biophys Res Commun* 177:343–349, 1991.

89. Bonanome A, Grundy SM. Effect of dietary stearic acid on plasma cholesterol and lipoprotein levels. *New Engl. J. Med.* 318:1244–1248, 1988.

90. Kris-Etherton PM, Derr J, Mitchell DC, Mustad VA, Russell ME, McDonnell ET, Salabsky D, Pearson TA, The role of fatty acid saturation on plasma lipids, lipoproteins and apolipoproteins. *Metabolism* 42:121–129, 1993.

91. Ahrens EH Jr, Hirsch J, Insull W Jr, Tsaltas TT, Blomstrand R, Peterson ML, The influence of dietary fats on serum-lipid levels in man. *Lancet* 1:943–953, 1957.

92. Hegsted DM, McGandy RB, Myers ML, Stare FJ, Quantitative effects of dietary fat on serum cholesterol in man. *Am J Clin Nutr* 17:281–295, 1965

93. Keys A, Anderson JT, Grande F, Serum cholesterol response to changes in the diet. IV. Particular saturated fatty acids in the diet. *Metabolism* 14:776–787, 1965,

94. Grundy SM, Influence of stearic acid on cholesterol metabolism relative to other long-chain fatty acids. *Am J Clin Nutr* 60:986S–990S, 1994.

95. Mustad VA, Kris-Etherton PM, Derr J, Reddy CC, Pearson TA, Comparison of the effects of diets rich in stearic acid versus myristic acid and lauric acid on platelet fatty acids and excretion of thromboxane A2 and PGI2 metabolites in healthy young men. *Metabolism* 42:463–469, 1993.

96. Hoak JC, Stearic acid, clotting, and thrombosis. *Am J Clin Nutr* 60:1050S-1053S, 1994.

6

LIPOXYGENASE METABOLITES AND CANCER METASTASIS

Keqin Tang and Kenneth V. Honn

Department of Radiation Oncology
Wayne State University School of Medicine
431 Chemistry
Detroit, Michigan 48202

INTRODUCTION

Metastasis

Cancer metastasis is a complicated process. For a transformed cell to form a successful metastatic colony, it must in general complete all or most of the well-defined steps that comprise the "metastatic cascade".[1-4] The first step is uncontrolled cell proliferation, characteristic of both benign and malignant tumor cells. Intrinsic or acquired genetic instability, together with various epigenetic factors, generate tumor cell variants that acquire unique phenotypic characteristics that dissociate them from the parent tumor population and thus allow these variants to escape from the "social" constraints imposed by the host. This step confers on these "mutated" tumor cells invasive or metastatic capabilities and is generally considered to be the first step leading to site-specific metastasis. In the next step, tumor cells, in response to various chemoattractants and cytokines derived from the host and/or tumor cells, migrate towards neighboring vasculature or intravasate into the vasculature of the tumor and thus enter the hematogenous or lymphatic circulation. Subsequently, tumor cells travel to and arrest in the microcirculation by specific adherence to the endothelial cells of the target organ. Thereafter, tumor cells induce endothelial cell retraction, exit from circulation (extravasation), interact with the organ-specific extracellular matrix (ECM), proliferate in response to local ("soil") growth factors, and finally form a metastatic colony. Failure at any one of these steps generally will abort the metastatic process. Completion of every step of the metastatic cascade is subject to a multitude of variable influences, an apparent example being the requirement of angiogenesis for the growth of both primary and secondary tumors.[5,6]

During the course of cancer metastasis, tumor cells have to undergo extensive cell-cell (such as tumor cell-tumor-cell, tumor cell-stromal cell, tumor cell-platelet, tumor cell-endothelial cell) and cell-ECM interactions.[7-9] These interactions are mediated by a wide spectrum of surface adhesion molecules present on both tumor cells and host cells, in particular target organ microvascular endothelial cells.[10-12] Specific tumor cell-endothelial cell interactions, to a large degree, determine the organ selectivity of metastasis.[13-15] Tumor cell interactions with other host cells such as platelets also significantly contribute to the formation of metastasis.[16-18] The ability of metastasizing tumor cells to invade, migrate, degrade, and to interact with host cells is modulated by a myriad of bioactive factors, among which are growth factors, cytokines, chemoattractants, motility factors, as well as lipid activators, derived from the host and/or tumor cells. A large body of evidence has demonstrated that eicosanoids play an important modulatory role in tumor cell-host interactions and thus in cancer metastasis.

Eicosanoids

The term eicosanoids refers to a group of oxygenated arachidonic acid (AA) metabolites, including prostaglandins, thromboxanes, leukotrienes, lipoxins, and various hydroperoxy and hydroxy fatty acids. Arachidonic acid (AA) is liberated from membrane phospholipids mainly by the action of phospholipase A_2 which also generates lysoglyceryl-phosphorylcholine which is further converted to platelet-activating factor (PAF). Arachidonic acid can also be released from the membrane phospholipids through the concerted actions of phospholipase C (PLC) and diacylglycerol lipase or of phospholipase D (PLD) and phospholipase A_2 (PLA_2). Once liberated, AA is metabolized through three major pathways, i.e., the cyclooxygenase (COX) pathway, which results in generation of prostacyclin (PGI_2), thromboxane (TXA_2), and various prostaglandins, the lipoxygenase (LOX) pathway, which gives rise to a variety of hydroxyeicosatetraenoic acids (HETEs), leukotrienes, and lipoxins; and the cytochrome P-450 monooxygenase pathway, from which various expoxyeicosatrienoic acids (EETs) are derived (for reviews see References 19–22). The first committed step in the biosynthesis of all eicosanoids is in the incorporation of molecular oxygen into polyunsaturated fatty acid and is catalyzed by COX, LOX or P-450-dependent oxygenase. These three key enzymes are expressed by most eukaryotic cells, but subsequent metabolism differs in different cell types. For example, COX is an endoplasmic reticulum-bound enzyme that occurs in virtually every cell type in the body, but arachidonic acid metabolism varies in different cells. In platelets, the COX pathway leads to TXA_2 synthesis, in endothelial cells it leads to PGI_2 synthesis, and in macrophages it leads mainly to synthesis of PGE_2. In contrast, mast cells primarily synthesize PGD_2. Another example is that all leukocytes possess 5-lipoxygenase, which adds a hydroperoxy group to C5 in AA, but in neutrophils subsequent metabolic steps lead to the biosynthesis of LTB_4, while in eosinophils, mast cells, basophils and macrophages, to the synthesis of cysteinyl-leukotrienes (i.e., LTC_4, LTD_4, LTE_4, and LTF_4).[20,21]

Arachidonic acid also can be derived from another polyunsaturated fatty acid, linoleic acid (LA). LA is generated from the intracellular triglyceride pool through the action of protein kinase A-activated lipase. LA is further converted, via the 15-LOX pathway, to 13-hydroxyoctadecadienoic acid (13-HODE).

Products of both COX and LOX pathways exert a variety of biological activities. Prostaglandin A, PGJ_2, and PGD_2 have been shown to possess antiproliferative and antiviral activities. Metabolites of AA and LA have been demonstrated to affect cell growth; release of collagenase IV, cathepsin B, and autocrine motility factor; hormone secretion, chloride transport, increased enzyme activity, altered immune response, and cell movement; CSF-1 gene transcription and nitric oxide synthesis. These oxygenated fatty acids are also involved in carcinogenesis, tumor promotion, radiation protection. EGF-induced cell signaling, even in mineralization by osteoblast cells (for a detailed account, see References 23 and 24 and references therein). In addition AA and LA metabolites have been demonstrated to affect synthesis or function of different cytokines and growth factors, and the COX/LOX products, cytokines, and growth factors are interregulated. Thus, EGF can enhance LOX activity and therefore production of HETE[25] and HODE[25], IL-1 induces the expression of COX;[27] TGF-α[28] as well as IL-1 and TNF[29] increases the activity of phospholipase A$_2$, a key enzyme in the release of AA from phospholipids. In the following discussion, we will focus on the effects of 12(S)-HETE, and 13-HODE, on tumor cell metastasis.

12(S)-HETE: INVOLVEMENT IN MULTIPLE STEPS OF CANCER METASTASIS

12(S)-HETE is the major AA metabolite of 12-lipoxygenase. Many different types of normal cells including platelets, neutrophils, macrophages, endothelial cells, and smooth muscle cells can synthesize 12(S)-HETE.[30–32] The physiological functions of this eicosanoid are not fully characterized, but accumulated data indicate that it is involved in a wide-spectrum of biological activities such as stimulating insulin secretion by pancreatic tissue, suppressing rennin production, chemoattracting leukocytes, and facilitating the attachment of macrophages to rat glomeruli during inflammation (reviewed in Reference 30). 12(S)-HETE also has been observed to reduce prostacyclin biosynthesis by vascular endothelial cells[33] and play a vital role in platelet activation and aggregation.[34] 12(S)-HETE is found to be the most prominent AA metabolite in menstrual blood, however, the biological significance of these findings is not yet clear.[35–36]

A variety of tumor cells including solid tumor cells express 12-lipoxygenase mRNA and protein.[37–42] RT-PCR together with immunoblotting revealed that cultured solid tumor cells from human, rat, and mouse express platelet-type 12-lipoxygenase mRNA and protein. Reverse phase HPLC identified 12(S)-HETE as the major metabolite of AA in these tumor cells which is confirmed by chiral phase HPLC and CG/MS analysis. Several lines of evidence suggest that the ability of tumor cells to synthesize 12(S)-HETE is a strong correlate of their metastatic potential. First, subpopulations of B16 amelanotic melanoma (B16a) have been isolated by centrifugal elutriation that demonstrate differential metastasis capabilities. The low metastatic B16a (i.e., LM180) cells biosynthesized similar amounts of 12(S)-HETE with only small quantities of 15-, 11-, and 5-HETEs. Furthermore, HM340 cells synthesized 4 times more 12(S)-HETE than LM180 cells when equal amounts of substrates were supplied.[43–45] The generation of higher amounts of 12(S)-HETE in HM340 cells appears to result from the presence of a higher level of 12-lipoxygenase mRNA in these cells[46]. Second, the correlation of 12(S)-HETE production and metastatic potential also was evaluated in several other tumor cell systems, i.e., Dunning rat prostate carcinoma (AT2.1 and GP9F3, low metastatic cell lines; MAT Lu and MLL; high metastatic lines), murine B16 melanoma (F1, low metastatic; F10 high metastatic), and murine K-1735 melanoma (C1–11, low metastatic clone; M1, high metastatic clone). In all these experiments it was observed that the high metastatic tumor cell lines generate significantly higher levels of 12(S)-HETE than low metastatic counterparts[46]. Third, tumor cell adhesion to fibronectin provokes a spike of 12(S)-HETE generation within 10 min indicating a late-type signaling activation of tumor cell 12-lipoxygenase[47]. Morphological studies demonstrated that tumor cell spreading on fibronectin is followed by 12-lipoxygenase translocation to the cell surface[47]. Fourth, adhesion of tumor cells to vascular endothelium is accompanied or immediately followed by a surge of 12(S)-HETE biosynthesis by tumor cells, which was correlated with tumor cell-induced endothelial cell retraction.[44,48] LM180 B16a cells generated little 12(S)-HETE upon adhesion and did not induce endothelial cell retraction. In contrast, HM340 B16a cells adhering to endothelium biosynthesized large amounts of 12(S)-HETE and induced prominent retraction of endothelial cell monolayers.[44] Fifth, pretreatment of tumor cells with a select platelet-type 12-lipoxygenase inhibitor, BHPP (N-benzyl-N-hydroxy-5-phenylpentanamide[34]) dose-dependently inhibited adhesion-induced tumor cell biosynthesis of 12(S)-HETE and endothelial cell retraction[44], platelet-enhanced tumor cell-induced endothelial cell retraction,[48] tumor cell (i.e., HM340 B16a cells) adhesion to endothelium and matrix,[39,43] and lung colonization by HM340 B16a cells[43]. Taken together, these data implicate 12-lipoxygenase and 12(S)-HETE production as important determining parameters of tumor cell metastasis. In the subsequent discussions, we will analyze in detail the versatile modulatory role of 12(S)-HETE in various intermediate steps of metastasis.

Effect of 12(S)-HETE on Tumor Cell Interactions with Extracellular Matrix (ECM)

Tumor cell-ECM interactions are mediated by adhesion receptors such as integrins, non-integrins and proteoglycans where integrins are considered to be the predominant matrix receptors.[49–51] Tumor cell-ECM interactions are key and multiple events during tumor progression, i.e., in the release of tumor cells from primary site, in the locomotion and invasion of dissociated tumor cells in interstitium, in intravasation and extravasation, and in the eventual establishment of the metastatic nodule. It is conceivable that the regulation of the surface expression of these matrix receptors has a great impact on tumor cell adhesive potentials.

Integrin receptors, after synthesis, are stored in tumor cells as well as in normal cells intracellularly in so called adhesosomes[52] and tumor cells have a large pool of intracellular receptors.[53] However, in normal cells, freshly synthesized and glycosylated integrins are not competent for their full range of ligand binding and signaling functions. Adhesion process initiated by either soluble or solid surface-bound ligands triggers cell activation. Ligand binding to the surface exposed integrins induces conformational change of the receptors increasing specificity. In parallel, such physical alterations in the receptors induce signal transmission to the cell interior by activation of a receptor associated tyrosine kinase, FAK (focal adhesion kinase[54–56]). As a consequence, cytoskeletal proteins become associated with the cytoplasmic domains of the integrins and this process immobilizes and concentrates the receptors to the areas of cell-matrix contacts called focal adhesion plaques. During the cytoskeleton-dependent process the cell shape changes dramatically from round to a spreading type.

Several extracellular factors, exemplified by ECM-derived soluble ligands and ECM-bound cytokines (characteristically these are heparin-binding cytokines such as TGF-β, FGFs, VEGF, etc.) may regulate or alter the adhesive properties of tumor cells. In the case of tumor cell - vessel wall interactions other cellular participants such as platelets and endothelial cells modulate the adhesion process by releasing cytokines and metabolites of AA to the micromilieu of the tumor cells. Therefore, a possibility is raised that the metabolites of AA generated in the extracellular space of tumor cells may modulate the tumor adhesiveness and thus tumor cell adhesion to ECM. Systematic studies indicated that 12(S)-HETE is able to stimulate metastatic murine tumor cell adhesion to fibronectin and subendothelial matrix in a dose and time dependent fashion, with a maximal effect observed at $0.1\mu M$ of 12(S)-HETE 15 minutes after stimulation.[53,57] Analysis of the mechanism of the 12(S)-HETE enhanced tumor cell adhesion indicated an increased surface expression of integrin receptors αIIbβ3 after 12(S)-HETE treatment.[53,57] In the murine tumor cell lines studied (3LL and B16a) the increased adhesiveness was due to overexpression of αIIbβ3 cytoadhesion,[58,59] a promiscuous matrix receptor physiologically expressed by platelets and cells of the hematopoietic lineage. Interestingly, TPA, a direct activator of PKC, mimicked the effect of 12(S)-HETE[57] suggesting a PKC-mediated pathway for the exogenous 12(S)-HETE effect (see below).

The 12(S)-HETE effect on tumor cell integrin expression was not due to an increased transcription of the gene or increased protein translation. Rather it was due to a rapid receptor translocation to the cell surface from the intracellular pool.[53,60] These studies further indicated that the 12(S)-HETE effect on intregrin upregulation and enhanced adhesion both depend upon microfilaments and intermediate filaments.[53,60]

12(S)-HETE does not only stimulate tumor cell adhesion to fibronectin but also promotes tumor cell spreading on this matrix protein, by inducing re-establishment of the filamentous cytoskeleton system and enhancing the formation of focal adhesion plaques.[47] This process appears to be dependent on the function of PKC.[47] The dose and time-frame of the 12(S)-HETE effect on spreading followed those observed previously for adhesion.[47]

Effect of Exogenous 12(S)-HETE on Tumor Cell Motility

Tumor cell motility is regulated by para- and autocrine cytokines including SF (scatter factor), MF (motility factor) and AMF (autocrine motility factor).[48] These cytokines act through their corresponding signaling receptors expressed at the cell surface of tumor cells. AMF was isolated as the first autocrine motility cytokine produced by tumor cells.[49] A receptor AMF, gp78, was later identified and sequenced.[59] These studies revealed a transmembrane protein with considerable homology to p53.[59]

Hydroxy fatty acid metabolites of AA including 12(S)-HETE have been shown to be involved in chemotactic movement of leukocytes where they serve as chemoattractants.[30] Therefore there exists a possibility that 12(S)-HETE also may modulate tumor cell movement. Studies on murine melanoma cell lines with different metastatic potential (B16a, K1735-Cl.11 and K1735 M1) indicate that 12(S)-HETE induces tumor cell motility comparable to AMF itself.[61] Subsequent mechanistic studies demonstrated that 12(S)-HETE, similar to its effect on tumor cell integrin receptor expression, increases gp78 expression at the cell surface in a dose and time-dependent manner.[61] The increased gp78 expression is due to translocation of the receptor to the cell surface from an intracellular tubulovesicular (TVS) structure positive for lysosomal markers.[62] Studies also indicated that this TVS pool is closely associated with microtubules and intermediate filaments, therefore it is reasonable to assume that the increased AMF receptor expression after 12(S)-HETE stimulation is also a cytoskeleton-dependent process.

More recently 12(S)-HETE also was observed to increase the invasive potential of AT2.1 rat prostate tumor cells. The tumor cell motility (as assessed by phagokinetic track assay) and invasion were significantly augmented by 12(S)-HETE.[63] The 12(S)-HETE effect was dramatically inhibited by a selective PKC inhibitor calphostin C, as well as by a cell membrane-permeable Ca^{2+}-chelator BAPTA,[63] suggesting the involvement of PKC (see below). Whether the 12(S)-HETE-promoted motility involves gp78 and whether 12(S)-HETE-augmented invasion in the rat tumor cell system involves activation of proteolytic enzymes are at present unclear.

The reciprocal relationship between 12-lipoxygenase/12(S)-HETE and AMF gp78 appears to be far more complicated than presented above. 12(S)-HETE apparently promotes an AMF-comparable and gp78-mediated tumor cell motility. On the other hand, gp78 also has been observed to stimulate 12(S)-HETE production in highly metastatic murine melanoma cells.[61] More interestingly, AMF, which stimulated the motility of the high (M1) but not low-metastatic (C1.11) variants of the K1735 murine melanoma, increased expression of 12-lipoxygenase protein in M1 sublines exclusively. Also, only the M1 (but not C1.11) cells responded to AMF stimulation with increased endogenous 12(S)-HETE production and 12(S)-HETE, like AMF, only increased the motility of M1 cells. Taken together, these data suggest that 12-lipoxygenase and 12(S)-HETE may be important intermediate signaling elements in AMF-gp78 mediated tumor cell motility.

Effect of Exogenous 12(S)-HETE on the Release of Lysosomal Enzymes

Earlier studies on various normal cells indicated that 12(S)-HETE induces lysosomal enzyme or hormone secretion.[30] Furthermore, lipoxygenase metabolites have been shown to be involved in collagenase IV release from tumor cells (reviewed in Reference 64). Secretion of proteins involves a so-called secretory mechanism, where the glycosylated protein is transported to the plasma membrane by the secretory pathway involving transport vesicles. Lysosomal enzymes utilize a mechanism which directs these proteins to lysosomes by recognizing the M6P residues on protein (M6P-R). Lysosomal enzymes are released from these vesicles after intra-or extracellular signaling. Tumor cells produce a wide range of proteolytic enzymes by which they digest the surrounding ECM. These degradative enzymes include collagenase IV, interstitial collagenases, heparinases, urokinase and cathepsins with variable substrate specificities.[65,66] Secretion of these enzymes from tumor cells provides a means by which tumor cells can cross tissue boundaries. In normal cells, the majority of such proteolytic enzymes are stored intracellularly and released only after appropriate stimulation. However, in tumor cells lysosomal enzymes, especially cathepsins, were found to be associated constitutively with the cell surface.[67,68]

It is known that the translocation of lysosomal proteins is mediated by cytoskeletal elements, primarily by microtubules. The demonstrated involvement of 12(S)-HETE in regulation of receptor (such as integrin $\alpha IIb\beta 3$) translocations through cytoskeleton-dependent mechanisms prompted us to study its effect on the surface expression and release of cathepsin B from tumor cells. At a concentration which is optimal for increased adhesiveness and motility of tumor cells (i.e., 0.1µM), 12(S)-HETE was observed to induce cathepsin B release from highly malignant cells, i.e., B16a murine melanoma cells and MCF10AneoT human mammary carcinoma cells.[64,66,69] Exogenous 12(S)-HETE first triggers the release of mature cathepsin B and then later the immature form, suggesting that the mature form of the enzyme is at or near the cell surface while the immature form is stored intracellularly. Subsequent morphologic studies indicated that, following 12(S)-HETE stimulation, the intracellular cathepsin B-containing vesicles are directed to specialized cell surfaces areas, i.e., lamellipodia and filopodia.[64] Analysis of the release process induced by 12(S)-HETE indicated that, similarly to its effect on tumor cell adhesion and motility, the stimulated release of cathepsin B is also a PKC-dependent process.[64]

Effect of Exogenous 12(S)-HETE on the Tumor Cell Infrastructure and Cystoskeleton

It is clear from the above discussion that 12(S)-HETE has a regulatory effect on complex cellular processes associated with cytoskeletal functions, such as adhesion to and spreading on ECM proteins, motility, lysosomal enzyme release and receptor translocation to the cell surface. Therefore, we have analyzed the effect of exogenous 12(S)-HETE on cytoskeleton in tumor cells.[70,71] These studies indicate that 12(S)-HETE induces translocation of organelles from the perinuclear space to the cell periphery paralleled by a reversible rearrangement of cytoskeletal filaments.[70] The earliest event post 12(S)-HETE treatment is a decreased labeling of cytoplasmic stress fibers with a concomitantly enhanced cortical actin labeling. These alterations in microfilaments are accompanied or immediately followed by microtubule polymerization and vimentin intermediate filament bundling and also by rearrangements of the focal adhesion protein vinculin.[70,71] The 12(S)-HETE effect on tumor cell cytoskeleton is PKC-dependent since pretreatment of tumor cells with a selective PKC inhibitor, i.e., calphostin C, abolished the 12(S)-HETE effects.[70] 12(S)-HETE-induced cytoskeletal alterations generally abate by ~30 min. However, simultaneous treatment of tumor cells with 12(S)-HETE and okadaic acid, a general serine/threonine protein phosphatase inhibitor, prevented disappearance of the 12(S)-HETE effects by 30 minutes.[71] These observations thus raise the possibility that a phosphatase(s), activated due to either the direct effect of 12(S)-HETE or homeostatic mechanisms, is responsible for dampening the 12(S)-HETE-triggered, PKC dependent, and phosphorylation-mediated cytoskeleton rearrangement.

12(S)-HETE as a Signaling Molecule

A growing body of recent evidence suggests that 12(S)-HETE may act as a second messenger in stimulus-response coupling in some cells.[72–74] For example, angiotensin II stimulates 12(S)-HETE biosynthesis and aldosterone secretion by adrenal cortical cells. Pharmacologic inhibitors of 12(S)-HETE biosynthesis suppress angiotensin II-induced aldosterone secretion, and exogenous 12(S)-HETE increases aldosterone secretion.[73] A similar series of observations suggests the participation of 12(S)-HETE or its precursor in neurotransmitter peptide-induced hyperpolarization by *Aplysia* neuronal cells.[72] 12(S)-HETE promotes epidermal proliferation, as does leukotriene B_4 (LTB_4).[74] On the other hand, treatment of rat renal cortical slices with 12-HPETE or 12(S)-HETE, but not LTB_4 or 5-HPETE, blocks PGI_2 (prostacyclin) or Iloprost-induced rennin secretion.[75] In addition, 12-HETE significantly increases intracellular levels of cyclic adenosine monophoshate (cAMP) in human arterial smooth muscle cells.[76]

We investigated a possible role of 12(S)-HETE in regulating expression of the 12-LOX gene, integrin genes (α_{IIb} and β_3), tumor suppressor genes (deleted in colon carcinoma [DCC], p53, retinoblastoma [Rb], adenomatous polyposis coli [APC], mutated in colon carcinoma [MCC]), and oncogene MDM-2 in human colon carcinoma (clone A) cells.[77] We found that 12(S)-HETE upregulates 12-LOX α_{IIb}, β_3, and MDM-2 but downregulates levels of DCC mRNA (unpublished data). Preliminary results indicate that 12(S)-HETE also increases 12-LOX and MDM-2 at the protein level. The effects of 12(S)-HETE were antagonized by the PKC inhibitor calphostin C. No alteration was observed for p53, Rb, or MCC genes.

The mechanism of mRNA regulation by 12(S)-HETE may be at the transcriptional level, because actinomycin D can abolish the effect of 12(S)-HETE on gene expression (unpublished data). When clone A cells are treated with BHPP, a 12-LOX selective inhibitor, the expression of the DCC gene is stimulated, whereas the expression of 12-LOX and of the MDM-2, α_{IIb}, β_3 genes is suppressed. These findings suggest that endogenous 12(S)-HETE in clone A cells may also regulate gene expression. When an endothelial cell line was treated with 12(S)-HETE, PKC-dependent transcriptional activation of the α_v integrin gene was found.[77] The ability of 12(S)-HETE to upregulate 12-LOX, integrins, and oncogenes and to down regulate tumor suppressor genes may contribute to its ability to increase metastasis.

12(S)-HETE activates PKC. Early experimental observations that 12(S)-HETE mimics the tumor promoter phorbol myristate acetate (PMA) in enhancing tumor cell integrin expression and adhesion[57] suggested that 12(S)-HETE may work by activating PKC. Supportive evidence was established when we observed that, in rat W256 cells, 12(S)-HETE induces a 100% - increase in membrane-associated PKC activity and that downregulation of PKC by prolonged treatment with PMA abolishes 12(S)-HETE-enhanced W256 cell adhesion to endothelium.[42] Subsequently, our group and others observed that essentially all 12(S)-HETE-evoked biologic responses encompassing rearrangement of cytoskeleton, promotion of cell adhesion and spreading, induction of endothelial cell retraction, release of proteolytic enzymes, enhancement of cell motility, and regulation of gene expression can be abrogated by selective PKC inhibitors.[42,78] Experiments by others also showed that 12(S)-HETE and some other HETEs (e.g., 15-HETE) activate PKC.[40] 12(S)-HETE mimics TPA (12–0-tetradecanoyl-phorbol-13-acetate) in activating PKC, and its effects can be blocked by 13(S)-HODE.[40]

PKC has been considered one of the most important phospholipid/CA^{2+}-dependent serine-threonine protein kinases positioned in the center of various signaling pathways.[79] The precise role of PKC in multistep tumorigenesis and cancer metastasis remains to be established; yet, generally it is agreed that abnormal expression of specific PKC isoforms is closely associated with cellular transformation and progression.[80] Precisely how 12(S)-HETE activates PKC has not been defined.

It appears that 12(S)-HETE in different cell types activates different isoforms of PKC. In murine B16a melanoma cells, PKCα is the only isoform detected by immunoblotting using isoform-specific antipeptide antibodies.[81] 12(S)-HETE stimulation leads to the increased translocation of PKC from the diffuse cytoplasmic pool to the plasma membrane. In rat AT2.1 prostate carcinoma cells, two isoforms of PKC, α and δ, were identified.[63] It is interesting that 12(S)-HETE preferentially activates PKCα by increasing the membrane association of this isoform, and the result is upregulation of tumor cell motility and invasive potential.[63] Treating tumor cells with thymelea toxin, a selective activator of PKCα, also promotes tumor cell motility and invasion,[63] a finding confirming the preferential activation of PKCα by 12(S)-HETE in this tumor cell line.

In murine microvessel endothelial (CD3) cells, at least two isoforms of PKC are expressed (unpublished data). Which isoform of PKC or whether both isoforms of PKC are activated by 12(S)-HETE is not known. In rat adrenal glomerulosa cells, the α and ε isoforms of PKC are expressed. In contrast to the situation in tumor cells, as described above, 12(S)-HETE and 15(S)-HETE alike predominantly activate PKCε, although angiotensin II increases membrane-bound levels of both PKC α and PKCε.

While the mechanism by which 12(S)-HETE activates PKC remains unclear, it is known that this fatty acid does not directly activate PKC as assessed by PKC activity assays in which 12(S)-HETE is included in the reaction mixtures.[80] But, 12(S)-HETE stimulation of tumor cells is followed by a rapid accumulation of diacylgycerol (DAG) and inositol 1,4,5-triphosphate (IP_3),[80] suggesting that phosphatidylinositol bis 4, 5-phosphate (PIP_2) is metabolized through the action of phospholipase C (PLC). Supporting this hypothesis are the findings that blocking the functions of G proteins by pertussis toxin in B16a cells and inhibiting the PLC activity by specific antagonists also suppress the effects of exogenous 12(S)-HETE.[80] These results indicate that 12(S)-HETE may function by activating an upstream G protein.

Alternatively, DAG may accumulate after 12(S)-HETE treatment as a result of downregulation of DAG kinase activity, as proposed by Setty and colleagues.[82] In the phosphatidylinositol cycle, DAG derived from various sources undergoes rapid phospho-rylation by DAG kinase to phosphatidic acid.[83] In bovine aortic endothelial cells, 12(S)-HETE as well as 15(S)-HETE inhibits DAG kinase activity. As a result, DAG accumulates in these cells.[82]

Certainly, it is possible that both mechanisms - the activation of upstream G protein and PLC and the inhibition of downstream DAG kinase - exists in cells and that together they eventually induce an increase in DAG and then activation of one or more specific PKC isoforms. However, it is unlikely that 12(S)-HETE inhibits DAG kinase activity in tumor cells because 12(S)-HETE causes accumulation of both DAG and IP_3 in a similar time-dependent manner[80] instead of DAG alone.

In various experimental settings, we observed that the effects of 12(S)-HETE (inte-grin expression, adhesion, spreading, retraction, etc) on tumor cells and vascular endothe-lial cells are antagonized by 13(S)-HODE, an eicosanoid derived from linoleic acid metabolism by 15-LOX.[83] The 13(S)-HODE effects appear to be mediated through inhibi-tion of PKC activation by 12(S)-HETE. However, as in the case of 12(S)-HETE, 13(S)-HODE does not directly suppress PKC activation. 13(S)-HODE does, however, block the cytoplasm-to-membrane translocation of PKC induced by 12(S)-HETE.[80-83] In epidermis where 13(S)-HODE is the major metabolite of linoleic acid, Cho and Ziboh[84] found that this fatty acid is incorporated in 13-HODE-containing DAG (13-HODE-DAG). Interest-ingly, they also showed that 13-HODE-DAG selectively inhibits PKCβ and subsequently inhibits its activity in epidermal cells.[84] Nevertheless, whether 13-HODE-DAG directly binds to PKCβ and subsequently inhibits its activity or inhibits PKC via some other sig-naling intermediates is unknown.

Accumulating experimental data suggests that 12(S)-HETE also may have one or more cell surface receptor or binding sites. Binding studies using human epidermal cell line SCL-II demonstrate that these cells have a single class of binding sites for 12(S)-HETE.[82,85] In human platelets, 12(S)-HETE was shown to block PGH_2/TXA_2-induced platelet aggregation.[86] Later, part of the inhibitory effect of 12(S)-HETE on PGH_2/TXA_2-induced platelet aggregation was found to be due to inhibition of PGH_2 and TXA_2 binding to their common receptor sites.[87] These findings suggest that 12(S)-HETE may interact with or bind to PGH_2/TXA_2 receptors. Recently, Herbertsson and Hammarstrom[88] found 12(S)-HETE binding sites in LLC cells. It is intriguing that subsequent subcellular frac-tionation studies identified potential 12(S)-HETE binding sites in multiple subcellular compartments.[89]

The cytosolic binding sites for 12(S)-HETE appear to be a large macromolecular ag-gregate with an approximate molecular weight of 669 kd that is sensitive to trypsin or chy-motrypsin.[89] Recenty, we showed that high-affinity 12(S)-HETE binding sites correlate with increased invasive potential of human tumor cells (unpublished data). Tumor cells with high-affinity 12(S)-HETE binding sites (PC-3 [human prostatic carcinoma] cells and U251P [human glioblastoma] cells) responded to 12(S)-HETE treatment with twofold to threefold invasiveness (unpublished data). Cells with low-affinity binding sites (DU 145 [human prostatic carcinoma] cells and U251N cells) could not be stimulated. We are cur-rently purifying and cloning the putative 12(S)-HETE receptor.

Hypothesis for Role of 12(S)-HETE in Metastasis

Based on the above findings, we offer the following hypothesis to explain the in-volvement of 12(S)-HETE in regulating tumor metastasis. In tumor cells, external 12(S)-HETE, through binding to the cell surface receptor, activates a pertussis toxin-sensitive G protein[80] which in turn activates PLC. Stimulation of PLC activates phosphatidylinositol metabolism, which generates both DAG and IP_3 from PIP_2. Together, DAG and IP_3 activate PKC. Functionally activated PKC mediates the phosphorylation of protein (eg, cytoskeletal elements), resulting in modulation of a variety of phenotypic properties of metastasizing tumor cells such as adhesion, spreading, and motility. 13(S)-HODE antagonizes the effects of 12(S)-HETE by binding to either the same site(s) for 12(S)-HETE or different site(s) that negatively affect 12(S)-HETE receptor function.

Adhesion of tumor cells to endothelial cells activates 12-LOX, leading to arachidonic acid metabolism to 12(S)-HETE. 12(S)-HETE, by mechanisms that are unknown, activates PKC and translocates PKC to the plasma membrane. PKC subsequently phosphorylates and reorganizes the cytoskeletal elements, which mobilize $\alpha_{IIb}\beta_3$ integrin-containing vesicles to the plasma membrane. Cytoskeletal rearrangements induced by 12(S)-HETE also are involved in modulation of tumor cell spreading on matrix and tumor cell motility.

Protein kinases (Raf1, Mos, and Mekk [Map kinase-kinase-kinase]) also may be components of the cellular signaling pathway(s) that are transduced from 12(S)-HETE, since they could be activated by PKC and thus lie downstream of PKC in the signaling pathway. These protein kinases can activate Map kinases (mitogen-activated protein kinases) or ERKs (extracellular signal-regulated kinases), which in turn phosphorylate transcriptional factors regulating gene expression and may eventually change cell behaviors such as proliferation, adhesion, migration, invasion and metastasis.

Since 14–3–3 proteins (a family of modulators of signaling proteins) activate both PKC and Raf1, it is possible that 14–3–3 proteins may amplify the signal transduced by 12(S)-HETE. Tumor cell-derived 12(S)-HETE likewise acts on endothelial cells, and this action results in enhanced $\alpha_v\beta_3$ integrin surface expression and tumor cell adhesion and in PKC-dependent phosphorylation and reorganization of cytoskeleton and endothelial cell reaction. In addition, 12(S)-HETE promotes the tumor cell release of cathepsin B to degrade the subendothelial matrix. In vivo, tumor cell interactions with platelets (through $\alpha_{IIb}\beta_3$ integrins and many other receptors) leads to platelet aggregation, activation, and release reactions. Platelet and tumor cell-derived 12(S)-HETE can similarly act on endothelial cells to initiate these diverse biologic responses and thus contribute to tumor cell metastasis.

ACKNOWLEDGMENT

This work was performed in my laboratory through support by NIH (CA29997 and 47115).

REFERENCES

1. Honn, K.V. and Sloane, B.F. Hemostatic Mechanisms and Metastasis. Martinus Nijhoff, Norwell, MA. 1984.
2. Weiss, L., Orr, F.W. and Honn, K.V. Interactions of cancer cells with the microvasculature during metastasis. FASEB J. 2, 12, 1988.
3. Honn, K.V., Powers, W.E., and Sloane, B.F., Mechanisms of Cancer Metastasis: Potential Therapeutic Implications. Martinus Nijhoff, Norwell, MA, 1986.
4. Filder, I.J. and Hart, I.R., Biological diversity in metastatic neoplasms: origins and implications. Science, 217, 998, 1982.

5. Folkman, J., Watson, K., Ingber, D., and Hanahan, D., Induction of angiogenesis during the transition from hyperplasia to neoplasia, Nature, 339, 58, 1989.

6. Liotta, L.A., Steeg, P.S., and Stetler-Stevenson, W.G., Cancer metastasis and angiogenesis: an imbalance of positive and negative regulation, Cell, 64, 327, 1991.

7. Weiss, L., Orr, F.W., and Honn, K.V., Interactions between cancer and the microvasculature: a rate-regulator for metastasis, Clin. Expl. Metastasis, 7, 127, 1989.

8. Hart, I.R., Goode, N.T., and Wilson, R.E., Molecular aspects of the metastatic cascade, Biochim. Biophys, Acta., 989, 65, 1989.

9. Liotta, L.A., Tumor invasion and metastasis: role of the extracellular matrix, Cancer Res., 46, 1, 1986.

10. Pauli, B.U. and Lee, C.L. Organ preference of metastasis: the role of organ- specifically modulated endothelial cells. Lab. Invest., 58, 379, 1988.

11. Belloni, P.N. and Tressler, R.J., Microvascular endothelial cell heterogeneity: interactions with leukocytes and tumor cells, Cancer Metastasis Rev., 8, 353, 1990.

12. Pauli, B.U., Augustin-Voss, H.G., El-Sabban, M.E., Johnson, R.C., and Hammar, D.A., Organ preference of metastasis: the role of endothelial cell adhesion molecules. Cancer Metastasis Rev., 9, 175, 1990.

13. Nicolson, G.L., Organ specificity of tumor metastasis: role of preferential adhesion, invasion, and of malignant cells at specific secondary sites, Cancer Metastasis Rev., 7, 143, 1988.

14. Auerbach, R., Pattern of tumor metastasis: organ selectivity in the spread of cancer cells, Lab. Invest., 58, 361, 1988.

15. Nicolson, G.L., Tumor and host molecules important in the organ preference of metastasis. Semin. Cancer Biol., 2, 143, 1991.

16. Honn, K.V., Grossi, I.M., Timar, J., Chopra, H., and Taylor, J.D., Platelets and cancer metastasis, in Microcirculation in Cancer Metastasis, Orr, F.W., Buchanan, M., and Weiss, L., Eds., CRC Press, Boca Raton, FL, 1991, 93.

17. Honn, K.V., Tang, D.G., and Chen, Y.Q. Platelets and cancer metastasis: more than an epiphenomenon. Semin. Thromb. Hemost., 18, 390, 1992.

18. Honn, K.V., Tang, D.G., and Crissman, J.D., Platelets and cancer metastasis: a casual relationship? Cancer Metastasis Rev., 11, 325, 1992.

19. Samuelsson, B., Goldyne, M., Granstrom, E., Hamberg, M., Hammarstrom, S., and Malmstern, C., Prostaglandins and thromboxanes, Annu. Rev. Biochem., 47, 997, 1978.

20. Needleman, P., Turk, J., Jakschik, B.A., Morrison, A.R., and Lefkowith, J.B., Arachidonic acid metabolism, Annu. Rev. Biochem., 55, 69, 1986.

21. Piper, P.J. and Samhoun, M.D., Leukotrienes, Br. Med. Bull., 43, 297, 1987.

22. Spector, A.A., Gordon, J.A., and Moore, S.A., Hydroxyeicosatetraenoic acids (HETEs). Prog. Lipid Res., 27, 271, 1988.

23. Honn, K.V. and Chen, Y.Q., Prostacyclin, hydroxy fatty acids and tumor metastasis, in Prostacyclin: New Perspectives in Basic Research and Novel Therapeutic Indications, Rubanyi, G.M. and Vane, J.R., Eds., Elsevier, Amsterdam, 1995.

24. Chen, Y.Q., Liu, B., Tang, D.G., and Honn, K.V., Fatty acid modulation of tumor cell-platelet-vessel wall interaction, Cancer Metastasis Rev., 11, 389, 1992.

25. Chang, W.C., Ning, C.C., Lin, M.T., and Huang, J.D., Epidermal growth factor enhances a microsomal 12-lipoxygenase activity in A431 cells, in Eicosanoids and Other Bioactive Lipids in Cancer, Inflammation and Radiation Injury, Nigam, S., Honn, K.V., Marnett, L.J., and Walter, T., Eds., Kluwer Academic, Norwell, MA, 1992, 463.

26. Glasgow, W.C. and Eling, T.E., Epidermal growth factor regulation of linoleic acid metabolism in Syrian hamster embryo fibroblasts, in Eicosanoids and Other Bioactive Lipids in Cancer, Inflammation and Radiation Injury, Nigam, S., Honn, K.V., Marnett, L.J., and Walden T., Eds., Kluwer Academic, Norwell, MA, 1992, p. 467.

27. Marier, J.A.M., Hla, T., and Maciag, T., Cyclooxygenase is an immediate early gene induced by interleukin-1 in human endothelial cells. J. Biol. Chem., 265, 10805, 1990.

28. Kast, R., Furstenberger, G., and Marks, F., Transforming growth factor alpha stimulated phospholipase A2 activity in mouse keratinocytes, in Eicosanoids and Other Bioactive Lipids in Cancer, Inflammation and Radiation Injury, Nigam, S., Honn, K.V., Marnett, L.J., and Walden, T., Eds., Kluwer Academic, Norwell, MA, 1992, 459.

29. Arita, H., Cytokine-induced phospholipase A2 and its possible relationship to eicosanoid formation, in Eicosanoids and Other Bioactive Lipids in Cancer, Inflammation and Radiation Injury., Nigam S., Honn, K.V., Marnett, L.J., and Walden, T., Eds, Kluwer Academic, Norwell, MA, 1992, p. 491.

30. Spector, A.A., Gordon, J.A., Moore, S.A., Hydroxyeicosatetraenoic acids (HETEs)., Prog. Lipid Res. 27, 271, 1988.

31. Natarajan, R., Gu J, L., Rossi, J., Gonzales, N., Lanting, L., XU, I., and Nadler, J., Elevated glucose and angiotensin II increase 12-lipoxygenase activity and expression in porcine aortic smooth muscle cells., Proc. Natl. Acad. Sci. USA, 90, 4947, 1993.

32. Kim, J.A., Gu, J., Natarajian, R., Berliner, J.A., and Nadler, J., Evidence that a leukocyte type of 12-lipoxygenase is expressed in normal human vascular and mononuclear cells., Clin. Res., 41, 148A, 1983.

33. Hadjiangapiou, C., and Spector, A.A., 12-hydroxyeicosatetraenoic acid reduces prostacyclin production by endothelial cells., Prostaglandins 31, 1135, 1986.

34. Sekiya, F., Takagi, J., Usui, T., Kawajiri, K., Kobayashi, Y., Sato, F., and Saito, Y., 12-hydroxyeicosatetraenoic acid plays a central role in the regulation of platelet activation. Biochem. Biophys. Res. Commun. 179, 345, 1991.

35. Hofer, G., Bieglmayer, C.H., Kopp, B., Janish, H., Measurement of eicosanoids in menstrual fluid by the combined use of high pressure chromatography and radio immunoassay., Prostaglandins, 45, 413, 1993.

36. Wetzka, B., Schafer, W., Scheibel, M., Nusing, R., Zahradnik, H.D., Eicosanoid production by intrauterine tissues before and after labor in short-term tissue culture., Prostaglandins, 45, 571, 1993.

37. Chang, W.C., Liu, Y.W., Ning, C.C., Suzuki, H., Yoshimoto, T., Yamamoto, S., Induction of arachidonate 12-lipoxygenase mRNA by epidermal growth factor in A431 cells., J. Biol. Chem., 268, 18734, 1993.

38. Chang, W.C., Ning, C.C, Lin, M.T., Huang, J.D., Epidermal growth factor enhances a microsomal 12-lipoxygenase activity in A431 cells., J. Biol. Chem., 267, 3657, 1992.

39. Chen, Y.Q., Duniec, Z.M., Liu, B., Hagmann, W., Gao, X., Shimoji, K ., Marnett, L.J., Johnson, C.R., Honn, K.V., Endogenous 12(S)-HETE production by tumor cells and its role in metastasis. Cancer Res., 54, 1574, 1994.

40. Hagmann, W., Kagawa, D., Renaud, C., Honn, K.V., Activity and protein distribution of 12-lipoxygenase in HEL cells: Induction of membrane-association by phorbol ester TPA, modulation of activity by glutathione and 13-HPODE, and Ca^{2+}-dependent translocation to membranes., Prostaglandins, 46, 471, 1993.

41. Hagmann, W., Maher, R., Honn, K.V., Intracellular distribution, activity, and Ca^{2+}-dependent translocation of 12-lipoxygenase in Lewis lung tumor cells. In: Eicosanoids and Other Bioactive Lipids in Cancer, Inflammation, and Radiation Injury, Honn, K.V., Marnett, L.J., and Nigam S., Eds., Plenum Publishing, in press, 1997.

42. Liu, B., Timar, J., Howlett, J., Diglio, C.A., Honn, K.V., Lipoxygenase metabolites of arachidonic and linoleic acids modulate the adhesion of tumor cells to endothelium via regulation of protein kinase C., Cell Regul. 2, 1045, 1991.

43. Liu, B., Marnett, L.J., Chaudhary, A., Ji, C., Blair, I.A., Johnson, C.R., Diglio, C.A., and Honn, K.V., Biosynthesis of 12(S)-hydroxyeicosatetraenoic acid by B16 amelanotic melanoma cells is a determinant of their metastatic potential. Lab. Invest., 70, 314, 1994.

44. Honn, K.V., Tang, D.G., Grossi, I., Duniec, Z.M., Timar, J., Renaud, C., Leithauser, M., Blair, I., Johnson, C.R., Diglio C.A., Kimler, V.A., Taylor, J.D., and Marnett, L.J., Tumor cell-derived 12(S)-hydroxyeicosatetraenoic acid induces microvascular endothelial cell retraction., Cancer Res., 54, 565, 1994.

45. Honn, K.V., Nelson, K.K., Renaud, C., Bazaz, R., Diglio, C.A., Timar, J., Fatty acid modulation of tumor cell adhesion to microvessel endothelium and experimental metastasis., Prostaglandins, 44, 413, 1992.

46. Tang, D.G., Honn, K.V., 12-Lipoxygenase, 12(S)-HETE, and cancer metastasis. Annals New York Acad. Sci., 744, 199, 194.

47. Timar, J., Chen, Y.Q., Liu, B, Bazaz, R., Taylor, J.D., Honn, K.V., The lipoxygenase metabolite 12(S)-HETE promotes αIIbβ3 integrin-mediated tumor cell spreading on fibronectin, Int. J. Cancer, 52, 594, 1992.

48. Honn, K.V., Tang, D.G., Grossi, I.M., Renaud, C., Duniec, Z.M., Johnson, C.R., Diglio, C.A., Enhanced endothelial cell retraction mediated by 12(S)-HETE: A proposed mechanism for the role of platelets in tumor cell metastasis. Exptl. Cell. Res., 210, 1, 1994.

49. Liotta , L.A., Mandler, R., Murano, G., Katz, D.A., Gordon, R.K., Chiang, P.K., Schiffman, E., Tumor-cell autocrine motility factor. Proc. Natl. Acad. Sci. USA, 83, 3302, 1986.

50. Watanabe, H., Carmi, P., Hogan, V., Raz, T., Silletti, S., Nabi, I.R., and Raz, A., Purification of human tumor cell autocrine motility factor and molecular cloning of its receptor., J. Biol. Chem., 226, 13442, 1991.

51. Ruoslahti, E., Giancotti, F.G., Integrins and tumor cell dissemination., Cancer Cells 4, 119, 1989.

52. Singer, I.J., Scott, S., Kawaka, D.W., Kazazis, D.M., Adhesosomes: specific granules containing receptors for laminin, c3bi, fibrinogen, fibronectin and vitronectin in human polymorphonuclear leukocytes and monocytes., J. Cell. Biol. 109, 3169, 1989.

53. Chopra, H., Timar, J., Chen, Y.Q., Rong, X.H., Grossi, I.M., Fitzgerald, L.A., Taylor, J.D., and Honn, K.V., The lipoxygenase metabolite 12(S)-HETE induces a cytoskeleton-dependent increase in surface expression of integrin αIIbβ3 on melanoma cells., Int. J. Cancer, 49, 774, 1991.

54. Zachary, I., Rozengurt, E., Focal adhesion kinase (p125^{FAK}): A point of convergence in the action of neuropeptides, integrins, and oncogenes., Cell 71, 891, 1992.
55. Shaller, M.D., Borman, C.A., Cobb, B.S., Vines, R.R., Reynolds, A.B., and Parsons, T.J., pp125^{FAK}, a structurally distinctive protein-tyrosine kinase associated with focal adhesions, Proc. Natl. Acad. Sci. USA, 89, 5192, 1992.
56. Guan, J.L., and Shalloway, D., Regulation of focal adhesion-as-associated protein tyrosine kinase by both cellular adhesion on oncogenic transformation., Nature, 358, 690, 1992.
57. Grossi, I.M., Fitzgerald, L.A., Unbarger, L.A., Nelson, K.D., Diglio, C.A., Taylor, J.D., and Honn, K.V., Bi-directional control of membrane expression and/or activation of the tumor cell IRGpIIB/IIIa receptor and tumor cell adhesion by lipoxygenase products of arachidonic and linoleic acid., Cancer Res. 49, 1029, 1989.
58. Chang, Y.S., Chen, Y.Q., Timar, J., Grossi, I.M., Fitzgerald, L.A., Diglio, C.A., and Honn, K.V., Increased expression of $\alpha IIb\beta 3$ integrin in subpopulations of murine melanoma cells with high lung-colonizing ability., Int. J.. Cancer, 51, 445, 1992.
59. Chen, Y.Q., Gao, X., Timar, J., Tang, D.G., Grossi, I.M., Chelladurai, M., Kunicki, T.J., Fligiel, S.E.G., Taylor, J.D., and Honn, K.V., Identification of the $\alpha IIb\beta 3$ integrin in murine tumor cells., J. Biol. Chem., 267, 17314, 1992.
60. Chopra, H., Timar, J., Rong, X., Grossi, I.M., Hatfield, J.S., Fligiel, S.E.G., Finch, C.A., Taylor, J.D., and Honn, K.V., Is there a role for the tumor cell integrin $\alpha IIb\beta 3$ and cytoskeleton in tumor cell-platelet interaction?, Clin. Exptl. Metastasis, 10, 125, 1992.
61. Timar, J., Silletti, S., Bazaz, R., Raz, A., and Honn, K.V., Regulation of melanoma-cell motility by the lipoxygenase metabolite 12(S)-HETE., Int. J. Cancer, 55, 1003, 1993.
62. Silletti, S., and Raz, A., Autocrine motility factor (AMF) is a growth factor., Biochem. Biophys. Res. Commun., 1994, 446, 1993.
63. Liu, B., Maher, R.J., Hannun, Y.A., Porter, A.T., and Honn, K.V., 12(S)-HETE increases in invasive potential of prostate tumor cells through selective activation of PKCα., J. Natl. Cancer Inst., 86, 1145, 1994.
64. Honn, K.V., Timar, J., Rozhin, J., Bazaz, R., Sameni, M., Ziegler, G., and Sloane, B.F., A lipoxygenase metabolite, 12(S)-HETE, stimulates protein kinase C-mediated release of cathespin B from malignant cells., Exptl. Cell. Res., 214, 120, 1994.
65. Schmitt, M. and Graff, J.H., Tumor-associated proteases., Fibrinolysis, 6,3, 1992.
66. Sloane, B.F., Moin, K., Sameni, M., Tait, L.R., Rozhin, J., and Ziegler, J., Membrane association of cathepsin B can be induced by transfection of human breast epithelial cells with c-Ha-ras oncogene., J. Cell Sci., 107, 373, 1994.
67. Sloane, B.F., Moin, K., Krepela, E. and Rozhin, J., Cathepsin B and its endogenous inhibitors: the role in tumor cell malignancy., Cancer Metastasis Rev., 9, 333, 1990.
68. Sloane, B.F., Rozhin, J., Johnson, K., Taylor, J., Crissman, J.D., and Honn, K.V., Cathepsin B: association with plasma membrane in metastatic tumors., Proc. Natl. Acad. Sci. USA, 83, 2483, 1986.
69. Sloane, B.F., Rozhin, J.R., Gomez, A.P., Grossi, I.M., and Honn, K.V., Effects of 12-hydroxyeicosatetraenoic acid on release of cathepsin B and cysteine proteinase inhibitors from malignant melanoma cells. In: Honn, K.V., Marnett, L.J., Nigam, S., Walden, T.L., Eds., Eicosanoids and Other Bioactive Lipids in Cancer and Radiation Injury., Kluwer, Boston, MA, p. 373, 1991.
70. Timar, J., Tang, D.G., Bazaz, R., Haddard, M.M., Kimler, V.A., Taylor, J.D., T and Honn, K.V., PKC mediates 12(S)-HETE-induced cytoskeletal rearrangement in B16a melanoma cells., Cell Motil. Cytoskel., 26, 49, 1993.
71. Tang, D.G., and Honn, K.V., Role of protein kinase C and phosphatases in 12(S)-HETE-induced tumor cell cytoskeletal reorganization. In: Honn, K.V., Marnett, L.J., Nigam, S., Eds., Eicosanoids and Other Bioactive Lipids in Cancer, Inflammation and Radiation Injury., Kluwer, Boston, MA, in press, 1997.
72. Piomelli, D., Volterra, A., Dale, N., Lipoxygenase metabolites of arachidonic acid as second messengers for presynaptic inhibition of aplysia sensory cells., Nature, 328, 38, 1987.
73. Nadler, J.L., Natarajan, R., and Stern, N., Specific action of the lipoxygenase pathway in mediating angiotension II-induced aldosterone synthesis in isolated adrenal glomerula cells, J. Clin. Invest., 80, 1763, 1987.
74. Chan, C.C., Duhamel, L., and Ford-Hutchinson, A., Leukotriene B4 and 12- hydroxyeicosatetraenoic acid stimulate epidermal proliferation in vivo in the guinea pig., J. Invest. Dermatol., 85, 333, 1985.
75. Antonipillai, I., 12-Lipoxygenase products are potent inhibitors of prostacyclin- induced renin release., Proc. Soc. Exp. Biol. Med., 194, 224, 1990.
76. Etingin, O.R., and Hajjar, D.P., Evidence for cytokine regulation of cholesterol metabolism in herpes viral-infected arterial cells by the lipoxygenase pathway., J. Lipid Res., 31, 299, 1990.

77. Tang, D.G., Diglio, C.A., and Honn, K.V., Transcriptional activation of endothelial cell integrin alpha$_v$ by protein kinase C activator 12(S)-HETE., J. Cell. Sci., 108, 2679, 1995.
78. Honn, K.V., Tang, D., Grossi, I.M., Duniec, Z.M., Timar, J., Renaud, C., Leithauser, M., Blair, I., Diglio, C.A., Taylor, J.D., and Marnett, L.J., Tumor cell-derived 12(S)-hydroxyeicosatetraenoic acid induces microvascular endothelial cell retraction., Cancer Res., 54, 565, 1994.
79. O'Brian, C.A., and Ward, N.E., Biology of protein kinase C family., Cancer Metastasis Rev., 8, 199, 1989.
80. Liu, B., Khan, W.A., Hannun, Y.A., Timar, J., Taylor, J.D., Lundy, S., Butovich, I., and Honn, K.V., 12(S)-HETE and 13(S)-HODE regulation of protein kinase Cα in melanoma cells: role of receptor mediated hydrolysis of inositol phospholipids., Proc. Natl. Acad. Sci. USA, 92, 9323, 1995.
81. Honn, K.V., Tang, D.G., Gao, Z., Butovich, I.A., Liu, B., Timar, J., and Hagmann, W., 12-Lipoxygenases and 12(S)HETE: role in cancer metastasis., Cancer Metastasis Rev., 13, 365, 1994.
82. Liu, B., Renaud, C., Nelson, K.K., Chen, Y.Q., Bazaz, R., Kowynia, J., Timar, J., Diglio C.A., and Honn, K.V., Protein kinase C inhibitor calphostin C reduces B16 amelanotic melanoma cell adhesion to endothelium and lung colonization., Int. J. Cancer, 52, 147, 1992.
83. Berridge, M.J. and Irvine, R.F., Inositol phosphates and cell signaling., Nature, 341, 197, 1989.
84. Cho, Y., and Ziboh, V.A., 13-Hydroxyocatadecaenoic acid reverse epidermal hyperproliferaiton via selective inhibition of protein kinase C-b activity. Biochem. Biophys. Res. Commun., 201, 257, 1994.
85. Gross E., Ruzicka, T., Restorff, B.V., Stolz, W., and Klotz, K.N., High-affinity binding and lack of growth-promoting activity of 12(S)hydroxyeicosatatraenoic acid (12[S]HETE) in a human epidermal cell lines., J. Invest. Dermatol., 94, 446, 1990.
86. Croset, M., and Lagarde, M., Stereospecific inhibition of PGH$_2$-induced aggregation by lipoxygenase products of eicosaenoic acids., Biochem. Biophys. Res. Commun., 112, 878, 1983.
87. Fonlupt, P., Croset, M., and Lagarde, M., 12(S)-HETE inhibits the binding of PGH$_2$/TXA$_2$ receptor ligands in human platelets, Thromb. Res., 63, 239, 1991.
88. Herbertsson, H. and Hammarstrom, S., High-affinity binding sites for 12(S)hydroxy-5,8,10,14-eicosatetraenoic acid (12(S)HETE) in carcinoma cells, FEBS Lett., 298.249, 1992.
89. Herbertsson, H. and Hammarstrom, S., Cytosolic 12(S)hydroxy-5,8,10,14- eicosatetraenoic acid binding sites in carcinoma cells. In: Honn, K.V., Marnett, L.J., Nigam, S., Eds., Eicosanoids and Other Bioactive Lipids in Cancer, Inflammation and Radiation Injury., Kluwer Acad. Publ., Norwell, MA, 1994.

MODULATION OF INTRACELLULAR SECOND MESSENGERS BY DIETARY FAT DURING COLONIC TUMOR DEVELOPMENT[*]

Robert S. Chapkin,[†] Yi-Hai Jiang, Laurie A. Davidson, and Joanne R. Lupton

Faculty of Nutrition and Molecular and Cell Biology Group
Texas A&M University
College Station, Texas 77843–2471

DIETARY FACTORS AND COLON CANCER

Marked differences in the incidence of colon cancer in populations migrating from low to high risk areas suggest that environmental factors, specifically dietary factors, play an important role in the etiology of colon cancer.[1,2] Dietary fat has received considerable attention as a possible risk factor in the etiology of colon cancer. A proposed hypothesis regarding the induction of colon cancer suggests that fats increase the amount of bile acids in the colon, which in turn are metabolized to secondary carcinogens to promote the growth of adenomas and subsequently invasive cancer.[3] Moreover, the amount and type of fat in the diet directly affects the concentration of free fatty acids, eicosanoids and diacylglycerols in the colonic lumen which can influence colonic intracellular second messengers and modify the risk of colon cancer development.[4,5,26,93] In comparison, dietary fiber is postulated to reduce the adverse effects of certain fats through several mechanisms, including dilution of carcinogens, decrease in transit time, reduction in fecal pH, chemical binding, and alteration in microflora metabolism[6,7]. However, experiments in animal models using a variety of fats and fibers have produced conflicting results. In some studies, fiber failed to protect,[8,9] and in others, fat failed to promote colon carcinogenesis.[10] The interpretation of this data is made even more complicated by the fact that different types of dietary fiber and fat may affect colon carcinogenesis in an interactive site-specific manner.[11]

[*] This work was supported in part by the American Institute for Cancer Research and NIH CA59034.
[†] Address Correspondence to: Dr. Robert S. Chapkin, 442 Kleberg Center, Texas A&M University, College Station, Texas 77843–2471. Phone: (409) 845–0419; Fax: (409) 862–2662; e-mail: chapkin@zeus.tamu.edu.

Dietary Fat and Colon Cancer

Since 1969, when Wynder et al.[12] first suggested that dietary fat might play a role in the etiology of colon cancer in humans, a considerable number of epidemiological studies have examined this hypothesis. There is a general consensus that dietary fat is promotive of colon cancer. Although the majority of correlations and case-control studies support this hypothesis,[13–15] a few studies show no positive relationship.[16–18] The interpretation of some epidemiological studies has been complicated by the inherent problems in testing a dietary hypothesis because of issues related to reliability, validity, and sensitivity to reveal narrow, but biologically significant differences. For example, a major difficulty has been the lack of measurement of fat consumption, i.e., types of saturated and classes of unsaturated fat in food of the population being studied. The importance of different types of fat with different fatty acid compositions, rather than total fat, cannot be discounted, because both epidemiological and experimental studies indicate that the colon tumor-promoting effect of dietary fat is dependent on the type of fat.[19,20] In addition, several other studies in humans have not taken into consideration confounding variables, such as dietary fiber content, since in several populations consuming diets high in fat, a high level of dietary fiber has been shown to reduce the risk for colon cancer.[19,20]

With regard to experimental colon carcinogenesis, numerous studies have shown that, irrespective of the carcinogen used, rats and mice fed high fat diets have a greater incidence of colon tumors than do animals fed low fat diets.[10,21] The promoting effects of dietary fat are highly dependent on the type of fat.[10,22–24] It has been established that dietary n-9 fatty acids (found in olive oil) do not promote carcinogen-induced colonic tumors when compared to n-6 polyunsaturated fatty acid-enriched diets (found in corn oil and safflower oil).[10] The varied effects of different classes of dietary fatty acids on colon cancer suggest that fatty acid composition is an important determining factor in colon tumor development. These data are particularly relevant because despite advancement in the treatment of colon cancer, the 5 year mortality rate has remained at 50% for almost 4 decades.[25] Therefore, chemopreventive dietary strategies must be developed in order to decrease the risk of colon cancer.

Dietary N-3 Polyunsaturated Fatty Acids and Colon Cancer

Among dietary factors, there is strong epidemiological, clinical and experimental data indicating a protective effect of n-3 polyunsaturated fatty acids (eicosapentaenoic acid, 20:5n-3 and docosahexaenoic acid, 22:6n-3) on colon cancer.[19,24,26–29] The average American normally consumes approximately 100 mg/day of 20:5n-3 and 22:6n-3 in the form of fish, versus 8000 mg/day in clinical trials and 6,000 - 12,000 mg/day by Greenland Eskimos (Table 1). Compared to diets rich in n-6 polyunsaturated fatty acids, which enhance the development of colon tumors,[10,30–32] n-3 fatty acid enriched diets reduce colon cancer incidence.[31–33] This protective effect is exerted at both the initiation and post-initiation stages of carcinogenesis.[10,33] We have shown that the balance between colonic epithelial cell proliferation and apoptosis can be favorably modulated by dietary n-3 polyunsaturated fatty acids, conferring resistance to toxic carcinogenic agents.[34] However, the underlying molecular mechanisms by which this class (n-3) of dietary fatty acids exerts its effect is not known. Recently, we demonstrated that n-3 polyunsaturated fatty acid supplementation can blunt carcinogen-induced ras mutation in the colon.[35] This is significant because ras gene mutations are a consistent somatic alteration in colorectal tumors. Approximately 50% of human colorectal carcinomas and a similar percentage of ade-

Table 1. Comparison of fatty acid consumption among epidemiological, clinical and experimental studies

Fatty Acid	Greenland Eskimo	Japanese	Canadian[d]	United States[d]	Clinical Trials	Mouse
% Energy as Fat	39[a]	30	39	38	40	7
Fat*	130[a]	70	80-110	80	88	0.150
18:2n-6*	2-3[b]	2-5[b]	14-19	14.7	14.7	0.080[c]
18:3n-3*	13.7[a]	0.4[b]	1.4-2.8	1.6	1.6	0
20:4n-6*	1.3	2.6	0.25	0.1	0.1	0
20:5n-3* and 22:6n-3*	6-12[b]	1-3[b]	0.15	0.1	8[e]	0.048[c]

*Values represent grams per day
[a]From ref. 96
[b]From ref. 97
[c]From ref. 98
[d]From ref. 99
[e]From ref. 100

nomas have K-ras mutations.[36] In comparision, almost 90% of axoxymethane-induced colonic tumors exhibit point mutations in K- and H-ras genes.[37,38] The acquisition of a mutant ras gene is a relatively early step in disease development[39] and leads to abnormal activation of ras.[40] This is noteworthy because the prolonged activation of ras is associated with altered cellular growth and malignant transformation,[41,42] reduced susceptibility to apoptosis,[43] altered glycerolipid (diacylglycerol) and sphingolipid-derived (ceramide) second messenger kinetics,[44] and activation and down-regulation of protein kinase C (PKC).[45–47] These data offer insight into mechanisms by which dietary n-3 polyunsaturated fatty acids may modify the risk of colon cancer development.

EXPERIMENTAL CARCINOGENESIS MODEL

The azoxymethane (AOM)-induced rat colon tumor model is a valuable model for studying the interaction between mediators of intracellular signal transduction and environmental factors. Evidence to support use of this model includes: (a) utilization of AOM provides a clear distinction between tumor initiation and promotion;[33] (b) AOM-induced ras expression and activation occurs as an early event in colon carcinogenesis;[37,48,49] (c) the development of tumors is responsive to the amount and type of dietary fat[28,33,50] and, AOM-induced colon tumors closely parallel human colonic neoplasia in pathologic features.[51] We have utilized this model system in order to investigate the mechanisms by which dietary fat modulates intracellular second messengers during colonic tumor development.[50]

THE ROLE OF DIACYLGLYCEROL AND PHOSPHOLIPASE C-γ 1 IN COLON CANCER

Modulation of intracellular lipid second messengers and deregulation of cell signaling pathways play a critical role in tumorigenesis.[52,53] Several reports demonstrate that transformation of cells by activated ras oncogene is associated with the elevation of diacylglycerol levels, consistent with the elevated activity of phospholipase C.[54–56] This is significant because diacylglycerol, an intracellular activator of PKC,[57] in part mediates colonic cell proliferation,[58-60] malignant transformation,[57,61] metastatic phenotype,[62] and opposes the induction of apoptosis.[63] The ability of diacylglycerol to attenuate apoptosis is relevant because normal colonic tissue ontogeny and function is the product of mechanisms regulating cell proliferation, differentiation and programmed cell death.[64] Paradoxically, several reports have demonstrated that diacylglycerol mass and PKC activity are actually lower in colonic tumors compared to adjacent uninvolved mucosa.[65-70] As aforementioned, this observation is inconsistent with previous studies indicating that the oncogenic transformation of cells is associated with the accumulation of diacylglycerol.[71,72] Unfortunately, the mechanisms responsible for the reduction of diacylglycerol mass and PKC activity in colonic tumors have not been elucidated.

The PLC-γ 1 signaling pathway has been implicated in the regulation of colonic cell growth and tumor development.[73,74] Overexpression of PLC-γ 1 has been shown to alter cell growth and oncogenic potential.[75] Moreover, PLC-γ 1 is highly expressed in human tissues obtained from individuals with familial adenomatous polyposis,[74] colorectal cancer,[73] breast cancer,[76] and hyperproliferative epidermal disease.[77] In contrast, PLC-γ 1 protein expression and activity are markedly decreased upon cell differentiation.[57] These results support the view that PLC-γ 1 plays an important role in controlling colonic cell proliferation and differentiation. Recently, using a 2 x 2 factorial design consisting of 2 fats (corn oil vs fish oil) in carcinogen-treated or control rats, we demonstrated that dietary fish oil (11.5 g/100 g diet), significantly decreased (P < 0.001) colonic PLC-γ 1 expression and diacylglycerol mass at 15 weeks post carcinogen injection.[50] These effects were consistent with a reduction in tumor incidence determined at 37 weeks post carcinogen injection.[50] We have recently demonstrated that AOM injected rats fed n-6 fatty acid enriched diets have significantly higher steady-state diacylglycerol levels compared with AOM injected animals fed n-3 fatty acid enriched diets and saline injected rats on all diets[50] (Table 2). The chronic increase in endogenous diacylglycerol could contribute to the clonal expansion of cells containing an activated ras oncogene.[61]

Our data suggest that oncogenic transformation of the colon is associated with an initial time-dependent accumulation of intracellular diacylglycerol,[44,50,71,72] which may be in part responsible for the observed down-regulation of colonic PKC expression and activity.[65-70] The presence of these changes in the colonic mucosa prior to overt neoplasia suggests that dietary n-3 polyunsaturated fatty acids are capable of influencing key molecular events which play a role in the early stages of malignant transformation.

Table 2. Effect of dietary fat and carcinogen on colonic DAG mass at 15 weeks post-AOM treatment

Fat / Carcinogen	Colonic DAG Mass
Corn/AOM	0.522 ± 0.049^a
Corn/Saline	0.253 ± 0.054^b
Fish/AOM	0.199 ± 0.035^b
Fish/Saline	0.237 ± 0.035^b

Colonic mucosal DAG mass is expressed as nmol   DAG / 100 nmol of total phospholipid (mol %). Dietary treatments were initiated one week prior to carcinogen injection. Mean $\pm$ SEM (n = 20). Values not sharing a common superscript are significantly different (p < 0.05).

Protein Kinase C Activation

Physiologically, most PKC isozymes are activated by the concerted action of phospholipids and the second messenger, diacylglycerol. In addition, some isozymes (α, βI, βII, γ) are also calcium-dependent. The interactions between PKC, Ca^{2+}, phospholipids and diacylglycerol have been elucidated. It has been determined that one molecule of diacylglycerol interacts with one molecule of PKC to cause activation.[78] Also, the structure and stereospecificity of diacylglycerol are critical for its ability to activate PKC. Thus, only sn-1,2-diacylglycerol is the active species; 1,3-diacylglycerol and sn-2,3-diacylglycerol are unable to activate PKC.[79] In addition to activation by calcium, phospholipids, and diacylcerol or phorbol esters, PKC can also be activated by cis-unsaturated fatty acids such as oleic acid (18:1n-9), arachidonic acid (20:4n-6), 20:5n-3 and 22:6n-3.[57,80] Fatty acid alone is inactive unless diacylglycerol or phorbol ester is present[81]. These observations allow for a model of PKC activation to be formulated where binding of ligands to appropriate receptors leads to activation of phospholipase C, phospholipase D and/or A$_2$, generating inositol trisphosphate (IP$_3$), diacylglycerol and/or fatty acid second messengers.[81]

Role of Protein Kinase C in Tumor Promotion

In 1982, Nishizuka and coworkers[82] reported that PKC was directly activated by tumor-promoting phorbol esters. In addition, other studies have shown that PKC is the major intracellular receptor for phorbol esters. Phorbol esters activate PKC in a manner analogous to the endogenous activator, diacylglycerol, except that phorbol ester activation can be maintained for prolonged periods of time due to their metabolic stability. Thus, prolonged stimulation of PKC has been proposed to be the mechanism for the tumor promoting action of the phorbol esters.[82] However, the action of phorbol ester on PKC results not only in activation, but also in translocation of PKC to various subcellular compartments, thereby regulating signaling events, gene expression, and PKC isozyme biological half-life. This raises the possibility that tumor promotion may be a consequence of either prolonged activation or the subsequent down-regulation (inactivation) of PKC by phorbol esters. In any case, the effect of phorbol esters on PKC was the first indication that prolonged activation or inhibition of PKC activity could play a critical role in cell regulation. Recently, other reports have indicated that PKC is regulated by many other tumor promoters such as bryostatins, unsaturated free fatty acids, and possibly by additional compounds, such as benzene and carbon tetrachloride, firming the link between PKC and tumor promotion.[83]

Role of Protein Kinase C in Colon Cancer

Human and rat colonic mucosa express both calcium-dependent (classical) and calcium-independent (novel and atypical) PKC isoforms with distinct subcellular distributions for each.[84] Alterations in PKC isozyme signaling have been directly linked to the pathogenesis of colon cancer.[68,83–86] This is not surprising in view of the fact that the PKC signaling pathway plays an important role in colonic cell proliferation, differentiation and apoptosis.[59,64,70,87–89] Previous studies have shown that decreased total PKC activity is a feature of colorectal cancer[67,85] and that dietary fish oil, containing n-3 polyunsaturated fatty acids, enhances PKC activation (membrane/cytosol activity ratio) in the colon.[59] In addition, dietary fat composition can alter the steady-state expression of individual PKC isozymes in the colon.[92,93]

Table 3. Effect of dietary fat and carcinogen on
atypical PKC expression in the preneoplastic
rat colon. Colonic cytosolic PKC λ–ζ
protein content was quantitated by densitometry of immunoblots and expressed a
band intensity (O.D. x band area) relative to a
constant amount of rat brain homogenate.
Scraped colonic mucosa was collected 15 weeks
after the second injection of AOM

Fat Source	15 Weeks
Corn Oil	0.86 ± 0.28 [b]
Fish Oil	2.65 ± 0.20 [a]
Carcinogen	15 Weeks
Azoxymethane	1.31 ± 0.23 [b]
Saline	2.20 ± 0.25 [a]

Mean ± SEM (n = 32–40) in a vertical column not sharing a
common superscript are significantly different (p < 0.05).

There is strong evidence that carcinogen/ras-induced elevation in the steady-state
levels of colonic diacylglycerol contributes to a decrease in select PKC isozymes (Table 3)[90], an important early event in multistage tumorigenesis.[83,91] Using the AOM-rat tumor model (refer to Section 3), we recently demonstrated that dietary n-3 polyunsaturated
fatty acids block the carcinogen-induced decrease in the steady-state levels of select colonic PKC isozymes[46] (Table 3). This may explain why this fat source protects against colon
cancer development. A model detailing the protective effects of n-3 polyunsaturated fatty
acids on colonic PKC-dependent signal transduction is summarized in Figure 1.

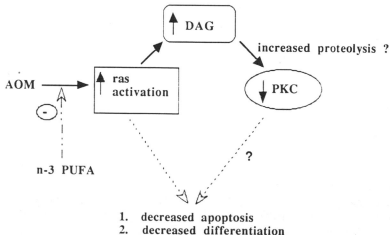

Figure 1. Putative antagonism of azoxymethane-induced ras activation by n-3 polyunsaturated fatty acids.

Noninvasive Detection of Protein Kinase C Isozymes as a Biomarker for Colon Cancer Using Fecal Messenger RNA: Clinical Implications

We have developed a noninvasive method utilizing feces containing exfoliated colonocytes as a sensitive technique for quantitating luminal mRNAs.[94] By incorporating the exquisite sensitivity of reverse transcriptase polymerase chain reaction (RT-PCR), we detected mRNAs for PKC-α, PKC-δ, PKC-ϵ and PKC-ζ isoforms in rat feces containing sloughed colonocytes. These data are consistent with immunoblot PKC protein expression data using scraped colonic mucosa.[84] In addition, using rapid competitive PCR,[95] we have quantitatively measured fecal PKC mRNAs isolated from control and carcinogen treated groups (Figure 2). Although there is no simple way to determine the effect of dietary fat on colonic disease activity, the ability to use fecal material containing exfoliated colonic epithelial cells as a means to monitor PKC expression may have predictive value in terms of monitoring the neoplastic process.

SUMMARY

In conclusion, dietary n-3 polyunsaturated fatty acids found in fish oil are capable of suppressing carcinogen-induced ras activation in the colon prior to overt neoplasia. This in turn blocks the oncogene driven increase in colonic diacylglycerol mass, preventing the persistent activation and chronic down-regulation of PKC isozymes, thereby maintaining tissue PKC levels. Since the maintenance of crypt PKC levels may sustain the homeostatic balance between cell proliferation, differentiation and apoptosis, the ability of dietary n-3 polyunsaturated fatty acids to block the carcinogen-induced decrease in steady-state levels of colonic mucosal PKC may in part explain why these fatty acids protect against colon tumorigenesis. Additional studies are required in order to elucidate the *mechanisms* by which select dietary lipids reduce colonic tumor incidence. This research focus is absolutely essential, because if we do not know why a dietary component is protective or promotive of cancer, then we have no right to attempt to modify eating behaviors.

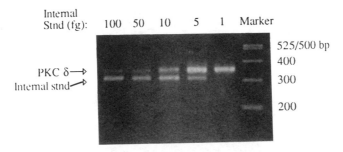

Figure 2. Semiquantitative competitive PCR of PKC-δ using fecal messenger RNA. Fecal poly (A+) RNA concentration was determined after densitometric scanning of a colonic mucosal total RNA standard curve, assuming colonic RNA contains 2% poly(A+) RNA. Decreasing amounts of internal standard were added to sequential tubes containing the same amount of reverse transcriptase reaction mixture. After PCR, products were separated by agarose gel electrophoresis and stained with ethidium bromide. *Far right lane,* DNA molecular size markers.

ACKNOWLEDGMENTS

We gratefully acknowledge Wen-Chi L. Chang, Dr. Harold Aukema and Dr. Deb Zoran, for contributing data cited in this review. We also acknowledge the NIH Fish Oil Test Material Program and Mr. Sid Tracy, Traco Labs, for providing the fish oil and corn oil used in our studies.

REFERENCES

1. Doll, R. (1976). Strategy for detection of cancer hazards to man. Nature 256:589–596.
2. Doll, R. and Peto, R. (1981). The causes of cancer: quantitative estimates of avoidable risks of cancer in the United States today. J. Natl. Cancer Inst. 66:1191–1238.
3. Reddy, B.S., Watanabe, K., Weisburger, J.H. and Wynder, E.L. (1977). Promoting effect of bile acids in colon carcinogenesis in germ-free and conventional F344 rats. Cancer Res. 47:644–648.
4. Chang, W-C., Lupton, J.R., Frolich, W., Schoeffler, G.L., Peterson, M.L. and Chen, X-Q. (1994). A very low intake of fat is required to decrease fecal bile acid concentrations in rats. J. Nutr. 124:181–187.
5. Pickering, J.S., Lupton, J.R. and Chapkin, R.S. (1995). Dietary fat, fiber, and carcinogen alter fecal diacylglycerol composition and mass. Cancer Res. 55:2293–2298.
6. Lupton, JR. (1995). Butyrate and colonic cytokinetics: differences between in vitro and in vivo studies. Eur. J. Cancer Prev. 4:373–378.
7. Lupton, J.R. (1995). Short-chain fatty acids and colon tumorigenesis: animal models. In: Physiological and clinical aspects of short-chain fatty acids. Cummings, J.H., Rombeau, J.L. and Sakata, T. (Eds). Cambridge University Press, New York, NY, pp.307–318.
8. Rogers, A.E. and Nauss, K.M. (1984). Contributions of laboratory animal studies of colon carcinogenesis. In: Large Bowel Cancer, A.J. Mastromarino and M.G. Brattain (eds). New York: Praeger, pp. 1–45.
9. Jacobs, L.R. and Lupton, J.R. (1986). Relationship between colonic luminal pH, cell proliferation, and colon carcinogenesis in 1,2-dimethylhydrazine-treated rats fed high fiber diets. Cancer Res. 46:1727–1734.
10. Reddy, B.S. (1993). Dietary fat, calories, and fiber in colon cancer. Prev. Med. 22:738–749.
11. Lee, D.Y., Chapkin, R.S. and Lupton, J.R. (1993). Dietary fat and fiber modulate colonic cell proliferation in an interactive site-specific manner. Nutr. Cancer 20:107–118.
12. Wynder, E.L., Kajitani, T., Ishikawa, S., Dodo, H. and Takano, A. (1969). Environmental factors of the colon and rectum. II. Japanese epidemiological data. Cancer 23: 1210–1220.
13. Kurihara, M., Aoki, K. and Tominaga, S. (1984). Cancer mortality stastistics in the world. Nagoya, Japan: Univ. of Nagoya Press.
14. Haenszel, W. (1982). Migrant studies. In Schottenfeld, D. and Fraumeni, J.F. (ed.) Cancer epidemiology and prevention. Philadelphia, PA: Saunders, pp. 194–207.
15. Weisburger, J.H. and Wynder, E.L. (1987). Etiology of colorectal cancer with emphasis on mechanism of action. In DeVita, Jr.V.T., Hellman,S., and Rosenberg, S.A. (ed.) Important advances in oncology. Philadelphia, PA: Lippincott, pp. 79–92.
16. Reddy, B.S. (1986). Diet and colon cancer: Evidence from human and animal model studies. In: Reddy, B.S. and Cohen, L.A. (eds.) Diet, nutrition and cancer: a critical evaluation: Boca Raton, FL: CRC Press, Vol. 1, pp. 47–66.
17. Reddy, B.S., Hedges, A.R., Laakso, K. and Wynder, E.L. (1978). Metabolic epidemiology of large bowel cancer: fecal bulk and constituents of high-risk North American and low-risk Finnish population. Cancer 42: 2832–2838.
18. Tuyns, A.J., Haelterman, M. and Kaaks, R. (1987). Colorectal cancer and the intake of nutrients: oligosaccharides are a risk factor, fats are not. A case-control study in Belgium. Nutr. Cancer 10: 181–196.
19. Potter, J.D., Slattery, M.L., Bostick, R.M. and Gapstur, S.M. (1993). Colon cancer: a review of the epidemiology. Epidemiol. Rev. 15:499–545.
20. Willet, W.C., Stampfer, M.J., Colditz, G.A., Rosner, B.A. and Speizer, F.E. (1990). Relations of meat, fat, and fiber intake to the risk of colon cancer in the prospective study among women. New Engl. J. Med. 323:1664–1672.
21. Nigro, N.D., Singh, D.V. and Pak, M.S. (1975). Effect of dietary beef fat on intestinal tumor formation by azoxymethane in rats. J. Natl. Cancer. Inst., 54: 439–442.
22. Reddy, B.S. and Maruyama, H. (1986). Effect of different levels of dietary corn oil and lard during the initiation phase of colon carcinogenesis in F344 rats. J. Natl. Cancer Inst. 77: 815–822.

23. Reddy, B.S. and Maeura, Y. (1984). Tumor promotion by dietary fat in azoxymethane-induced colon car-
    cinogenesis in female F344 rats: Influence of amount and source of dietary fat. J. Natl. Cancer Inst. 72:
    745–750.

24. Reddy, B.S. and Maruyama, H. (1986). Effect of dietary fish oil on azoxymethane-induced colon carcino-
    genesis in male F344 rats. Cancer Res. 46: 3367–3370.

25. Wingo, P.A., Tong, T. and Bolden, S. (1995). Cancer statistics, 1995. CA Cancer J. Clin. 45: 8–30.

26. Lee, D.Y., Lupton, J.R., Aukema, H.M. and Chapkin, R.S. (1993). Dietary fat and fiber alter rat colonic
    mucosal lipid mediators and cell proliferation. J. Nutr. 123:1808–1817.

27. Minoura, T., Takata, T., Sakaguchi, M., Takada, H., Yamamura, M. and Yamamoto, M. (1988). Effect of
    dietary eicosapentaenoic acid on azoxymethane-induced colon carcinogenesis in rats. Cancer Res.
    48:4790–4794.

28. Reddy, B.S. (1994). Chemoprevention of colon cancer by dietary fatty acids. Cancer Metas. Rev.
    13:285–302.

29. Anti, M., Armelao, F., Marra, G., Percesepe, A., Bartoli, G.M., Palozz, P., Parrella, P., Canetta, C., Gen-
    tiloni, N., De Vitis, I. and Gasbarrini, G. (1994). Effects of different doses of fish oil on rectal cell prolif-
    eration in patients with sporadic colonic adenomas. Gastroenterology 107:1709–1718.

30. Bull, A. W., Soullier, B. K., Wilson, P. S., Hayden, M. T., and Nigro, N. D. (1981). Promotion of
    azoxymethane-induced intestinal cancer by high fat diets in rats. Cancer Res. 41:3700–3705.

31. Deschner, E.E., Lytle, J., Wong, G., Ruperto, J. and Newmark, H.L. (1990). The effect of dietary omega-3
    fatty acids (fish oil) on azoxymethanol-induced focal areas of dysplasia and colon tumor incidence. Cancer
    66:2350–2356.

32. Minoura, T., Takata, T., Sakaguchi, M., Takada, H., Yamamura, M., and Yamamoto, M. (1988). Effect of
    dietary eicosapentaenoic acids on azoxymethane-induced colon carcinogenesis in rat. Cancer Res.
    48:4790–4794.

33. Reddy, B. S., Burill, C. and Rigotty, J. (1991). Effect of diets high in ω-3 and ω-6 fatty acids on initiation
    and postinitiation stages of colon carcinogenesis. Cancer Res. 51:487–491.

34. Chang, W.C., Jiang, Y.H., Chapkin, R.S. and Lupton, J.R.. (1995). Predictive value of proliferation, differ-
    entiation and apoptosis as intermediate markers for colon tumorigenesis. FASEB J. 9:A869.

35. Jiang, Y.H., Lupton, J.R. and Chapkin, R.S. (1995). Modulation of intermediate biomarkers of colon cancer
    by dietary fat, fiber and carcinogen in rat colonocytes. FASEB J. 9:A868.

36. Bos, J.L. (1988). The ras gene family and human carcinogenesis. Mutation Res. 195:255–271.

37. Singh, J., Kelloff, G. and Reddy, B.S. (1993). Intermediate biomarkers of colon cancer: modulation of ex-
    pression of ras oncogene by chemopreventive agents during azoxymethane induced colon carcinogenesis.
    Carcinogenesis 14:699–704.

38. Singh, J., Kulkarni, N., Kelloff, G. and Reddy, B.S. (1994). Modulation of axoxymethane-induced muta-
    tional activation of ras protooncogenes by chemopreventive agents in colon carcinogenesis. Carcinogenesis
    15:1317–1323.

39. Fearon, E. R., and Vogelstein, B. (1990). A genetic model for colorectal tumorigenesis. Cell 61:759–767.

40. White, MA., Nicollete, C., Minden, A., Polverino, A., Van Aeist, L., Karin, M. and Wigler, M.H. (1995).
    Multiple ras functions can contribute to mammalian cell transformation. Cell 80:533–541.

41. Lowry, D.R. and Willumsen, B.M. (1993). Function and regulation of ras. Annu. Rev. Biochem.
    62:851–891.

42. Shirasawa, S., Furuse, M., Yokoyama, N. and Sasazuki, T. (1993). Altered growth of human colon cancer
    cell lines disrupted at activated Ki-ras. Science 260:85–88.

43. Chen, C.Y. and Faller, D.V. (1995). Direction of p21ras-generated signals towards cell growth or apoptosis
    by protein kinase C and bcl-2. Oncogene 11:1487–1498.

44. Laurenz, J.C., Gunn, J.M., Jolly, C.A. and Chapkin, R.S. (1996). Alteration of glycerolipid and sphin-
    golipid-derived second messenger kinetics in ras transformed 3T3 cells. Biochim. Biophys. Acta
    1299:146–154.

45. Haliotis, T., Trimble, W., Chow, S., Bull, S., Mills, G., Girard, P., Kuo, J.F. and Hozumi, N. (1990). Expres-
    sion of ras oncogene leads to down-regulation of protein kinase C. Int. J. Cancer 45:1177–1183.

46. Jiang, Y.H., Lupton, J.R. and Chapkin, R.S. (1996). Select dietary fats and fibers block carcinogen-induced
    post-transcriptional down-regulation of colonic protein kinase C. FASEB J. 10:A494.

47. Colapietro, A.M., Goodell, A.L. and Smart, R.C. (1993). Characterization of benzo[a]pyrene-initiated
    mouse skin papillomas for Ha-ras mutations and protein kinase C levels. Carcinogenesis 14:2289–2295.

48. Stopera, S.A., Murphy, L.C. and Bird, R.P. (1992). Evidence for a ras gene mutation in azoxymethane-in-
    duced colonic aberrant crypts in Sprague-Dawley rats: earliest recognizable precursor lesions of experi-
    mental colon cancer. Carcinogenesis 13:2081–2085.

49. Vivona, A.A., Shpitz, B., Medline, A., Bruce, W.R., Hay, K., Ward, M.A., Stern, H.S. and Gallinger, S. (1993). K-ras mutations in aberrant crypt foci, adenomas and adenocarcinomas during azoxymethane-induced colon carcinogenesis. Carcinogenesis 14:1777–1781.

50. Jiang, Y.H., Lupton, J.R., Jolly, C.A., Davidson, L.A., Aukema, H.M. and Chapkin, R.S. (1996). Dietary fat and fiber differentially regulate intracellular second messengers during tumor development in rat colon. Carcinogenesis 17:1227–1233.

51. Ahnen, D.J. (1985). Are animal models of colon cancer relevant to human diseases? Dig. Dis. Sci. 30:103s–106s.

52. Hunter,T. (1991). Cooperation between oncogenes. Cell 64: 249–272.

53. Powis, G. and Phil, D. (1994). Inhibitors of phosphatidylinositol signalling as antiproliferative agents. Cancer Metas. Rev. 13: 91–103.

54. Lacal, J.C., Moscat, J. and Aaronson, S.A. (1987). Novel source of 1, 2-diacylglycerol elevated in cells transformed by Ha-ras oncogene. Nature 330: 269–272.

55. Wolfman, A. and Macara, I.G. (1987). Elevated levels of diacylglycerol increased phorbol ester sensitivity in ras-transformed fibroblasts. Nature 325: 359–361.

56. Preiss, J.E., Loomis, C.R., Bishop, W.R. and Bell, R.M. (1986). Quantitative measurement of *sn*-1, 2-diacylglycerols present in platelets, hepatocytes, and *ras* and *sis*-transformed normal rat kidney cells. J. Biol. Chem. 261: 8597–8600.

57. Nishizuka, Y.(1992). Intracellular signaling by hydrolysis of phospholipids and activation of protein kinase C. Science 258: 607–614.

58. Choi, P.M., Tchou-Wong, K-M. and Weinstein, I.B. (1990). Overexpression of protein kinase C in HT29 colon cancer cells causes growth inhibition and tumor suppression. Mol. Cell. Biol. 10:4650–4657.

59. Chapkin, R.S., Gao, J., Lee, D-Y.K. and Lupton, J.R. (1993). Dietary fibers and fats alter rat colonic protein kinase C activity: correlation to cell proliferation. J. Nutr. 123: 649–655.

60. DeRubertis, F.R. and Craven, P.A. (1987). Relationship of bile salt stimulation of colonic epithelial phospholipid turnover and proliferative activity: role of activation of protein kinase C. Prev. Med. 16:572–579.

61. Mills, K.J., Reynolds, S.H. and Smart, R.C. (1993). Diacylglycerol is an effector of the clonal expansion of cells containing activated Ha-ras genes. Carcinogenesis 14:2645–2648.

62. Herbert, J.M. (1993). Protein kinase C: A key factor in the regulation of tumor cell adhesion to the endothelium. Biochem. Pharmacol. 45: 527–537.

63. Hannun, Y.A. and Obeid, L.M. (1995). Ceramide: an intracellular signal for apoptosis. TIBS 20: 73–77.

64. Saxon, M. L., Zhao, X., and Black, J. P. (1994). Activation of protein kinase C isozymes is associated with post-mitotic events in intestinal epithelial cells in situ. J. Cell Biol. 126: 747–763.

65. Kopp, R., Noelke, B., Sauter, G., Schildberg, F.W., Paumgartner, G. and Pfeiffer, A. (1991). Altered protein kinase C activity in biopsies of human colonic adenomas and carcinomas. Cancer Res. 51:205–210.

66. Guillem, J.G., O'Brien, C.A., Fitzer, C.J., Forde, K.A., Logerfor, P., Treat, M. and Weinstein, I.B. (1987). Altered levels of protein kinase C and Ca^{2+} dependent protein kinase in human colon carcinomas. Cancer Res. 47:2036–2039.

67. Baum, C.L., Wali, R.K., Sitrin, M.D., Bolt, G.J.G. and Brasitus, T.A. (1990). 1,2-Dimethylhydrazine-induced alterations in protein kinase C activity in the rat preneoplastic colon. Cancer Res. 50:3915–3920.

68. Wali, R.K., Baum, C.L., Bolt, M.J.G., Dudeja, P.K., Sitrin, M.D. and Brasitus, T.A. (1991). Down-regulation of protein kinase C activity in 1,2-dimethylhydrazine-induced rat colonic tumors. Biochim. Biophys. Acta 1092:119–123.

69. Levy, M. F., Pocsidio, J., Guillem, J. G., Forde, K., LoGerfo, P., and Weinstein, I. B. (1993). Decreased levels of protein kinase C enzyme activity and protein kinase C mRNA in primary colon tumors. Dis. Colon Rectum 36: 913–921.

70. Doi, S., Goldstein, D., Hug, H., and Weinstein, I. B. (1994). Expression of multiple isoforms of protein kinase C in normal human colon mucosa and colon tumors and decreased levels of protein kinase C β and η mRNAs in the tumors. Molec. Carcinogenesis 11: 197–203.

71. Chiarugi, V., Bruni, P., Pasquali, F., Magnelli, L., Basi, G., Ruggiero, M. and Farnararo, M. (1989). Synthesis of diacylglycerol de novo is responsible for permanent activation and down-regulation of protein kinase C in transformed cells. Biochem. Biophys. Res. Comm. 164:816–823.

72. Price, B.D., Morris, J.D.H., Marshall, C.J. and Hall, A. (1989). Stimulation of phosphatidylcholine hydrolysis, diacyglycerol release, and arachidonic acid production by oncogenic ras is a consequence of protein kinase C activation. J. Biol. Chem. 264:16638–16643.

73. Noh, D.Y., Lee, Y.H., Kim, S.S., Kim, Y.I. Ryu, S.H., Suh, P.G. and Park, J.G. (1994). Elevated content of phopholipase C-γ 1 in colorectal cancer tissues. Cancer 73:36–41.

74. Park, J.G., Lee, Y.H., Kim, S.S., Park, K.J., Noh, D.Y., Ryu, S.H. and Suh, P.G. (1994). Overexpression of phospholipase C-γ 1 in familial adenomatous polyposis. Cancer Res. 54:2240–2244.

75. Smith, M.R., Ryu, S.H., Suh, P.G., Rhee, S.G. and Kung, H.F. (1989). S-phase induction and transformation of quiescent NIH 3T3 cells by microinjection of phospholipase C. Proc. Natl. Acad. Sci. USA 86:3659–3663.

76. Arteaga, C.L., Johnson, M.D., Todderud, G., Coffey, R.J., Carpenter, G. and Page, D.L. (1991). Elevated content of tyrosine kinase substrate phospholipase C-γ 1 in primary human breast carcinomas. Proc. Natl. Acad. Sci. USA 88:10435–10439.

77. Nanney, L.B., Gates, R.E., Todderud, G., King, L.E. and Carpenter, G. (1992). Altered distribution of phospholipase C-γ 1 in benign hyperproliferative epidermal disease. Cell Growth Differ. 3:233–239.

78. Hannun, Y.A., Loomis, C.R. and Bell, R.M. (1986). Protein kinase C activation in mixed micelles: mechanistic implications of phospholipid, diacylglycerol, and calcium interdependencies. J. Biol. Chem. 261: 7184–7190.

79. Bell, R.M. and Burns, D.J. (1991). Lipid activation of protein kinase C. J. Biol. Chem. 266: 4661–4664.

80. Murakami, K., Chan, S.Y. and Routtenberg, A. (1986). Protein kinase C activation by cis-fatty acid in the absence of Ca^{2+} and phospholipid. J. Biol. Chem. 261: 15424–15429.

81. Asaoka, Y., Tsujishita, Y. and Nishizuka, Y. (1996). Lipid signaling for protein kinase C activation. In: Handbook of lipid research, Volume 8: Lipid second messengers, R.M. Bell, Ed., Plenum Press, New York, NY, pp. 59–74.

82. Castagna, M., Takai, Y., Kaibuchi, K., Sano, K., Kikkawa, U. and Nishizuka, Y. (1982). Direct activation of calcium-activated, phospholipid-dependent protein kinase by tumor-promoting phorbol esters. J. Biol. Chem. 257: 7847–7851.

83. Blobe, G.C., Obeid, L.M. and Hannun, Y.A. (1994). Regulation of protein kinase C and role in cancer biology. Cancer Metas. Rev. 13:411–431.

84. Davidson, L. A., Jiang, Y. H., Derr, J. N., Aukema, H. M., Lupton, J. R., and Chapkin, R. S. (1994). Protein kinase C isoforms in human and rat colonic mucosa. Arch. Biochem. Biophys. 312: 547–553.

85. Sakanode, Y., Hatada, T., Kusunoki, M., Yanagi, H., Yamamura, T., and Utsunomiya, J. (1991). Protein kinase C activity as a marker for colorectal cancer. Int. J. Cancer 48: 803–806.

86. Wali, R. K., Frawley, B. P., Hartmann, S., Roy, H. K., Khare, S., Scaglione-Sewell, B. A., Earnest, D. L., Sitrin, M. D., Brasitus, T. A., and Bissonnette, M. (1995). Mechanism of action of chemoprotective ursode oxycholate in the azoxymethane model of rat colonic carcinogenesis: potential roles of protein kinase C-α, -β_{II}, and -ζ. Cancer Res. 55: 5257–5264.

87. Jiang, Y. H., Aukema, H. M., Davidson, L. A., Lupton, J. R., and Chapkin, R. S. (1995). Localization of protein kinase C in rat colon. Cell Growth Differ. 6: 1381–1386.

88. Craven, P. A., and DeRubertis, F. R. (1987). Subcellular distribution of protein kinase C in rat colonic epithelial cells with different proliferative activities. Cancer Res. 47: 3434–3438.

89. Craven, P. A., Pfanstiel, J., and DeRubertis, F. R. (1987). Role of activation of protein kinase C in the stimulation of colonic epithelial proliferation and reactive oxygen formation by bile acids. J. Clin. Invest. 79: 532–541.

90. Haliotis, T., Trimble, W., Chow, S., Bull, S., Mills, G., Girard, P., Kuo, J.F. and Hozumi, N. (1990). Expression of ras oncogene leads to down-regulation of protein kinase C. Int. J. Cancer 45:1177–1183.

91. Clemens, M. J., Trayner, I., and Menaya, J. (1992). The role of protein kinase C isoenzymes in the regulation of cell proliferation and differentiation. J. Cell Sci. 103: 881–887.

92. Davidson, L. A., Lupton, J. R., Jiang, Y. H., Chang, W. C., Aukema, H. M., and Chapkin, R. S. (1995). Dietary fat and fiber alter rat colonic protein kinase C isozyme expression. J. Nutr. 125: 49–56.

93. Lee, D.Y., Lupton, J.R. and Chapkin, R.S. (1992). Prostaglandin profile and synthetic capacity of the colon: Comparison of tissue sources and subcellular fractions. Prostaglandins 43:143–164.

94. Davidson, L.A., Jiang, Y.H., Lupton, J.R. and Chapkin, R.S. (1995). Noninvasive detection of putative biomarkers for colon cancer using fecal messenger RNA. Cancer Epidemiol. Biomarkers Prev. 4:643–647.

95. Jiang, Y.H., Davidson, L.A., Lupton, J.R. and Chapkin, R.S. (1996). Rapid competitive PCR determination of relative gene expression in limiting tissue samples. Clin. Chem. 42:227–231.

96. Bang, H.O., Dyerberg, J. and Sinclair, H.M. (1980). The composition of the Eskimo food in north western Greenland. Am. J. Clin. Nutr. 33:2657–2661.

97. Kinsella, J.E. (1987) Dietary fats and cardiovascular disease. In: Seafoods and fish oils in human health and disease. J.E. Kinsella, Ed., Marcel Dekker, New York, NY, p. 14.

98. Hosack-Fowler, K., Chapkin, R.S. and McMurray, D.N. (1993). Effects of purified dietary ethyl esters on murine T-lymphocyte function. J. Immunol. 151:5186–5197.

99. Jonnalagadda, S.S., Egan, S.K., Heimbach, J.T., Harris, S.S. and Kris-Etherton, P.M. (1995). Fatty acid consumption pattern of Americans: 1987–1988 USDA nationwide food consumption survey. Nutr. Res. 15:1767.

100. Kremer, J.M. (1991). Clinical studies of omega-3 fatty acid supplementation in patients who have rheuma-
      toid arthritis. Rheum. Dis. Clin. North America 17:391–402.

DIET, APOPTOSIS, AND CARCINOGENESIS

Craig D. Albright,[*] Rong Liu, Mai-Heng Mar, Ok-Ho Shin,
Angelica S. Vrablic, Rudolf I. Salganik, and Steven H. Zeisel

Department of Nutrition
School of Public Health and School of Medicine
The Universtiy of North Carolina at Chapel Hill
Chapel Hill, North Carolina 27599

ABSTRACT

It is known that long-term withdrawal of choline from the diet induces hepatocellular carcinomas in animal models in the absence of known carcinogens. We hypothesize that a choline deficient diet (CD) alters the balance of cell growth and cell death in hepatocytes and thus promotes the survival of clones of cells capable of malignant transformation. When grown in CD medium (5 μM or 0 μM choline) CWSV-1 rat hepatocytes immortalized with SV40 large T-antigen underwent p53-independent apoptosis (terminal dUTP end-labeling of fragmented DNA; laddering of DNA in agarose gel). CWSV-1 cells which were adapted to survive in 5 μM choline acquired resistance to CD-induced apoptosis and were able to form hepatocellular carcinomas in nude mice. These adapted CWSV-1 cells express higher amounts of both the 32 kDa membrane-bound and 6 kDa mature form of TGFα compared to cells made acutely CD. Control (70 μM choline) and adapted cells, but not acutely deficient hepatocytes, could be induced to undergo apoptosis by neutralization of secreted TGFα. Protein tyrosine phosphorylation is known to protect against apoptosis. We found decreased EGF receptor tyrosine phosphorylation in acutely choline deficient CWSV-1 cells. TGFß1 is an important growth-regulator in the liver. CWSV-1 cells express TGFß1 receptors and this peptide induced cell detachment and death in control and acutely deficient cells. Hepatocytes adapted to survive in low choline were also resistant to TGFß1, although TGFß1 receptors and protein could be detected in the cytoplasm of these cells. The non-essential nutrient choline is important in maintaining plasma membrane structure and function, and in intracellular signaling. Our results indicate that acute withdrawal of choline induces p53-independent programmed cell death in hepatocytes, whereas cells adapted to survive in low choline are resistant to this form of apoptosis, as well as to cell death induced by TGFß1. Our results also suggest that CD

* Phone: (919) 966–0131; Fax: (919) 966–7216.

may induce alterations (mutations?) in growth factor signaling pathways which may enhance cell survival and malignant transformation.

INTRODUCTION

In humans, dietary factors are implicated in the genesis of cancer in several major organs, including the colon, breast, liver, lung, prostate, and uterus. Classical carcinogenesis paradigms link repeated episodes of cell injury and death, and alterations in cell proliferation to the establishment of altered foci of cells (*e.g.,* epithelial growth variants, enzyme altered foci, dysplasia) which are often clonal in nature, and which have an increased potential to undergo malignant transformation when confronted with the appropriate stress factors. Although knowledge of the regulatory mechanisms of multistep carcinogenesis is rapidly evolving, the role of specific dietary and nutritional factors at the cellular and molecular biological level in malignant transformation is not well understood.

In animal models withdrawal of the nutrient choline causes liver cancer in the absence of known carcinogens (1,2), and can increase the incidence of hepatocellular carcinomas in animals treated with a chemical carcinogen (3,4). Animals which are made choline deficient undergo cell injury, cell death, and compensatory proliferation in the liver prior to the appearance of altered preneoplastic foci and carcinomas (5,6). Thus this model is well-suited to investigate the role of nutrients in multistep carcinogenesis. We have developed a cell culture model of rat hepatocytes to investigate how the withdrawal of choline pushes immortalized, nontumorigenic hepatocytes to undergo malignant transformation (7–9). Our research focuses on the possible role of cell-cell signaling pathways and intracellular mediators in p53-independent programmed cell death and malignant transformation.

Diet and Carcinogenesis

Two principle mechanisms appear to link dietary factors to increased cancer risk: (1) deficiencies of micronutrients increase the induction of oxidative DNA modifications, and (2) nutrient factors which increase mitogenesis increase the risk for mutations or gene amplification, permitting the evolution of malignant phenotypes (10). We believe that in addition to premutagenic DNA modifications *per se*, factors in the diet may induce unrepairable DNA modifications resulting in apoptotic cell death. Apoptosis may protect against the establishment of permanent genomic alterations in dividing cells. The ability of cells to resist apoptotic triggers may be a crucial event in carcinogenesis.

It is known that several factors in the diet are able to enhance the process of carcinogenesis by direct interaction with DNA. For example, food-derived heterocyclic amines, such as PhIP (2-amino-1-methyl-6-phenylimidazo [4,5-b]pyridine) found in grilled meat, are linked to an increased risk of breast cancer in humans, and have been shown to form DNA adducts in the rat mammary gland (11). Similary, feeding rats a riboflavin-deficient diet increased the number of DNA single strand breaks in aflatoxin-treated animals (12). Dietary factors may target other components of the carcinogenesis process as well. In laboratory animals, feeding a western-style high fat diet increased the risk for cancers of the breast (13) and pancreas (14), whereas feeding a low fat diet inhibited promotion of two-stage carcinogenesis (15). Studies show that the omega-6 fatty acids may provide a growth/survival advantage to transformed cells. High dietary corn oil rich in omega-6 fatty acids increased the activities of PI-PLC and PLA2 associated with PKC-dependent transmembrane signal transduction and cell proliferation in azoxymethane-induced colon carcinomas compared to adjacent normal colonic mucosa (16). In mice, feeding a diet high in omega-6 fatty acids caused immune suppression, delaying the rejection of transplanted tumors (17).

Enhancement of carcinogenesis may also be due to interactions between dietary and endogenous factors. For example, sodium deoxycholate, the predominant bile salt present in the colon, caused apoptosis in normal colonic goblet cells, but not in normal appearing cells of the colonic mucosa in cancer patients (18). Thus a high fat diet and the chronic presence of bile salts could select for cells resistant to apoptosis, thereby increasing the risk for malignant transformation. In contrast, feeding animals a high protein diet resulted in overexpression of PKCß in the particulate fraction and selective promotion of the growth of carcinogen-altered GGT(+) hepatocytes (19). In our laboratory, prolonged activation of PKC in association with liver carcinogenesis is also observed in animals fed a choline deficient diet (20, 21). The role of caloric levels in carcinogenesis may not be so straightforward. Additional studies show that malnutrition causes reductions in tissue levels of antioxidants (*e.g.,* GSH) (22), capable of scavenging reactive oxygen species and thus inhibiting the activitation of mediators of apoptosis (*e.g.,* NF-κB) (23). Thus it is possible that a low protein diet could inhibit the activation of oxidant-dependent apoptotic triggers, thereby promoting the survival of transformed cells. Studies are needed which focus on the regulatory role of individual dietary nutrients in carcinogenesis.

Choline Deficiency and Apoptosis

The possible role of choline in membrane structure, signal transduction and carcinogenesis has been extensively reviewed (24,25). Withdrawal of the nutrient choline from the diet induces accumulation of lipids, cell injury and cell death, compensatory proliferation and carcinogenesis in rat liver. These alterations reproduce many of the features of human liver cancer, making the choline deficiency (CD) model well-suited for investigations of mechanisms of nutrient regulation of multistep carcinogenesis (26).

Apoptosis is a form of physiological cell death which can be distinguished from necrosis by the absence of inflammation in the latter, and on the basis of morphological features (27,28). We have established a cell culture model of choline deficiency (7,8) using SV40 large T antigen immortalized hepatocytes derived from male F344 rats (a generous gift from Dr. Harriet Isom) (29,30). When these cells are acutely deprived of choline in serum-free defined medium, they undergo apoptosis as measured by TUNEL labeling (Figure 1) (see (7) for additional features of apoptosis in CWSV-1 hepatocytes). Apoptosis is often characterized at the molecular level by the activation of a Ca^{2+} - Mg^{2+} dependent endonuclease and the generation of 180 bp DNA fragments (31). We observed that acute CD induced a characteristic fragmentation of DNA typical of an apoptotic process (Figure 1 inset). In contrast, cells which were adapted to survive in low choline (5 μM) were resistant to CD apoptosis. Scavenging of reactive oxygen species (ROS) using *N*-acetylcysteine, and chelation of copper ions using neocuproine prevented apoptosis in response to acute CD (32). We are aware that the availability of choline can be influenced by the levels of folate, methionine and vitamin B12 (33). Choline deficiency may result in folate deficiency (34). We observed that CD apoptosis was not reversed by the addition of extra methionine or betaine, bypassing the folate requirement, or by the actual addition of extra folate and vitamin B12 (8) (Figure 2). This it appears that CD apoptosis is due to the choline moiety itself, and to the generation of ROS, and not to the availability of methyl-donors.

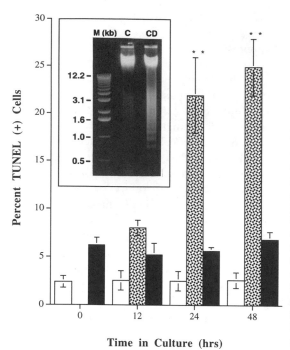

Figure 1. Choline deficiency induces DNA strand breaks in CWSV-1 hepatocytes. Cells were maintained for 4 days in control (70 μM choline) medium then switched acutely for 2 days to experimental medium (70 μM or 5 μM choline) or gradually adapted to survive in choline deficient (5 μM choline) medium. DNA strand breaks were detected using a TUNEL method. Increased numbers of TUNEL(+) cells were detected in acute choline deficient cultures (stippled bars), but not in cells adapted to survive in CD medium (black bars), compared to control cells (70 μM choline) (open bars). Mean ± SD; ** = p, .01 (t-test). Inset: cleavage of DNA into Å 200 bp fragments characteristic of apoptosis was observed in acute CD cells. Isolated DNA was solubilized and resolved in a 1.8% agarose gel as described previously (7).

Choline Deficiency and p53 Expression

In cells capable of expressing a wild-type p53 gene product, a wide variety of agents which damage DNA or inhibit DNA replication appear to arrest cells at a late point in G1, the so-called p53 restriction point (35), leading to cell cycle arrest or p53-dependent apoptosis (36,37). Induction of the cyclin dependent kinase inhibitor p21$^{\text{Cip1/Waf1}}$ (38) by DNA damaging agents is p53-dependent (39), and correlates with the induction of apoptosis in response to γ-irradiation (40). We found that lethal doses of γ-irradiation induced the expression of p53 protein and p21$^{\text{Cip1/Waf1}}$ in primary cultures of rat hepatoctyes, but not in CWSV-1 cells (Figure 3). Our studies suggest that induction of apoptosis by choline deficiency in CWSV-1 cells occurs independent of p53 (32). CWSV-1 cells which are adapted

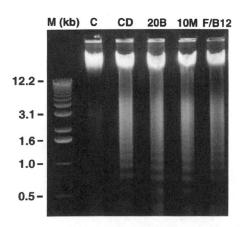

Figure 2. Choline deficiency-apoptosis is not prevented by methyl donors. Control and acute CD hepatocytes were treated with extra methionine, betaine, folate or vitamin B12 and isolated DNA was resolved in a 1.8% agarose gel (8). The addition of methyl donors to acute CD CWSV-1 cells did not prevent the induction by acute CD of DNA fragmentation typical of apoptosis. No cytotoxic effects of methyl donors were observed in controls (8).

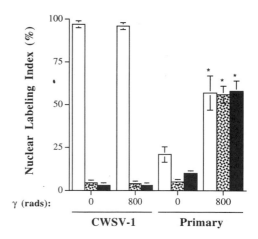

Figure 3. Induction of p53-dependent apoptosis in CWSV-1 hepatocytes. CWSV-1 cells and cultures of primary rat hepatocytes were treated with a lethal dose (800 rads) of γ-irradiation, and 6 hrs later were assayed for the nuclear expression of p53 (open bars) or p21$^{Cip1/Waf1}$ (stippled bars) by an immunocytochemical method (32) and for the induction of DNA strand breaks using a TUNEL method (black bars) (7). In CWSV-1 cells SV40 large T antigen binds and inactivates p53 protein accounting for the prominent nuclear labeling observed in these cells. Primary hepatocytes, but not CWSV-1 cells, showed an increased nuclear expression of p53-dependent mediators and apoptosis in response to γ-irradiation, a p53-dependent trigger of apoptosis (see (32) for review). Mean ± SD; * = p < .01 (t-test).

to survive in low choline exhibit increased anchorage-independent growth and they form hepatocellular carcinomas in nude mice (41). Our studies suggest that loss of alternative, p53-independent pathways in cells with a defective p53-dependent primary apoptosis pathway is an important risk factor for completion of carcinogenesis.

Choline Deficiency and TGFα/EGF Signaling

Growth factors and cell-cell communication (42), and interactions with extracellular matrix proteins (43) provide positive and negative signals for cell survival. In rats elevated levels of TGFα mRNA were detectable in the liver after four weeks on a CD diet and persisted as tumors developed (44). Increased amounts of 30 kDa TGFα and intermediate sized protein, representing the membrane-bound and partially cleaved forms (Å 20 kDa) which generate the mature 6 kDa form of TGFα (45,46), were observed in adapted CWSV1 hepatocytes compared to acute deficient cells (Figure 4). TGFα and epidermal growth factor (EGF) share a receptor which has intrinsic tyrosine kinase activity. Expression of EGF receptor protein was detected in CWSV-1 hepatocytes (Figure 5). Antibody neutralization of secreted TGFα resulted in a triggering of apoptosis in control and adapted CWSV-1 cells, but not in acutely deficient hepatocytes (Table 1). Previous reports show that feeding a CD diet for 1 week has no effect on the level of TGFα mRNA

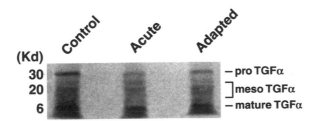

Figure 4. Expression of TGFα by CWSV-1 hepatocytes. Comparison of the expression of TGFα protein in control, acute choline deficient and adapted CWSV1 hepatocytes. Cells were maintained as described in the legend to Figure 1. Cell lysates were prepared in the presence of protease and phosphatase inhibitors as decribed previously (45). Equivalent amounts of protein were subjected to SDS-PAGE and Western blotting using anti-TGFα antibody (Santa Cruz Biotechnology) and an ECL method.

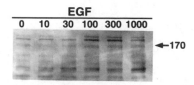

Figure 5. Epidermal growth factor (EGF) receptor expression in CWSV-1 cells. Control (70 μM choline) CWSV1 hepatocytes were maintained as described in Figure 1. Western blotting was performed using anti-EGF receptor antibody, which recognizes the cytoplasmic domain in the EGF relevant to exons 15–18; the antibody reportedly does not inhibit the tyrosine kinase activity of the EGF receptor (Upstate Biotechology).

(44). Our results suggest that CD may decrease the expression of TGFα in acutely deficient cells, whereas in adapted CWSV-1 cells elevated levels of this EGF receptor ligand may be an important survival factor. To confirm the effects of antibody neutralization we showed that addition of EGF also suppressed apoptosis in CWSV-1 hepatocytes (Figure 6). These results suggest that acute CD hepatocytes express sufficient levels of EGF receptor to permit rescue from apoptosis by stimulation of this pathway.

Choline Deficiency and TGFß1 Signaling

Although not typically found in normal hepatocytes, TGFß1 protein has been identified in hepatocytes undergoing apoptosis, mainly in regressing liver (47). In nonregressing hepatocytes, it is not clear whether the primary role of TGFß1 is to suppress cell growth in preparation for apoptosis, as has been suggested by some (48), or whether TGFß1 is an actual autocrine-paracrine trigger of apoptosis in hepatocytes. Most types of cells release TGFß1 as a latent complex in a molecular size range of Å 65–114 kDa, depending upon the extent of glycosylation (49, 50).

We assayed for the expression of TGFß1 in control (70 μM choline), acute CD hepatocytes and cells adapted to survive in 5 μM choline. Similar levels of expression of a Å65 kDa protein, consistent with a latent form of TGFß1, were observed in all three classes of CWSV-1 cells (Figure 7). Once activated, binding of TGFß1 to heterodimeric type I and II TGFß1 receptors is required for signal transduction (51). Expression of p27^{Kip1}, a cyclin-dependent kinase inhibitor with some sequence similarity to p21$^{Cip1/Waf1}$ (52), is an essential component linking TGFß1 expression to cell cycle arrest (53). Our data shows that CWSV-1 hepatocytes express TGFß1 type I and II receptors; p27^{Kip1} is expressed in the

Table 1. Endogenous TGFα and choline deficiency-induced apoptosis

| | Percent TUNEL (±) Cells | | |
Anti-TGFα (μl/ml)	Control	Deficient	Adapted
0	2.67 ± 0.69	20.77 ± 3.12	5.60 ± 1.13
0.1	10.67 ± 1.94*	18.04 ± 3.32	9.30 ± 2.76*
1.0	14.08 ± 0.31*	19.94 ± 0.48	18.50 ± 1.07*
5.0	16.30 ± 3.88*	22.82 ± 1.53	16.00 ± 3.42*

Induction of DNA strand breaks in control, acute deficient and adapted CWSV1 hepatocytes following neutralization of secreted TGFα. Cells were maintained for 2 days in experimental media as described in Figure 1, with or without the addition of anti-TGFα antibody (Santa Cruz Biotechnology) to neutralize endogenous TGFα. Mean ± SD, n = 2-4/group; * = p < .01 vs.ANOVA.

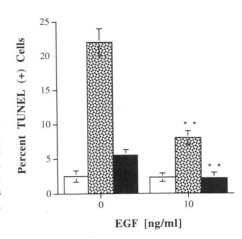

Figure 6. EGF signaling prevents apoptosis in choline deficient hepatocytes. Control (open bars) and acute CD (stippled bars), and adapted hepatocytes (black bars) were treated
for 2 days with or without EGF (Sigma) and the occurrence
of DNA strand breaks was assayed using a TUNEL method.
A significant reduction in the number of TUNEL (+) cells
undergoing apoptosis was observed in EGF treated cultures.
Mean ± SD; ** = $p < .01$ (t-testd).

cytoplasm of normal appearing cells, and in acute CD is found in the nucleus in morphologically apoptotic cells (Table 2). Although TGFß1-related proteins were detected immunocytochemically in cells adapted to survive in low choline, nuclear expression of $p27^{Kip1}$
was rarely seen in these cells. This change may be an impediment to the induction of CD
apoptosis in adapted CWSV-1 cells.

Cell detachment is a characteristic feature of TGFß1-induced apoptosis (54). When
treated with an active form of TGFß1, control and acute CD CWSV-1 cells, but not cells
adapted to low choline, showed a high rate of detachment from the monolayer (Figure 8)
and exhibited morphologic features of apoptosis (not shown). In animals, TGFß1 mRNA
synthesis increases by one week on a CD diet and persists for up to 6 weeks (44). Our results suggest that choline deficiency, perhaps by altering membrane composition in the vi-
cinity of apoptosis-transducing receptors, or by inducing the expression and/or activation
of latent triggers of apoptosis may induce apoptotic cell death independent of p53 expression in hepatocytes.

SUMMARY

Choline deficiency induces a p53-independent form of apoptosis in immortalized
nontumorigenic rat hepatocytes. Loss of responsiveness to this form of apoptosis occurs in
hepatocytes adapted to survive in low choline; such cells express increased levels of the

Figure 7. Expression of TGFß1 by CWSV-1 hepatocytes. Comparison of the expression of TGFß1
protein in control, acute deficient and adapted
CWSV-1 hepatocytes. Cells were maintained as
described in the legend to Figure 1. Equivalent
amounts of protein from cell lysates were subjected to SDS-PAGE and Western blotting using
anti-TGFGß1 antibody (Clone V, Santa Cruz
Biotechnology) and an ECL method. CWSV-1
cells express a Å 65 kDa protein consistent with a
latent form of TGFß1 (49, 50).

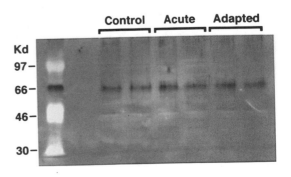

Table 2. Expression of TGFß1-related proteins by CWSV-1 hepatocytes

Antibody Probe	Control	Deficient	Adapted
TGFß1 Protein	+	++	++
Type I Receptor	+	+	+
Type II Receptor	+	+	+
p27^{Kip1} (Nuclear)	+	++/+++	+/±

Expression of TGFß1-related proteins in control, acute deficient and adapted CWSV-1 hepatocytes. Cells were maintained for 2 days in experimental media as described in Figure 1, and then probed using affinity purified antibodies (Santa Cruz Biotechnology) raised against TGFß1 protein, its cognate receptor types I and II, which together transduce the inhibitory effects of TGFß1, and p27^{Kip1}, a cyclin dependent kinase inhibitor implicated in growth arrest and apoptosis. Cells which were made acutely choline deficient, or ones adapted to survive in 5 µM choline-containing medium, exhibit a more intense immunocytochemical labeling for TGFß1 protein; whereas the intensity of labeling with the anti-TGFß1 receptor antibodies appeared to be similar in all three cell classes. Acutely deficient CWSV-1 hepatocytes, but not adapted hepatocytes, showed a pronounced increase in the number of nuclei labeled with the anti-p27^{Kip1} antibody.

mitogen TGFα, and are tumorigenic. Alterations to these signaling pathways may be important factors in multistep carcinogenesis induced by withdrawal of the nutrient choline from the diet.

ACKNOWLEDGMENT

This work was supported by a grant from the University of North Carolina University Research Council and the Institute of Nutrition, and by a grant from the American Institute for Cancer Research.

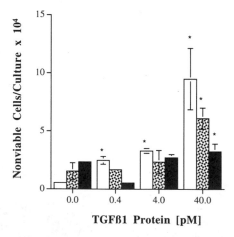

Figure 8. TGFß1 and cell detachment. Kinetics of detachment and cell death (trypan blue exclusion) in CWSV-1 hepatocytes in response to TGFß1. When treated for 2 days with hrTGFß1 (R & D Systems), control (70 µM choline) (open bars) and acute CD (5 µM choline) (stippled bars) hepatocytes, but not CWSV-1 cells adapted to low choline (black bars), exhibit a high rate of cell detachment, a characteristic feature of TGFß1 induced apoptosis (54). Mean ± SD; * = p < .01 (t-test).

REFERENCES

1. Ghoshal, A.K. and Farber, E. (1984). The induction of liver cancer by dietary deficiency of choline and methionine without added carcinogens. *Carcinogenesis* **5**, 1367–1370.
2. Chandar, N. and Lombardi, B. (1988). Liver cell proliferation and incidence of hepatocellular carcinomas in rats fed consecutively a choline-devoid and choline- supplemented diet. *Carcinogenesis* **9**, 259–263.
3. Rogers, A.E. (1995). Methyl donors in the diet and responses to chemical carcinogens. *Am J Clin Nutrition* **61**,659S–665S.
4. Yokoyama, S., Sells, M.A., Reddy, T.V. and Lombardi, B. (1985). Hepatocarcinogenic and promoting action of a choline-devoid diet in the rat. *Cancer Res* **45**, 2834–2842.
5. Abonabi, S.E., Lombardi, B. and Shinozuka, H. (1982). Possible mechanism(s) of action of a choline devoid diet and phenobarbital as promoters of carcinogenesis. *Cancer Res* **42**, 412–415.
6. Ghoshal, A.K., Ahluwalia, M. and Farber, E. (1983). The rapid induction of liver cell death in rats fed choline-deficient methionine low diet. *Am J Pathol* **113**, 309–314.
7. Albright, C.D., Liu, R., Bethea, T.C., da Costa, K.-A., Salganik, R.I. and Zeisel, S.H. (1996). Choline deficiency induces apoptosis in SV40-immortalized CWSV-1 rat hepatocytes in culture. *FASEB J* **10**, 510–516.
8. Shin, Ok-Ho, Mar, M.-H., Albright, C.D., Citarella, M.T., da Costa, K.-A. and Zeisel, S.H. (1996). Methyl-group donors cannot prevent apoptotic death of rat hepatocytes induced by choline deficiency. *J Cell Biochem* (in press).
9. Zeisel, S.H., da Costa, K.-A., Albright, C.D. and Shin, O.-K. (1995). Choline and hepatocarcinogenesis in the rat. *Adv Exp Biol Med* **375**, 65–74.
10. Shigenaga, M.K. and Ames, B.N. (1993). Oxidants and mitogenesis as causes of mutation and cancer: the influence of diet. *Basic Life Sciences* **61**, 419–436.
11. Snyderwine, E.G. (1994). Some perspectives on the nutritional aspects of breast cancer research. Food-derived heterocyclic amines as etiologic agents in human mammary cancer. *Cancer* **74 (3 Suppl)**, 1070–1077.
12. Webster, R.P., Gawde, M.D. and Bhattacharya, R.K. (1996). Modulation of carcinogen- induced DNA damage and repair enzyme activity by dietary riboflavin. *Cancer Lett.* **98**, 129–135.
13. Khan, N., Yang, K., Newmark, H., Wong, G., Telang, N., Rivlin, R. and Lipkin, M. (1994). Mammary ductal epithelial cell hyperproliferation and hyperplasia induced by a nutritional stress diet containing four components of a western-style diet. *Carcinogenesis* **15**, 2645–2648.
14. Appel, M.J. and Woutersen, R.A. (1994). Modulation of growth and cell turnover of preneoplastic lesions and of prostaglandin levels in rat pancreas by dietary fish oil. *Carcinogenesis* **15**, 2107–2112.
15. Birt, D.F., Pelling, J.C., Anderson, J. and Barnett, T. (1994). Consumption of reduced- energy/low-fat diet or constant-energy/high-fat diet during mezerin treatment inhibited mouse skin tumor promotion. *Carcinogenesis* **15**, 2341–2345.
16. Rao, C.V., Simi, B., Wynn, T.T., Garr, K. and Reddy, B.S. (1996). Modulating effect of amount and types of dietary fat on colonic mucosal phospholipase A2, phosphatidylinositol-specific metabolite formation during different stages of colon tumor promotion in male F344 rats. *Cancer Res* **56**, 532–537.
17. Black, H.S., Okotie-Eboh, G., Gerguis, J., Urban, J.I. and Thornby, J.I. (1995). Dietary fat modulates immunoresponsiveness in UV-irradiated mice. *Photochem Photobiol* **62**, 964–969.
18. Payne, C.M., Bernstein, H., Bernstein, C. and Garewal, H. (1995). Role of apoptosis in biology and pathology: resistance to apoptosis in colon carcinogenesis. *Ultrastructural Biol* **19**, 221–248.
19. La Porta, C.A. and Comolli, R. (1995). Over-expression of protein kinase C delta is associated with a delay in preneoplastic lesion development in diethylnitrosamine-induced rat hepatocarcinogenesis. *Carcinogenesis* **16**, 1233–1238.
20. da Costa, K.-A., Cochary, E.F., Blusztajn, J.K., Garner, S.C. and Zeisel, S.H. (1993). Accumulation of 1,2-sn-diradylglycerol with increased nuclear membrane-associated protein kinase C may be the mechanism for spontaneous hepatocarcinogenesis in choline deficient rats. *J Biol Chem* **268**, 2100–2105.
21. da Costa, K.-A., Garner, S.C., Chang, J. and Zeisel, S.H. (1995). Effects of prolonged (1 year) choline deficiency and subsequent refeeding of choline on 1,2-sn-diradylglycerol, fatty acids and protein kinase C in rat liver. *Carcinogenesis* **16**, 327–334.
22. Enwonwu, C.O and Meeks, V.I. (1995). Bionutrition and oral cancer in humans. *Critical Rev Oral Biol* **6**, 5–17.
23. Anderson, M.T., Staal, F.J.T., Gitler, C., Herzenberg, L.A. and Herzenberg, L.A. (1994). Separation of oxidant-initiated and redox-regulated steps in the NF-κB signal transduction pathway. *Proc Natl Acad Sci USA* **91**, 11527–11531.
24. Zeisel, S.H. (1981). Dietary choline: biochemistry, physiology, and pharmacology. *Annu Rev Nutr* **1**, 95–121.

25. Zeisel, S.H. (1993). Choline phospholipids: signal transduction and carcinogenesis. *FASEB J* **7**, 551–557.
26. Ghoshal, A.K. and Farber, E. (1993). Choline deficiency, lipotrope deficiency and the development of liver disease including liver cancer: a new perspective. *Lab Investigation* **68**, 255–260.
27. Columbano, A. (1995). Cell death: current difficulties in discriminating apoptosis from necrosis in the context of pathological processes in vivo. *J Cell Biochem* **58**,181–190.
28. Wyllie, A.H. (1987). Cell death. *Int J Cytol* **17 (Suppl)**, 755–785.
29. Woodworth, C.D., Secott, T. and Isom, H.C. (1986). Transformation of rat hepatocytes by transfection with simian virus 40 DNA to yield proliferating differentiated cells. *Cancer Res* **46**, 4018–4026.
30. Woodworth, C.D., Krieder, J.W., Mengel, L., Miller, T., Meng, Y. and Isom, H.C. (1988). Tumorigenicity of simian virus 40-hepatocyte cell lines: effect of in vitro and in vivo passage on expression of liver-specific genes and oncogenes. *Molec Cell Biol* **8**, 4492–4501.
31. Arends, M.J., McGregor, A.H., Toft, N.J., Brown, E.J.H. and Wyllie, A.H. (1993). Susceptibility to apoptosis is differentially regulated by *c-myc* and mutated *Ha-ras* oncogenes and is associated with endonuclease availability. *Br J Cancer* **68**, 1127–1133.
32. Albright, C.D., Salganik, R.I., Kaufmann, W.K. and Zeisel, S.H. (1996). Apoptosis is induced independent of p53 by choline deficiency in SV40-immortalized hepatocytes and a human hepatoma cell line. (*Submitted*).
33. Zeisel, S.H. and Blusztajn, J.K. (1994). Choline and human nutrition. *Annu Rev Nutr* **14**, 269–296.
34. Kim, Y.-L., Miller, J.W., da Costa K.-A., Nadeau, M., Smith, D., Selhub, J., Zeisel, S.H. and Mason, J.B. (1995). Folate deficiency causes secondary depletion of choline and phosphocholine in liver. *J Nutr* **124**, 2197–2203.
35. Meikrantz, W. and Schlegel, R. (1995). Apoptosis and the cell cycle. *J Cell Biochem* **58**, 160–174.
36. Sen, S. and D'Incalci, M. (1992). Apoptosis: biochemical events and relevance to cancer chemotherapy. *FEBS Lett* **307**,122–127.
37. Symonds, H., Krall, L., Remington, L., Saenz-Robles, M., Lowe, S., Jacks, T. and Van Dyke, T. (1994). p53-dependent apoptosis suppresses tumor growth and progression in vivo. *Cell* **78**, 703–711.
38. Harper, J.M., Adami, G.R., Wei, N., Keyomarsi, K., Elledge, S.J. (1993). The p21 Cdk-interacting protein Cip 1 is a potent inhibitor of G1 cyclin-dependent kinases. *Cell* **75**, 805–816.
39. El-Deiry, W.S., Tokino, T., Velculescu, V.E., Levy, D.B., Parsons, R., Trent, J.M., Lin, D., Mercer, W.E., Kinzler, K.W. and Vogelstein, B. (1993). *WAF1*, a potential mediator of p53 tumor suppression. *Cell* **75**, 817–825.
40. El-Deiry, W.S., Harper, J.W., O'Connor, P.M., Velculescu, V.E., Canman, C.E., Jackman, J., Peitenpol, J.A., Burrell, M., Hill, D.E., Wang, Y., Wiman, K.G., Mercer, W.E., Kastan, M.B., Kohn, K.W., Elledge, S.J., Kinzler, K.W. and Vogelstein, B. (1994). *WAF1/CIP1* is induced in *p53*-mediated G1 arrest and apoptosis. *Cancer Res* **54**, 1169–1174.
41. Zeisel, S.H., Albright, C.D., Mar, M.-H., Salganik, R.I., Shin, Ok-Ho and da Costa, K,- A. (1996). Choline deficiency selects for resistance to apoptosis and causes tumorigenic transformation of rat hepatocytes. (*Submitted*).
42. Albright, C.D., Jones, R.T., Hudson, E.A., Fontana, J.A. and Resau, J.H. (1990). Intercellular communication in bronchial epithelial cells: review of evidence for a possible role in lung carcinogenesis. *Toxicol Pathol* **6**, 379–398.
43. Boudreau, N., Sympson, C.J., Werb, Z. and Bissell, M.J. (1995). Suppression of ICE and apoptosis in mammary epithelial cells by extracellular matrix. *Science* **267**, 891–893.
44. Shinozuka, H., Masuhara, M., Kubo, Y. and Katyal, S.L. (1993). Growth factor and receptor modulations in rat liver by choline-methionine deficiency. *J Nutr Biochem* **4**, 610–617.
45. Luetteke, N.C., Michalopoulos, G.K., Teixido, J., Gilmore, R., Massagué, J. and Lee, D.C. (1988). Characterization of high molecular weight transforming growth factor α produced by rat hepatocellular carcinoma cells. *Biochemistry* **27**, 6487–6494.
46. Paniella, A. and Massagué, J. (1991). Multiple signals activate cleavage of the membrane transforming growth factor-α precursor. *J Biol Chem* **266**, 5769–5773.
47. Bursch, W., Oberhammer, F., Jirtle, R.L., Askari, M., Sedivy, R., Grasl-Kraupp, B., Purchio, A.F. and Schulte-Hermann, R. (1993). Transforming growth factor-ß1 as a signal for induction of cell death by apoptosis. *Br J Cancer* **67**, 531–536.
48. Santoni-Rugiu, Nagy, P., Jensen, M.R., Factor, V.M. and Thorgeirsson, S.S. (1996). Evolution of neoplastic development in the liver of transgenic mice co-expressing *c-myc* and transforming growth factor-α. *Am J Pathol* **149**, 407–428.
49. Horimoto, M., Kato, J., Takimoto, R., Terui, T., Mogi, Y. and Niitsu, Y. (1995). Identification of a transforming growth factor beta-1 activator derived from a human gastric cancer cell line. *Br J Cancer* **72**, 676–682.

50. McMahon, G.A., Dignam, J.D. and Gentry, L.E. (1996). Structural characterization of the latent complex between transforming growth factor beta 1 and beta 1-latency-associated peptide. *Biochem J* **313**, 343–351.

51. Franzen, P., ten Dijke, P., Ichijo, H., Yamashita, H., Schultz, P., Heldin, C.H. and Miyazono, K. (1993). Cloning of TGFß type I receptor that forms a heteromeric complex with the TGFß type II receptor. *Cell* **75**, 681–692.

52. Polyak, K., Lee, M.H., Erdjument-Bromage, H., Koff, A., Roberts, J.M., Tempst, P. and Massagué, J. (1994). Cloning of p27Kip1, a cyclin-dependent kinase inhibitor and a potential mediator of extracellular antimitogenic signals. *Cell* **78**, 59–66.

53. Coats, S., Flanagan, W.M., Nourse, J. and Roberts, J.M. (1996). Requirement of p27^{Kip1} for restriction point control of the fibroblast cell cycle. *Science* **272**, 877–880.

54. Oberhammer, F., Wilson, J.W., Dive, C., Morris, I.D., Hickman, J.A., Wakeling, A.E., Walker, P.R. and Sikorska, M. (1993). Apoptotic death in epithelial cells: cleavage of DNA to 300 and/or 50 kb fragments prior to or in the absence of internucleosomal fragmentation. *EMBO J* **12**, 3679–3684.

THE ROLE OF PEROXISOME PROLIFERATOR ACTIVATED RECEPTOR α IN PEROXISOME PROLIFERATION, PHYSIOLOGICAL HOMEOSTASIS, AND CHEMICAL CARCINOGENESIS

Frank J. Gonzalez[*]

Laboratory of Metabolism
Division of Basic Sciences
National Cancer Institute
National Institutes of Health
Bethesda, Maryland 20892

INTRODUCTION

Peroxisomes and Peroxisome Proliferation

Peroxisomes are single membrane-bound organelles, found in all cells, that contain a variety of enzymes involved in a number of metabolic processes[1]. The most well characterized reactions carried out by peroxisomes are those involved in fatty acid β-oxidation. The peroxisomal β-oxidation system metabolizes very long chain and long chain fatty acids and cannot metabolize short chain fatty acids (<6 units) whereas the mitochondrial system most efficiently oxidizes long, medium and short chain fatty acids down to two carbon units. The peroxisomal acyl-CoA oxidase enzyme generates H_2O_2 while the corresponding mitochondrial enzymes lead to production of NADH. Since plants lack mitochondria, peroxisomes are solely responsible for their fatty acid β-oxidation. The production of H_2O_2 by peroxisomal acyl-CoA oxidase has been used as a cytological marker for the organelle through staining with diaminobenzidine and historically accounts for the name "peroxisomes"[1]. Peroxisome proliferation can result in an excess of H_2O_2 that can potentially result in toxicity as discussed below. Under usual circumstances H_2O_2 is decomposed to molecular oxygen and water by catalase and glutathione peroxidase.

*Correspondence: Frank J. Gonzalez, Building 37, Room 3E-24, National Institutes of Health, Bethesda, Maryland 20892. Phone: (301) 496–9067; Fax: (301) 496–8419; Email fjgonz@helix.nih,gov.

Dietary Fat and Cancer, edited by AICR
Plenum Press, New York, 1997

Human genetic deficiencies in peroxisome biogenesis and individual peroxisomal enzymes have been described that result in accumulation of long chain fatty acids and multiple functional deficiencies.[2,3] The most severe of the peroxisome deficiencies cause neurological abnormalities resulting in neonatal death.

Peroxisomes also carry out the β-oxidation of the side chain of cholesterol during the synthesis of bile acids, and participate in the biosynthesis of cholesterol,[3] of ether glycolipids and dolichols. Catabolism of purines, polyamines, glyoxylate and certain amino acids have been attributed to peroxisome-localized enzymes. Thus, peroxisomes are essential organelles for maintaining cellular and organismal homeostasis.

The number of peroxisomes can be increased by treatment of rodents with high fat diets, starvation, cold temperature, ACTH and certain chemicals generically termed peroxisome proliferators.[1] Peroxisome proliferation has also been demonstrated in yeast incubated with fatty acids and methanol and in plants during germination and light exposure. Peroxisome proliferators include a structurally diverse group of chemicals that include 1) hypolipidemic drugs used in lowering serum triglycerides and cholesterol including clofibrate, gemfibrozil, fenofibrate, benzofibrate and etofibrate; 2) the azole antifungal compounds such as bifenazole; 3) leukotriene D_4 antagonists; 4) herbicides; 5) pesticides; 6) phthalate esters used in the plastics industry like di-(2-ethylhexyl)phthalate and related compounds; 7) simple solvents including trichloroethylene; and 8) the endogenous chemicals phenylacetate and the steroid dehydroepiandosterone and its sulfate conjugate. Among the most potent peroxisome proliferators is the experimental drug WY-14,643. Peroxisome proliferators are structurally diverse and share only a carbon backbone and carboxylic acid or other acidic group, except for those compounds that require conversion to an active metabolite by the xenobiotic-metabolizing carboxylesterases.

Peroxisome proliferation is most pronounced in liver, kidney and to a lesser extent in heart. In liver, the number of peroxisomes increase from about 500–600 per cell to over 5,000 per cell after two to three weeks of exposure to peroxisome proliferators in the diet[1]. This is accompanied by an increase in cell volume resulting in hepatomegaly. In addition, a transient hepatocellular hyperplasia occurs during early stages of treatment.

PPARα

The Peroxisome Proliferator-Activated Receptor

Coincident with an increase in number of peroxisomes, several peroxisomal enzymes are induced. The transcription, mRNA and protein levels of the key β-oxidation enzymes acyl-CoA oxidase and enoyl-CoA hydratase/3-hydroxyacyl-CoA dehydrogenase (bifunctional enzyme) are markedly elevated by peroxisome proliferator treatment.[4] Genes encoding the microsomal cytochrome P450s in the CYP4A family are also activated by these agents.[5,6] These data were the first to suggest that a nuclear receptor-mediated mechanism is responsible for the pleiotropic effects of peroxisome proliferators A receptor was then cloned from mouse and found to be responsive to peroxisome proliferators in a *trans*-activation transfection assay.[7,8] This receptor, named the peroxisome proliferator-activated receptor (PPAR), was found to be a member of the nuclear receptor superfamily that includes the estrogen, progesterone, thyroid, glucocorticoid, retinoic acid and a growing number of "orphan" receptors that exhibit the domain structure of the nuclear receptors but their natural ligands are not known. The conserved modular domain structure of this superfamily allowed the construction of chimeric receptors with the estrogen receptor[7] and glucocorticoid receptor DNA-binding domains[9] with the ligand-binding domain of PPARα which aided in the initial characterization of the ligand dependence of the original mouse PPAR.

Three members of the PPAR family designated PPARα, β and γ were also cloned from *Xenopus leavis*.[10] The frog PPARα exhibited the highest amino acid sequence similarity with the original mouse PPAR. Subsequently, several additional receptors related to PPARα were identified in mammals. PPARγ[11,12] and PPARγ2[13] share highest similarity with frog PPARγ and appear to be derived by alternative splicing of the same gene. The mammalian form most closely related to the frog PPARβ, has been called NUC1[14,15,16] and FAAR.[17] Thus, there appear to be three members of the PPAR family, although others may yet be discovered.

Function of PPARs

Recent evidence suggest that the three PPARs may have different functions. The tissue distribution of each receptor is quite distinct. PPARα mRNA has been detected at highest abundance in the liver, kidney and heart, all of which are tissues that exhibit peroxisome proliferation.[15] PPARα is also found at low levels in adipose tissue and the gastrointestinal tract.[15,18] All tissues expressing PPARα are known to have high rates of lipid metabolism. Expression of PPARβ is found in most tissues analyzed.[15,19] In tissues expressing PPARα, PPARβ mRNA is present at a markedly lower abundance. PPARβ is expressed in adipose tissue, small intestine, skeletal muscle, lung, heart and brain. PPARγ and PPARγ2 are found predominantly in adipose tissue and spleen. Comparative analysis of the tissue distribution of the PPAR's expression suggest that the different receptors fulfill specific functions in the adult animal and possibly distinct functions during development.

The tissue distribution of PPARγ2 suggests an involvement in adipose cell gene expression which was confirmed by a series of experiments.[20,21] 1) PPARγ2 in cooperation with the retinoic acid X receptor α (RXRα) can bind to the adipocyte P2 gene enhancer. The enhancer is also stimulated to respond in *trans*-activation transfection assays in the presence of hypolipidemic agents, certain fatty acids and 9-*cis* retinoic acid. 2) Expression of PPARγ2 is induced when fibroblasts spontaneously differentiate to adipocytes in the presence of insulin and serum. 3) Forced expression of PPARγ2 in fibroblast by a retroviral expression vector stimulates adipocyte differentiation. This effect is promoted by activators of the receptor including clofibric and linoleic acid, the arachidonic acid analog ETYA and the leukotriene antagonist LY-171883. Further evidence for a critical role of PPARγ in adipocyte differentiation was suggested by the identification of ligands that were previously known to have marked adipogenic effects[22] (PPARγ not PPARγ2 was used in these experiments).

Thiazolidinediones, derivatives of antidiabetic agents that increase the insulin sensitivity of target tissues in animal models, were shown to be activators of PPARγ2 in a *trans*-activation assay and bind to recombinant receptor expressed in bacteria.[22] These data indicate that PPARγ may modulate glucose and insulin levels either through an indirect effect mediated by serum lipids or by a direct action at target tissues. A recent report has revealed that a peroxisome proliferator can alter glucose transport, although the mechanism and role of PPARs in this effect was not addressed.[23] Taken together, these data establish a critical role for PPARγ (and γ2) in physiological homeostasis.

To date, there is no evidence for a physiological or developmental role for PPARβ, the receptor that exhibits a more ubiquitous expression pattern than the other two family members. In addition, no ligands or activators have been found for this receptor and no target genes have been identified. PPARβ is not responsive to several peroxisome proliferators but, in *trans* activation assays, is able to antagonize activation of gene promoters mediated by PPARα.[16,17]

A physiological role for PPARα is suggested by its role in peroxisome proliferation and activation of genes encoding fatty acid metabolizing enzymes, as noted earlier. Direct gene activation by PPARα has been suggested by *trans*-activation studies using chimeric receptor constructs and intact cDNA-expressed receptors.[24,25] In addition to the peroxisomal β-oxidation enzymes and microsomal fatty acid hydroxylase P450s, liver fatty acid binding protein[26] and the genes encoding mitochondrial enzymes medium-chain acyl-CoA dehydrogenase[27] 3-hydroxy-3-methylglutaryl-CoA synthase[28] and malic enzymes[29] are also activated by PPARα as indicated by *trans*-activation assays. It was suggested that induction of the malic enzyme by peroxisome proliferators is a paradox since it is not directly involved in oxidation of fatty acids and is generally considered a lipogenic enzyme.[29] However, malic enzyme could play a role in mobilizing the acetyl-CoA formed during fatty acid catabolism to gluconeogenesis. Collectively, the battery of enzymes under control of PPARα suggest a critical role for the receptor in modulating lipid catabolism and perhaps. under certain conditions, even lipid synthesis.

The ligands for PPARα have not been demonstrated through direct binding experiments, and have only been inferred through *trans*-activation studies. In addition to all peroxisome proliferators tested to date, PPARα mediates *trans*-activation by fatty acids with the long chain dicarboxylic fatty acids being the most potent.[9,10,30] There also is an indirect relationship between the ability of a fatty acid or fatty acid derivatives such as sulfur substituted fatty acids to be metabolized and its potency to induce β-oxidation enzymes.[31,32] These data suggest that increases in cellular fatty acids lead to an activation of PPARα by a fatty acid or metabolite resulting in induction of enzymes responsible for fatty acid β-oxidation.

An indirect mechanism for induction of peroxisome proliferation was suggested in which xenobiotic peroxisome proliferators directly inhibit mitochondrial fatty acid β-oxidation resulting in accumulation of long chain fatty acids which, when metabolized to dicarboxylic acids, activate PPARα.[33] Inhibition of carnitine palmitoyltransferase I by oleic acid, which is not normally a peroxisome proliferator, results in induction of CYP4A1, acyl-CoA oxidase and liver fatty acid binding protein. Thus, fatty acids may be essential for the full induction process mediated by peroxisome proliferators. Induction studies with clofibrate have shown that activation of the *CYP4A* genes precedes activation of the acyl-CoA oxidase gene.[30,31,32] When the P450s are inhibited, the induction of acyl-CoA oxidase by clofibrate is blocked but the *CYP4A1* gene is induced.[30] These data support the hypothesis that long chain dicarboxylic acids produced through increased CYP4A1 ω-hydroxylation are essential for induction of the peroxisomal enzymes. These data would argue that the *CYP4A* genes and the genes encoding peroxisomal enzymes are controlled differently. Perhaps they are regulated by different PPARs or the *CYP4A* genes are more sensitive to weaker ligand-bound PPARα than are the acyl-CoA oxidase gene that requires ligands with higher affinities. However, the P450 and peroxisomal genes each have similar upstream DNA elements that interact with PPARα.[34]

Mechanism of Gene Activation by PPARα

PPARα activates gene expression through binding to a response element located upstream of target genes called the peroxisome proliferator-activated response element

(PPRE). The PPRE is also called a DR-1 motif that interacts with members of the nuclear receptor superfamily. It consists of two half sites of TGACCT separated by a one base spacer.[34] A PPRE is found upstream of all PPARα target genes examined to date. Binding to PPRE has been demonstrated using cellular extracts containing PPARα, cDNA-expressed PPARα and *in vitro* translated receptor. Binding does not require addition of a putative ligand to cell extracts or expressed PPARα. However, addition of another member of the nuclear receptor superfamily, the retinoid X receptor α (RXRα) markedly enhances *in vitro* binding and *trans*-activation by PPARα. Optimal *trans*-activation occurs in the presence of retinoic acid or the natural RXRα ligand 9-*cis* retinoic acid.[34-37] It is also noteworthy that the RXRα-PPARα complex can *trans*-activate the hepatitis B virus enhancer I by binding to a retinoic acid response element.[37] Thus, PPARα must form a functional dimer with RXRα for optimum gene activation potential and the complex can activate target genes through PPRE and potentially other regulatory elements.

PPARα also interacts with other members of the nuclear receptor superfamily. It was found to form heterodimers with the thyroid hormone receptor (TR) and to interact with a non-classical thyroid hormone response DR-2 element (TRRE).[38] PPARα behaves as a dominant-negative regulator of thyroid hormone action when complexed with TRα and a positive regulator with TRβ. With a different TRRE, PPARα inhibits activation with both TRs.

Another member of the nuclear steroid receptor superfamily, ARP-1 dimerizes with RXRα and activates transcription of genes encoding apolipoproteins (apo) A-I, A-II, B and C-III.[39,40] These genes are also affected by peroxisome proliferator fibrate drugs (see below). ARP-1/RXRα binds to a proximal element in the *CYP4A6* gene and blocks transcription mediated by PPARα/RXRα.[41] It is also of note that HNF-4, which is believed to be a member of the steroid receptor superfamily that acts in the absence of a ligand, represses PPARα-dependent activation of the acyl-CoA oxidase regulatory element.[42] In contrast, HNF-4 enhanced PPARα-dependent activation of the promoter controlling the bifunctional enzyme.[42] This factor also modulates the medium chains acyl-CoA dehydrogenase gene[43] and the genes encoding apoC-III.[44]

It should be emphasized that the studies showing interactions between various receptors and transcription factors and their regulatory elements were done using transfection assays and DNA binding. There is no evidence that these relationships are of functional significance in the intact cell or in tissues. Understanding receptor "cross talk" requires knowledge of the cellular concentrations of each receptor, their dimerization affinities with each other and the binding affinities of each dimer pair with the different target gene regulatory elements. Thus, the cellular content of the PPARα, TR and Arp-1 with RXRα and other receptors and their ligands might dictate gene expression patterns.

It remains a possibility that other receptors will be found that can influence the activity of the PPAR receptors in different cell types. It is essential to sort out the complexity of the signal transduction by the different members of the nuclear steroid receptor superfamily in order to understand the functions of PPARα in animal development and physiology and in the toxic and carcinogenic effects of peroxisome proliferators.

The fibrate class of drugs can also lead to suppression of gene expression. The apoA-I,[45] apoA-II,[46] apoC-III[44] and lecithin cholesterol acyltransferase[47] are all suppressed by treatment of animals with fibrate drugs. The alterations of these enzymes are thought to mediate the lipid and cholesterol lowering effects of hypolipidemic drugs. However, drugs can have different effects in human cells. Gemfibrozil actually increases apoA-1 expression whereas fenofibrate lowers its expression in the HepG2 cell line.[48,49] In clinical studies, fibrates are known to increase apoA-I and apoA-II levels.[50] Heterologous promoter *trans*-activation studies using the apo A-I gene upstream elements suggest that the effects of fibrates are independent of PPARα and act near the transcription start site.[51] The apo A-I gene does have an apparent PPRE upstream of the negative regulatory element that can eliminate or modulate the negative effects of fibrates. This study clearly does not totally explain the mechanism of gene suppression by fibrate drugs but it should be emphasized that the results were obtained using reporter gene constructs and HepG2 cell transfection assays and may not reflect the regulation of apoA-1 in an intact animal.

PEROXISOME PROLIFERATOR-INDUCED HEPATOCARCINOGENESIS

Chronic administration of peroxisome proliferators to rodents results in hepatic neoplasia.[52–56] Almost every peroxisome proliferator tested causes liver cancer in either mice or rats. The potent compounds WY-14,643 and LY-171883 cause 100% tumors at 2 years when rats are fed a level of 0.1% or less and the weaker chemicals di(2-ethylhexyl)phthalate and trichloroethylene yield hepatocellular carcinoma at 2 years when administered to rats at 1.2% to 2.5% and 1200 mg/kg, respectively.

Peroxisome proliferators are referred to as non-genotoxic carcinogens since they yield negative results in the common genotoxicity tests including the *Salmonella typhimurium*-based Ames test and eukaryote-based DNA damage test systems.[57–61] They are also unable to function as initiators of carcinogenesis in animals.[62,63] The mechanism by which these agents cause cancer is not currently understood. One hypothesis is that peroxisome proliferation causes genotoxicity by increasing intracellular H_2O_2.[55] This is based on the fact that peroxisome proliferators markedly increase fatty acid β-oxidation but only modestly elevate catalase, the peroxisomal enzyme responsible for metabolizing the H_2O_2 generated by acyl-CoA oxidase. Thus, excess H_2O_2 could escape from the organelle and react with cellular macromolecules. Indeed, peroxide-modified lipids have been found in hepatocytes of peroxisome proliferator-treated rats.[64,65] Damaged DNA in the form of 8-hydroxydeoxyguanosine has been detected in livers of rats chronically exposed to different peroxisome proliferators.[66,67] However, other studies reported no increases in deoxyguanosine modification in liver DNA of treated rats compared to controls.[68,69] It is noteworthy that in experiments where DNA damage was detected, total liver DNA was analyzed whereas in the negative studies, DNA was derived from purified nuclei. This suggests that the increased damage upon treatment with peroxisome proliferators was largely due to cytoplasmic mitochondrial DNA.

Further evidence supporting a role for H_2O_2 -mediated DNA damage and carcinogenesis was recently generated by showing that stable over-expression of acyl-CoA oxidase in NIH-3T3 cells produced DNA damage and cell transformation as assessed by cell growth in soft agar and nude mice.[70] These data establish that H_2O_2 can lead to cell transformation *in vitro*. However, peroxisomal proliferators cause transformation of cultured cells in the absence of apparent peroxisome proliferation[71,72] thus lending support to the notion that excess H_2O_2 is not involved in the genetic damage leading to carcinogenesis.

Another hypothesis that has been advanced through experimental data is that peroxisome proliferators are tumor promoters that promote spontaneously initiated hepatocytes.[73] Tumor promoters are thought to exert their effects, in part, through the stimulation of cell division.[74] Peroxisome proliferators, as noted, cause hepatomegaly and hypertrophy. The latter is due to a mitogenic effect that occurs soon after commencement of peroxisome proliferator feeding[75]. To address the relationship between DNA replication and carcinogenesis, a weak peroxisome proliferator diethylhexylpthalate (DEHP) was compared with the potent WY-14,643.[76] An initial burst of DNA replication was found for both compounds within a few days after initiation of feeding. After 18 days, livers from WY-14,643-treated animals showed a persistent increase in DNA labeling over DEHP-fed rats that ranged from 0.6 to 15-fold out to 369 days. The amount of peroxisome proliferation and increase in peroxisomal enzyme induction were only 10 to 20% higher for the WY-14,643-treated group. This led to the conclusion that peroxisome proliferation is not correlated with carcinogenicity and that persistent cell replication is an important factor. Is should be noted, however, that other studies have shown a correlation between peroxisome proliferation and hepatocarcinogenesis.[77,78]

A third hypothesis to account for the carcinogenic effect of peroxisome proliferators and perhaps other non-genotoxic carcinogens is that these agents suppress apoptosis or programmed cell death.[79] Incubation of rat hepatocytes with the potent peroxisome proliferator nafenopin resulted in maintenance of viability for at least four weeks in contrast to eight days with control cultures.[80,81] This was associated with a decrease in cells exhibiting signs of apoptosis such as condensed and fragmented nuclei. Apoptosis in control primary cultures could be increased by TGFβ1 and this increase is inhibited by nafenopin. A similar effect was seen in the Reuber hepatoma cell line. Nafenopin was also found to synergize with the growth factor EGF to promote clonal outgrowth of primary rat hepatocytes.[82] This was due to an alteration of balance between mitosis and apoptosis. Inhibition of apoptosis could be responsible for the carcinogenic effects of peroxisome proliferators if cells destined for programmed cell death have genetic damage and are more prone to develop into cancer cells (Figure 1).

In conclusion, the mechanism of hepatocarcinogenesis for peroxisome proliferators remains unclear. Perhaps a combination of the elements of all three hypotheses discussed above is relevant. Understanding the mechanism of action of peroxisome proliferators will aid in both human risk assessment and in a general comprehension of the processes involved in tumorigenesis.

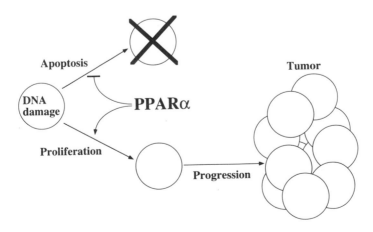

Figure 1.

SPECIES DIFFERENCES AND HUMAN RISK ASSESSMENT

Rats and mice are highly responsive to peroxisome proliferators and develop hepato-carcinomas upon long-term administration.[54] Hamsters also exhibit peroxisome proliferation albeit to a lesser degree than rats.[83] In contrast, guinea pigs and marmosets are not responsive, even to the potent agent nafenopin. Studies of Rhesus monkeys and humans have yielded different results and appear to suggest that these species are only weakly responsive or not responsive.[84]

The mechanisms by which humans and non-human primates are resistant to peroxisome proliferation are not known. Humans do possess the PPARα receptor indicating that lack of receptor is not responsible for the refractivity to peroxisome proliferation.[24] However, the cellular content of receptor in human liver has not been determined and compared with that in responsive species. The human acyl-CoA oxidase gene also has a PPRE that is responsive in *trans*-activation transfection assays, thus indicating that the response elements may not differ between humans and rodents.[85] Peroxisome proliferators do evoke a biological response in humans which is the basis for the therapeutic efficacy of the fibrate drugs which reduce serum lipid levels. Surprisingly, even though these drugs have been used in humans for more than 30 years, their mechanisms of action are not fully understood.

THE PPARα-NULL MOUSE

Isolation and Characterization of the PPARα cDNA

In an effort to determine the mechanism of regulation of cytochromes P450 in the CYP4A family,[5,6,86] the PPARα cDNA was cloned from rat and mouse liver cDNA libraries by use of polymerase chain reaction. The human PPARα cDNA was isolated from human liver cDNA libraries using the rat cDNA as a probe. All cDNAs were completely sequenced and shown to be active in *trans*-activation transfection assays using the rabbit *CYP4A6* and acyl-CoA oxidase gene's PPREs.[24] However, differences were noted between the rat and human PPARα cDNA in the ED_{50} for clofibrate, but these differences were not significant enough to explain the refractivity of humans to peroxisome proliferators.

Production of the PPARα-Null Mouse

The mouse PPARα cDNA was used to isolate genomic clones from mouse gene libraries derived from Balb/C and 129/Sv strains. Gene targeting constructs were produced from the 129/Sv library and used to electroporate embryonic stem cells. The construct contained exon's 7 and 8 of the PPARα gene; exon 8 was used to insert the phosphoribo-syltransferase II gene cassette. The latter allowed for selection of clones with G418 resistance and disruption of gene function.[87,88] Exon 8 encodes the putative ligand-binding/dimerization domain as estimated by comparison with the domain structures of other members of the steroid receptor superfamily.[89,90] There was initial concern that leaving the putative DNA-binding domain intact could result in production of a receptor that was truncated but retained the capacity to bind DNA. The truncated receptor could bind to PPREs and make interpretations of data obtained with the knockout mice difficult. However, analysis of homozygous knockout mice revealed that the mRNA from the targeted gene was unstable and no protein was detectable.

Mice containing two copies of the disrupted PPARα gene, designated PPARα-null, showed no apparent gross phenotypic abnormalities either externally or internally.[87] Administration of clofibrate and WY-14,643 to the PPARα-null by feeding the compounds for two weeks did not elicit a pleiotropic response seen in control mice. Increased peroxisomes, hepatomegaly and induction of several target genes were also not detected in null animals administered peroxisome proliferators. These data establish with certainty that the PPARα is the receptor responsible for the pleiotropic effects of peroxisome proliferators and imply that PPARβ and PPARγ have minor roles, if any, in peroxisome proliferation.

Lipid Metabolism in the PPARα-Null Mice

PPARα-null mice of 2 to 3 months of age treated with WY-14,643 showed microvesicles of fat in the liver.[87] As these animals get older, they tend to increase in weight as compared to control or heterozygous animals (unpublished results). The fat was found in the epididymal and omental regions. Analysis of livers from 8- month-old mice maintained on normal animal chow contained numerous microvesicles of fat in the liver. The number of fat droplets increased in animals fasted for 24 hours. These data indicate that lipid metabolism is altered in the PPARα-null mice. The mechanism for the altered lipid metabolism might be related to the ability of the mouse to respond to an increased lipid load by elevating the levels of peroxisomal and mitochondrial β-oxidation enzymes. The relative numbers of peroxisomes in PPARα-null mice are similar to untreated control mice. Our data suggest that in mice (and perhaps rats) the constitutive level of peroxisomes may not be sufficient to maintain lipid homeostasis and cope with lipid overload. Alternatively, the enzyme constituents of peroxisomes differ between the knockout and control animals.

The Role of PPARα in DHEA Induction of Peroxisome Proliferation

Dehydroepiandrosterone (DHEA) is an endogenous steroid of no known physiological function. It must be converted to its 3β-sulfate derivative DHEA-S in order to be active in peroxisome proliferation.[91] Administration of high levels of DHEA or DHEA-S to rats results in peroxisome proliferation. Similar to other peroxisome proliferators, DHEA causes an increase in peroxisomal fatty acid β-oxidation[92,93] and under chronic administration, results in hepatocarcinogenesis.[94] Paradoxically, DHEA has also demonstrated chemopreventive properties in experimental carcinogenesis systems[95] and by correlative studies in humans.[96] Controversy exists regarding whether DHEA or DHEA-S act through the PPARα.[97] In some studies, DHEA and DHEA-S were unable to stimulate *trans*-activation mediated by PPARα.[7,9] DHEA-S does differ in its biological effects from the more classical xenobiotic peroxisome proliferators; it does not induce gene expression in the kidney[96] and it does not stimulate growth arrest in cultured cell lines[98] but can cause inhibition of cell growth.[98-100] The latter property may be related to its chemoprevention properties.

To investigate the role of PPARα in peroxisome proliferation mediated by DHEA-S, the PPARα-null mice were examined.[101] PPARα-null mice were unresponsive to the induction of target genes encoding peroxisomal fatty acid β-oxidation enzymes and CYP4A P450s when administered DHEA-S, in contrast to normal mice that were readily induced by the steroid. *Trans*-activation assays using the intact rat PPARα also showed that DHEA-S was active.[96] These data establish that PPARα-mediated the response to DHEA-S and illustrate the utility of the PPARα-null mouse model in determining whether a specific compound is a classical peroxisome proliferator.

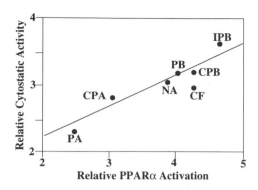

Figure 2. Relative cytostatic activity vs. PPARα activation.

PPARα AND GROWTH CONTROL

Mechanism of action of phenylacetate and its analogs. Phenylacetate is a naturally occurring plasma component currently in clinical trials as an antitumor agent.[102] Phenylacetate
and its analog phenylbutyrate cause cytostasis and differentiation of a number of tumor cell
lines grown *in vitro* and in animals.[103–107] Many of the properties of these compounds and their
effects on cells were similar to those of peroxisome proliferators suggesting that they may act
through the PPARα. This was examined by use of *trans*-activation assays using recombinant
hPPARα.[108] Phenylacetate and its derivatives were able to induce expression of reporter genes
through the hPPARα. The potency of activation was highly correlated with the potency of
growth arrest (Figure 2). In addition, expression of PPARα in cultured cells was also stimulated by these agents.[108] These results establish that PPARα is active in human cells and plays
a role in control of the cell cycle.

CONCLUSIONS AND FUTURE STUDIES

PPARα and Peroxisome Proliferation

Studies using *trans*-activation assays and the PPARα-null mouse model indicate that
PPARα mediates the biological action of peroxisome proliferators. To date all putative
target genes that are known to be induced by peroxisome proliferators, have been found to
have a PPRE in their promoter regions and their promoters respond to peroxisome proliferators in the presence of PPARα in *trans*-activation assays. The PPARα-null mice are totally resistant to peroxisome proliferation and target gene induction by peroxisome
proliferators. The endogenous chemicals DHEA-S and phenylacetate were also shown to
act through PPARα and it appears that PPARβ and PPARγ cannot substitute for loss of
PPARα in the null mouse model.

The endogenous functions of PPARα and PPARγ seem to be related to fatty acid homeostasis; the function of the more ubiquitously-expressed PPARβ is not known. An endogenous ligand for PPARγ has been identified as 15-deoxy-delta 12,14-prostaglandin
J_2.[109] Possible candidates for endogenous ligands for PPARα include the eicosanoid 8S-
HETE[110] and leukotriene B_4 (unpublished results).

PPARα and the Mechanism of Action of Non-Genotoxic Carcinogens

Studies using the PPARα-null mouse would predict that the PPARα mediates the hepatocarcinogenesis of peroxisome proliferators in susceptible species. Hepatomegaly, due to increase in cell number and size is not observed in the null mouse fed WY-14,643.[87] Preliminary data indicate that these animals do not exhibit the increase in DNA synthesis associated with the chronic administration of this compound (unpublished results). A carcinogenesis bioassay is underway to establish with certainty whether the PPARα-null is resistant to the induction of liver tumors. It is anticipated that the null mouse model will also provide a unique opportunity to dissect the mechanism of action of peroxisome proliferator-induced cancer. The target genes required for this process could be determined. Among the possibilities are genes encoding cell cycle control factors. Indeed, peroxisome proliferators have been shown to induce expression of a number of immediate early genes including c-*fos*, c-*jun*, *jun*B *erg*-1 and *NUP475*.[98] As noted above, peroxisome proliferators can stimulate DNA synthesis and inhibit apoptosis in hepatocytes, and can cause growth arrest and differentiation in tumor cell lines. Thus, depending on the cellular context, peroxisome proliferators can cause various perturbations in the cell cycle. This property probably plays a role carcinogenesis and tumor promotion, and the growth arrest and chemopreventive properties of peroxisome proliferators. Considerable work remains to determine how PPARα affects cell cycle control.

PPARα and Species Differences in Response to Peroxisome Proliferators

Peroxisome proliferation has not been demonstrated in humans. Hypolipidemic drugs do have biological effects in humans as they are effective serum lipid and cholesterol lowering agents. The mechanisms for this differential response are not known. One possibility is that humans have low hepatic levels of the receptor. This was suggested by the low frequency of PPARα cDNA clones in human liver cDNA libraries as compared to rat and mouse liver libraries.[24] The PPARα has been shown to be functional in *trans*-activation assays in humans although some quantitative differences in relative ligand response were noted between humans and rats.[24] One possibility that could account for the species differences is that the low levels of receptor in human hepatocytes are not sufficient to activate the full battery of genes that are activated in rodents including those involved in the process of peroxisome proliferation and cell cycle control (Figure 3). Perhaps the low receptor levels are sufficient to activate a subset of genes encoding enzymes or lipid transporters that result in the beneficial therapeutic effects of hypolipidemic drugs. Alternatively, the receptor expressed in extrahepatic tissues may be sufficient to cause the lowering of serum triglycerides by hypolipidemic drugs.

In order to test the hypothesis that low hepatic PPARα causes the species differences in response to peroxisome proliferators, gene replacements are being carried out. The human PPARα gene, cloned from a bacterial artificial chromosome library, will be introduced into the PPARα-null mice. It is anticipated, based on other studies on transgenic mice, that transgenic lines will be obtained which differ in the level of expression of the receptor. Based on the above hypothesis, lines in which the receptor is expressed at low levels comparable to that in human liver, would be expected to express a limited number of genes when challenged with peroxisome proliferators, while those that express high levels, as in a normal mouse, would exhibit all of the pleiotropic effects of peroxisome proliferators. These experiments will establish a mechanism for species differences

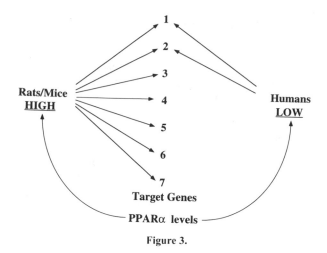

Figure 3.

in response to peroxisome proliferators and will allow a more rational basis for risk assessment decisions regarding these agents.

REFERENCES

1. Reddy, J.K. and Mannaerts, G.P. Peroxisomal lipid metabolism. Annu. Rev. Nutr., 14: 343–370, 1994.
2. Wanders, R.J., Schutgens, R.B., and Barth, P.G. Peroxisomal disorders: a review. J. Neuropathol. Exp. Neurol., 54: 726–739, 1995.
3. Wanders, R.J., Barth, P.G., Schutgens, R.B., and Tager, J.M. Clinical and biochemical characteristics of peroxisomal disorders: an update. Eur. J. Pediatr., 153: S44-S48, 1994.
4. Reddy, J.K., Goel, S.K., Nemali, M.R., Carrino, J.J., Laffler, T.G., Reddy, M.K., Sperbeck, S.J., Osumi, T., Hashimoto, T., and Lalwani, N.D. Transcription regulation of peroxisomal fatty acyl-CoA oxidase and enoyl-CoA hydratase/3-hydroxyacyl-CoA dehydrogenase in rat liver by peroxisome proliferators. Proc. Natl. Acad. Sci. USA., 83: 1747–1751, 1986.
5. Kimura, S., Hardwick, J.P., Kozak, C.A., and Gonzalez, F.J. The rat clofibrate-inducible CYP4A subfamily. II. cDNA sequence of IVA3, mapping of the Cyp4a locus to mouse chromosome 4, and coordinate and tissue-specific regulation of the CYP4A genes. DNA, 8: 517–525, 1989.
6. Kimura, S., Hanioka, N., Matsunaga, E., and Gonzalez, F.J. The rat clofibrate-inducible CYP4A gene subfamily. I. Complete intron and exon sequence of the CYP4A1 and CYP4A2 genes, unique exon organization, and identification of a conserved 19-bp upstream element. DNA, 8: 503–516, 1989.
7. Issemann, I., Prince, R.A., Tugwood, J.D., and Green, S. The peroxisome proliferator-activated receptor:retinoid X receptor heterodimer is activated by fatty acids and fibrate hypolipidaemic drugs. J. Mol. Endocrinol., 11: 37–47, 1993.
8. Green, S. PPAR: a mediator of peroxisome proliferator action. Mutat. Res., 333: 101–109, 1995.
9. Gottlicher, M., Widmark, E., Li, Q., and Gustafsson, J.A. Fatty acids activate a chimera of the clofibric acid-activated receptor and the glucocorticoid receptor. Proc. Natl. Acad. Sci. USA., 89: 4653–4657, 1992.
10. Dreyer, C., Krey, G., Keller, H., Givel, F., Helftenbein, G., and Wahli, W. Control of the peroxisomal beta-oxidation pathway by a novel family of nuclear hormone receptors. Cell, 68: 879–887, 1992.
11. Zhu, Y., Alvares, K., Huang, Q., Rao, M.S., and Reddy, J.K. Cloning of a new member of the peroxisome proliferator-activated receptor gene family from mouse liver. J. Biol. Chem., 268: 26817–26820, 1993.
12. Zhu, Y., Qi, C., Korenberg, J.R., Chen, X.N., Noya, D., Rao, M.S., and Reddy, J.K. Structural organization of mouse peroxisome proliferator-activated receptor gamma (mPPAR gamma) gene: alternative promoter use and different splicing yield two mPPAR gamma isoforms. Proc. Natl. Acad. Sci. USA., 92: 7921–7925, 1995.

13. Tontonoz, P., Hu, E., Graves, R.A., Budavari, A.I., and Spiegelman, B.M. mPPAR gamma 2: tissue-specific regulator of an adipocyte enhancer. Genes. Dev., 8: 1224–1234, 1994.

14. Chen, F., Law, S.W., and O'Malley, B.W. Identification of two mPPAR related receptors and evidence for the existence of five subfamily members. Biochem. Biophys. Res. Commun., 196: 671–677, 1993.

15. Kliewer, S.A., Forman, D.M., Blumberg, B., Ong, E.S., Borgmeyer, U., Mangelsdorf, D.J., Umesono, K., and Evans, R.M. Differential expression and activation of a family of murine peroxisome proliferator-activated receptors. Proc. Natl. Acad. Sci. USA., 91: 7355–7359, 1994.

16. Jow, L. and Mukherjee, R. The human peroxisome proliferator-activated receptor (PPAR) subtype NUC1 represses the activation of hPPAR alpha and thyroid hormone receptors. J. Biol. Chem., 270: 3836–3840, 1995.

17. Amri, E.Z., Bonino, F., Ailhaud, G., Abumrad, N.A., and Grimaldi, P.A. Cloning of a protein that mediates transcriptional effects of fatty acids in preadipocytes. Homology to peroxisome proliferator-activated receptors. J. Biol. Chem., 270: 2367–2371, 1995.

18. Lemberger, T., Staels, B., Saladin, R., Desvergne, B., Auwerx, J., and Wahli, W. Regulation of the peroxisome proliferator-activated receptor alpha gene by glucocorticoids. J. Biol. Chem., 269: 24527–24530, 1994.

19. Braissant, O., Foufelle, F., Scotto, C., Dauca, M., and Wahli, W. Differential expression of peroxisome proliferator-activated receptors (PPARs): tissue distribution of PPAR-alpha, -beta, and -gamma in the adult rat. Endocrinology, 137: 354–366, 1996.

20. Tontonoz, P., Hu, E., and Spiegelman, B.M. Stimulation of adipogenesis in fibroblasts by PPAR gamma 2, a lipid-activated transcription factor [published erratum appears in Cell 1995 Mar 24;80(6):following 957]. Cell, 79: 1147–1156, 1994.

21. Tontonoz, P., Hu, E., and Spiegelman, B.M. Regulation of adipocyte gene expression and differentiation by peroxisome proliferator activated receptor gamma. Curr. Opin. Genet. Dev., 5: 571–576, 1995.

22. Lehmann, J.M., Moore, L.B., Smith-Oliver, T.A., Wilkison, W.O., Willson, T.M., and Kliewer, S.A. An antidiabetic thiazolidinedione is a high affinity ligand for peroxisome proliferator-activated receptor gamma (PPAR gamma). J. Biol. Chem., 270: 12953–12956, 1995.

23. Goecke-Flora, C.M., Wyman, J.F., Jarnot, B.M., and Reo, N.V. Effect of the peroxisome proliferator perfluoro-n-decanoic acid on glucose transport in the isolated perfused rat liver. Chem. Res. Toxicol., 8: 77–81, 1995.

24. Sher, T., Yi, H.F., McBride, O.W., and Gonzalez, F.J. cDNA cloning, chromosomal mapping, and functional characterization of the human peroxisome proliferator activated receptor. Biochemistry, 32: 5598–5604, 1993.

25. Muerhoff, A.S., Griffin, K.J., and Johnson, E.F. The peroxisome proliferator-activated receptor mediates the induction of CYP4A6, a cytochrome P450 fatty acid omega-hydroxylase, by clofibric acid. J. Biol. Chem., 267: 19051–19053, 1992.

26. Brandes, R., Kaikaus, R.M., Lysenko, N., Ockner, R.K., and Bass, N.M. Induction of fatty acid binding protein by peroxisome proliferators in primary hepatocyte cultures and its relationship to the induction of peroxisomal beta-oxidation. Biochim. Biophys. Acta, 1034: 53–61, 1990.

27. Gulick, T., Cresci, S., Caira, T., Moore, D.D., and Kelly, D.P. The peroxisome proliferator-activated receptor regulates mitochondrial fatty acid oxidative enzyme gene expression. Proc. Natl. Acad. Sci. USA., 91: 11012–11016, 1994.

28. Rodriguez, J.C., Gil-Gomez, G., Hegardt, F.G., and Haro, D. Peroxisome proliferator-activated receptor mediates induction of the mitochondrial 3-hydroxy-3-methylglutaryl-CoA synthase gene by fatty acids. J. Biol. Chem., 269: 18767–18772, 1994.

29. Castelein, H., Gulick, T., Declercq, P.E., Mannaerts, G.P., Moore, D.D., and Baes, M.I. The peroxisome proliferator activated receptor regulates malic enzyme gene expression. J. Biol. Chem., 269: 26754–26758, 1994.

30. Kaikaus, R.M., Chan, W.K., Lysenko, N., Ray, R., Ortiz de Montellano, P.R., and Bass, N.M. Induction of peroxisomal fatty acid beta-oxidation and liver fatty acid-binding protein by peroxisome proliferators. Mediation via the cytochrome P-450IVA1 omega-hydroxylase pathway. J. Biol. Chem., 268: 9593–9603, 1993.

31. Demoz, A., Vaagenes, H., Aarsaether, N., Hvattum, E., Skorve, J., Gottlicher, M., Lillehaug, J.R., Gibson, G.G., Gustafsson, J.A., and Hood, S. Coordinate induction of hepatic fatty acyl-CoA oxidase and P4504A1 in rat after activation of the peroxisome proliferator-activated receptor (PPAR) by sulphur-substituted fatty acid analogues. Xenobiotica, 24: 943–956, 1994.

32. Berge, R.K., Aarsland, A., Kryvi, H., Bremer, J., and Aarsaether, N. Alkylthioacetic acid (3-thia fatty acids)—a new group of non-beta-oxidizable, peroxisome-inducing fatty acid analogues. I. A study on the

structural requirements for proliferation of peroxisomes and mitochondria in rat liver. Biochim. Biophys. Acta, 1004: 345–356, 1989.

33. Kaikaus, R.M., Sui, Z., Lysenko, N., Wu, N.Y., Ortiz de Montellano, P.R., Ockner, R.K., and Bass, N.M. Regulation of pathways of extramitochondrial fatty acid oxidation and liver fatty acid-binding protein by long-chain monocarboxylic fatty acids in hepatocytes. Effect of inhibition of carnitine palmitoyltransferase I. J. Biol. Chem., 268: 26866–26871, 1993.

34. Aldridge, T.C., Tugwood, J.D., and Green, S. Identification and characterization of DNA elements implicated in the regulation of CYP4A1 transcription. Biochem. J., 306: 473–479, 1995.

35. Kliewer, S.A., Umesono, K., Noonan, D.J., Heyman, R.A., and Evans, R.M. Convergence of 9-cis retinoic acid and peroxisome proliferator signalling pathways through heterodimer formation of their receptors. Nature, 358: 771–774, 1992.

36. Gearing, K.L., Gottlicher, M., Teboul, M., Widmark, E., and Gustafsson, J.A. Interaction of the peroxisome-proliferator-activated receptor and retinoid X receptor. Proc. Natl. Acad. Sci. USA., 90: 1440–1444, 1993.

37. Huan, B., Kosovsky, M.J., and Siddiqui, A. Retinoid X receptor alpha transactivates the hepatitis B virus enhancer 1 element by forming a heterodimeric complex with the peroxisome proliferator-activated receptor. J. Virol., 69: 547–551, 1995.

38. Bogazzi, F., Hudson, L.D., and Nikodem, V.M. A novel heterodimerization partner for thyroid hormone receptor. Peroxisome proliferator-activated receptor. J. Biol. Chem., 269: 11683–11686, 1994.

39. Ladias, J.A. and Karathanasis, S.K. Regulation of the apolipoprotein AI gene by ARP-1, a novel member of the steroid receptor superfamily. Science, 251: 561–565, 1991.

40. Ladias, J.A., Hadzopoulou-Cladaras, M., Kardassis, D., Cardot, P., Cheng, J., Zannis, V., and Cladaras, C. Transcriptional regulation of human apolipoprotein genes ApoB, ApoCIII, and ApoAII by members of the steroid hormone receptor superfamily HNF-4, ARP-1, EAR-2, and EAR-3. J. Biol. Chem., 267: 15849–15860, 1992.

41. Palmer, C.N., Hsu, M.H., Muerhoff, A.S., Griffin, K.J., and Johnson, E.F. Interaction of the peroxisome proliferator-activated receptor alpha with the retinoid X receptor alpha unmasks a cryptic peroxisome proliferator response element that overlaps an ARP-1-binding site in the CYP4A6 promoter. J. Biol. Chem., 269: 18083–18089, 1994.

42. Winrow, C.J., Marcus, S.L., Miyata, K.S., Zhang, B., Capone, J.P., and Rachubinski, R.A. Transactivation of the peroxisome proliferator-activated receptor is differentially modulated by hepatocyte nuclear factor-4. Gene Expr., 4: 53–62, 1994.

43. Carter, M.E., Gulick, T., Raisher, B.D., Caira, T., Ladias, J.A., Moore, D.D., and Kelly, D.P. Hepatocyte nuclear factor-4 activates medium chain acyl-CoA dehydrogenase gene transcription by interacting with a complex regulatory element. J. Biol. Chem., 268: 13805–13810, 1993.

44. Hertz, R., Bishara-Shieban, J., and Bar-Tana, J. Mode of action of peroxisome proliferators as hypolipidemic drugs. Suppression of apolipoprotein C-III. J. Biol. Chem., 270: 13470–13475, 1995.

45. Staels, B., Van Tol, A., Andreu, T., and Auwerx, J. Fibrates influence the expression of genes involved in lipoprotein metabolism in a tissue-selective manner in the rat. Arterioscler. Thromb., 12: 286–294, 1992.

46. Berthou, L., Saladin, R., Yaqoob, P., Calder, P., Fruchart, J.C., Denefle, P., Auwerx, J., and Staels, B. Regulation of rat liver apolipoprotein A-I, apolipoprotein A-II, and acyl-CoA oxidase gene expression by fibrates and dietary faty acids. Eur. J. Biochem., 232: 179–187, 1985.

47. Staels, B., Van Tol, A., Skretting, G., and Auwerx, J. Lecithin:cholesterol acyltransferase gene expression is regulated in a tissue-selective manner by fibrates. J. Lipid. Res., 33: 727–735, 1992.

48. Bovard-Houppermans, S., Ochoa, A., Fruchart, J.C., and Zakin, M.M. Fenofibric acid modulates the human apolipoprotein A-IV gene expression in HepG2 cells. Biochem. Biophys. Res. Commun., 198: 764–769, 1994.

49. Schmidt, A., Endo, N., Rutledge, S.J., Vogel, R., Shinar, D., and Rodan, G.A. Identification of a new member of the steroid hormone receptor superfamily that is activated by a peroxisome proliferator and fatty acids. Mol. Endocrinol., 6: 1634–1641, 1992.

50. Mellies, M.J., Stein, E.A>, Khoury, P., Lamkin, G., and Glueck, C.J. Effects of fenofibrate on lipids, lipoproteins and apolipoproteins in 33 subjects with primary hypercholesterolaemia. Atherosclerosis 78: 167–182, 1989.

51. Vu-Dac, N., Schoonjans, K., Laine, B., Fruchart, J.C., Auwerx, J., and Staels, B. Negative regulation of the human apolipoprotein A-I promoter by fibrates can be attenuated by the interaction of the peroxisome proliferator-activated receptor with its response element. J. Biol. Chem., 269: 31012–31018, 1994.

52. Reddy, J.K., Azarnoff, D.L., and Hignite, C.E. Hypolipidaemic hepatic peroxisome proliferators form a novel class of chemical carcinogens. Nature, 283: 397–398, 1980.

53. Reddy, J.K., Scarpelli, D.G., Subbarao, V., and Lalwani, N.D. Chemical carcinogens without mutagenic activity: peroxisome proliferators as a prototype. Toxicol. Pathol., 11: 172–180, 1983.

54. Reddy, J.K. and Lalwani, N.D. Carcinogenesis by hepatic peroxisome proliferators: evaluation of the risk of hypolipidemic drugs and industrial plasticizers to humans. Crit. Rev. Toxicol., 12: 1–58, 1993.

55. Reddy, J.K. and Rao, M.S. Peroxisome proliferators and cancer: mechanisms and implications. Trends. Pharmacol. Sci., 7: 631–636, 1996.

56. Rao, M.S., Kokkinakis, D.M., Subbarao, V., and Reddy, J.K. Peroxisome proliferator-induced hepatocarcinogenesis: levels of activating and detoxifying enzymes in hepatocellular carcinomas induced by ciprofibrate. Carcinogenesis, 8: 19–23, 1987.

57. Warren, J.R., Simmon, V.F., and Reddy, J.K. Properties of hypolipidemic peroxisome proliferators in the lymphocyte [3H]thymidine and Salmonella mutagenesis assays. Cancer Res., 40: 36–41, 1980.

58. Glauert, H.P., Reddy, J.K., Kennan, W.S., Sattler, G.L., Rao, V.S., and Pitot, H.C. Effect of hypolipidemic peroxisome proliferators on unscheduled DNA synthesis in cultured hepatocytes and on mutagenesis in Salmonella. Cancer Lett., 24: 147–156, 1984.

59. Yoshikawa, K., Tanaka, A., Yamaha, T., and Kurata, H. Mutagenicity study of nine monoalkyl phthalates and a dialkyl phthalate using Salmonella typhimurium and Escherichia coli. Food. Chem. Toxicol., 21: 221–223, 1983.

60. Agarwal, D.K., Lawrence, W.H., Nunez, L.J., and Autian, J. Mutagenicity evaluation of phthalic acid esters and metabolites in Salmonella typhimurium cultures. J. Toxicol. Environ. Health, 16: 61–69, 1985.

61. Cattley, R.C., Smith-Oliver, T., Butterworth, B.E., and Popp, J.A. Failure of the peroxisome proliferator WY-14,643 to induce unscheduled DNA synthesis in rat hepatocytes following in vivo treatment. Carcinogenesis, 9: 1179–1183, 1988.

62. Williams, G.M., Maruyama, H., and Tanaka, T. Lack of rapid initiating, promoting or sequential syncarcinogenic effects of di(2-ethylhexyl)phthalate in rat liver carcinogenesis. Carcinogenesis, 8: 875–880, 1987.

63. Cattley, R.C., Marsman, D.S., and Popp, J.A. Failure of the peroxisome proliferator WY-14,643 to initiate growth-selectable foci in rat liver. Toxicology., 56: 1–7, 1989.

64. Conway, J.G., Tomaszewski, K.E., Olson, M.J., Cattley, R.C., Marsman, D.S., and Popp, J.A. Relationship of oxidative damage to the hepatocarcinogenicity of the peroxisome proliferators di(2-ethylhexyl)phthalate and Wy-14,643. Carcinogenesis, 10: 513–519, 1989.

65. Marsman, D.S., Goldsworthy, T.L., and Popp, J.A. Contrasting hepatocytic peroxisome proliferation, lipofuscin accumulation and cell turnover for the hepatocarcinogens Wy-14,643 and clofibric acid. Carcinogenesis, 13: 1011–1017, 1992.

66. Kasai, H., Okada, Y., Nishimura, S., Rao, M.S., and Reddy, J.K. Formation of 8-hydroxydeoxyguanosine in liver DNA of rats following long-term exposure to a peroxisome proliferator. Cancer Res., 49: 2603–2605, 1989.

67. Takagi, A., Sai, K., Umemura, T., Hasegawa, R., and Kurokawa, Y. Relationship between hepatic peroxisome proliferation and 8-hydroxydeoxyguanosine formation in liver DNA of rats following long-term exposure to three peroxisome proliferators; di(2-ethylhexyl) phthalate, aluminium clofibrate and simfibrate. Cancer Lett., 53: 33–38, 1990.

68. Hegi, M.E., Ulrich, D., Sagelsdorff, P., Richter, C., and Lutz, W.K. No measurable increase in thymidine glycol or 8-hydroxydeoxyguanosine in liver DNA of rats treated with nafenopin or choline-devoid low-methionine diet. Mutat. Res., 238: 325–329, 1990.

69. Cattley, R.C. and Glover, S.E. Elevated 8-hydroxydeoxyguanosine in hepatic DNA of rats following exposure to peroxisome proliferators: relationship to carcinogenesis and nuclear localization. Carcinogenesis, 14: 2495–2499, 1993.

70. Chu, S., Huang, Q., Alvares, K., Yeldandi, A.V., Rao, M.S., and Reddy, J.K. Transformation of mammalian cells by overexpressing H_2O_2-generating peroxisomal fatty acyl-CoA oxidase. Proc. Natl. Acad. Sci. USA, 92: 7080–7084, 1995.

71. Mikalsen, S.O., Holen, I., and Sanner, T. Morphological transformation and catalase activity of Syrian hamster embryo cells treated with hepatic peroxisome proliferators, TPA and nickel sulphate. Cell Biol. Toxicol., 6: 1–13, 1990.

72. Tsutsui, T., Watanabe, E., and Barrett, J.C. Ability of peroxisome proliferators to induce cell transformation, chromosome aberrations and peroxisome proliferation in cultured Syrian hamster embryo cells. Carcinogenesis, 14: 611–618, 1993.

73. Cattley, R.C., Marsman, D.S., and Popp, J.A. Age-related susceptibility to the carcinogenic effect of the peroxisome proliferator WY-14,643 in rat liver. Carcinogenesis (in press), 1996.

74. Farber, E. Experimental induction of hepatocellular carcinoma as a paradigm for carcinogenesis. Clin. Physiol. Biochem., 5: 152–159, 1987.

75. Moody, D.E., Rao, M.S., and Reddy, J.K. Mitogenic effect in mouse liver induced by a hypolipidemic drug, nafenopin. Virchows Arch. B. Cell Pathol., 23: 291–296, 1977.

76. Marsman, D.S., Cattley, R.C., Conway, J.G., and Popp, J.A. Relationship of hepatic peroxisome proliferation and replicative DNA synthesis to the hepatocarcinogenicity of the peroxisome proliferators di(2-ethyl-hexyl)phthalate and [4-chloro-6-(2,3-xylidino)-2-pyrimidinylthio]acetic acid (Wy-14,643) in rats. Cancer Res., 48: 6739–6744, 1988.

77. Reddy, J.K., Reddy, M.K., Usman, M.I., Lalwani, N.D., and Rao, M.S. Comparison of hepatic peroxisome proliferative effect and its implication for hepatocarcinogenicity of phthalate esters, di(2-ethylhexyl) phthalate, and di(2-ethylhexyl) adipate with a hypolipidemic drug. Environ. Health Perspect., 65: 317–327, 1986.

78. Tomaszewski, K.E., Agarwal, D.K., and Melnick, R.L. In vitro steady-state levels of hydrogen peroxide after exposure of male F344 rats and female B6C3F1 mice to hepatic peroxisome proliferators. Carcinogenesis, 7: 1871–1876, 1986.

79. Barrett, J.C. Mechanisms for species differences in receptor-mediated carcinogenesis. Mutat. Res., 333: 189–202, 1995.

80. Bayly, A.C., Roberts, R.A., and Dive, C. Suppression of liver cell apoptosis in vitro by the non-genotoxic hepatocarcinogen and peroxisome proliferator nafenopin. J. Cell Biol., 125: 197–203, 1994.

81. Roberts, R.A., Soames, A.R., Gill, J.H., James, N.H., and Wheeldon, E.B. Non-genotoxic hepatocarcinogens stimulate DNA synthesis and their withdrawal induces apoptosis, but in different hepatocyte populations. Carcinogenesis, 16: 1693–1698, 1995.

82. James, N.H. and Roberts, R.A. The peroxisome proliferator class of non-genotoxic hepatocarcinogens synergize with epidermal growth factor to promote clonal expansion of initiated rat hepatocytes. Carcinogenesis, 15: 2687–2694, 1994.

83. Lake, B.G., Evans, J.G., Gray, T.J., Korosi, S.A., and North, C.J. Comparative studies on nafenopin-induced hepatic peroxisome proliferation in the rat, Syrian hamster, guinea pig, and marmoset. Toxicol. Appl. Pharmacol., 99: 148–160, 1989.

84. Reddy, J.K. and Rao, M.S. Peroxisome proliferation and hepatocarcinogenesis. 1992.

85. Varanasi, U., Chu, R., Huang, Q., Castellon, R., Yeldandi, A.V., and Reddy, J.K. Identification of a peroxisome proliferator-responsive element upstream of the human peroxisomal fatty acyl coenzyme A oxidase gene. J. Biol. Chem., 271: 2147–2155, 1996.

86. Hardwick, J.P., Song, B.J., Huberman, E., and Gonzalez, F.J. Isolation, complementary DNA sequence, and regulation of rat hepatic lauric acid omega-hydroxylase (cytochrome P-450LA omega). Identification of a new cytochrome P-450 gene family. J. Biol. Chem., 262: 801–810, 1987.

87. Lee, S.S., Pineau, T., Drago, J., Lee, E.J., Owens, J.W., Kroetz, D.L., Fernandez-Salguero, P.M., Westphal, H., and Gonzalez, F.J. Targeted disruption of the alpha isoform of the peroxisome proliferator-activated receptor gene in mice results in abolishment of the pleiotropic effects of peroxisome proliferators. Mol. Cell Biol., 15: 3012–3022, 1995.

88. Gonzalez, F.J., Fernandez-Salguero, P., Lee, S.S., Pineau, T., and Ward, J.M. Xenobiotic receptor knockout mice. Toxicol. Lett., 82–83: 117–121, 1995.

89. Lemberger, T., Saladin, R., Vazquez, M., Assimacopoulos, F., Staels, B., Desvergne, B., Wahli, W., and Auwerx, J. Expression of the peroxisome proliferator-activated receptor alpha gene is stimulated by stress and follows a diurnal rhythm. J. Biol. Chem., 271: 1764–1769, 1996.

90. Issemann, I. and Green, S. Cloning of novel members of the steroid hormone receptor superfamily. J. Steroid. Biochem. Mol. Biol., 40: 263–269, 1991.

91. Yamada, J., Sakuma, M., Ikeda, T., and Suga, T. Activation of dehydroepiandrosterone as a peroxisome proliferator by sulfate conjugation. Arch. Biochem. Biophys., 313: 379–381, 1994.

92. Yamada, J., Sakuma, M., Ikeda, T., Fukuda, K., and Suga, T. Characteristics of dehydroepiandrosterone as a peroxisome proliferator. Biochim. Biophys. Acta, 1092: 233–243, 1991.

93. Frenkel, R.A., Slaughter, C.A., Orth, K., Moomaw, C.R., Hicks, S.H., Snyder, J.M., Bennett, M., Prough, R.A., Putnam, R.S., and Milewich, L. Peroxisome proliferation and induction of peroxisomal enzymes in mouse and rat liver by dehydroepiandrosterone feeding. J. Steroid. Biochem., 35: 333–342, 1990.

94. Hayashi, F., Tamura, H., Yamada, J., Kasai, H., and Suga, T. Characteristics of the hepatocarcinogenesis caused by dehydroepiandrosterone, a peroxisome proliferator, in male F-344 rats. Carcinogenesis, 15: 2215–2219, 1994.

95. Shibata, M., Hasegawa, R., Imaida, K., Hagiwara, A., Ogawa, K., Hirose, M., Ito, N., and Shirai, T. Chemoprevention by dehydroepiandrosterone and indomethacin in a rat multiorgan carcinogenesis model. Cancer Res., 55: 4870–4874, 1995.

96. Gordon, G.B., Helzlsouer, K.J., and Comstock, G.W. Serum levels of dehydroepiandrosterone and its sulfate and the risk of developing bladder cancer. Cancer Res., 51: 1366–1369, 1991.

97. Waxman, D.J. Role of metabolism in the activation of dehydroepiandrosterone as a peroxisome prolifera-
     tor. Mol. Pharmacol., 50: 67–74, 1996.
98. Ledwith, B.J., Johnson, T.E., Wagner, L.K., Pauley, C.J., Manam, S., Galloway, S.M., and Nichols, W.W.
     Growth regulation by peroxisome proliferators:opposing activities in early and late G1. Cancer Res., 56:
     3257–3264, 1996.
99. Schulz, S. and Nyce, J.W. Inhibition of protein farnesyltransferase: a possible mechanism of tumor preven-
     tion by dehydroepiandrosterone sulfate [published erratum appears in Carcinogenesis 1995 Jan;16(1):149]
     [retracted by Nyce JW. In: Carcinogenesis 1995 May;16(5):1257]. Carcinogenesis, 15: 2649–2652, 1994.
100. Nyce, J.W. Inhibition of protein farnesyltransferase: a possible mechanism of tumor prevention for dehy-
     droepiandrosterone   sulfate   [retraction   of   Schulz   S,   Nyce   JW.   In: Carcinogenesis   1994
     Nov;15(11):2649–52]. Carcinogenesis, 16: 1257, 1995.
101. Peters, J.M., Zhou, Y.C., Ram, P.A., Lee, S.S.T., Gonzalez, F.J., and Waxman, D.J. Peroxisome prolifera-
     tor-activated receptor alpha required for gene induction by dehydroepiandrosterone-3-beta-sulfate. Mol.
     Pharmacol., 50: 67–74, 1996.
102. Thibault, A., Cooper, M.R., Figg, W.D., Venzon, D.J., Sartor, A.O., Tompkins, A.C., Weinberger, M.S.,
     Headlee, D.J., McCall, N.A., and Samid, D. A phase I and pharmacokinetic study of intravenous phenylacetate in
     patients with cancer. Cancer Res., 54: 1690–1694, 1994.
103. Samid, D., Shack, S., and Myers, C.E. Selective growth arrest and phenotypic reversion of prostate cancer
     cells in vitro by nontoxic pharmacological concentrations of phenylacetate. J. Clin. Invest., 91: 2288–2295,
     1993.
104. Samid, D., Ram, Z., Hudgins, W.R., Shack, S., Liu, L., Walbridge, S., Oldfield, E.H., and Myers, C.E. Se-
     lective activity of phenylacetate against malignant gliomas: resemblance to fetal brain damage in
     phenylketonuria. Cancer Res., 54: 891–895, 1994.
105. Liu, L., Shack, S., Stetler-Stevenson, W.G., Hudgins, W.R., and Samid, D. Differentiation of cultured hu-
     man melanoma cells induced by the aromatic fatty acids phenylacetate and phenylbutyrate. J. Invest. Der-
     matol., 103: 335–340, 1994.
106. Stockhammer, G., Manley, G.T., Johnson, R., Rosenblum, M.K., Samid, D., and Lieberman, F.S. Inhibition
     of proliferation and induction of differentiation in medulloblastoma- and astrocytoma-derived cell lines
     with phenylacetate. J. Neurosurg., 83: 672 681, 1995.
107. Lipschutz, J.H., Samid, D., and Cunha, G.R. Phenylacetate is an inhibitor of prostatic growth and develop-
     ment in organ culture. J. Urol., 155: 1762–1770, 1996.
108. Pineau, T., Hudgins, W.R., Liu, L., Chen, L.C., Sher, T., Gonzalez, F.J., and Samid, D. Activation of a hu-
     man peroxisome proliferator-activated receptor by the antitumor agent phenylacetate and its analogs. Bio-
     chem. Pharmacol., 52: 659–667, 1996.
109. Forman, B.M., Tontonoz, P., Chen, J., Brun, R.P., Spiegelman, B.M., and Evans, R.M. 15-Deoxy-delta 12,
     14-prostaglandin J2 is a ligand for the adipocyte determination factor PPAR gamma. Cell, 83: 803–812,
     1995.
110. Yu, K., Bayona, W., Kallen, C.B., Harding, H.P., Ravera, C.P., McMahon, G., Brown, M., and Lazar, M.A.
     Differential activation of peroxisome proliferator-activated receptors by eicosanoids. J. Biol. Chem., 270:
     23975–23983, 1995.

A HYPOTHETICAL MECHANISM FOR FAT-INDUCED RODENT HEPATOCARCINOGENESIS

Daniel J. Noonan[1,2] and Michelle L. O'Brien[2]

[1]Department of Biochemistry
[2]Graduate Center for Toxicology
University of Kentucky
800 Rose Street, Lexington, Kentucky 40536

PPAR, A FAMILY OF TRANSCRIPTION FACTORS REGULATING FAT METABOLISM

The regulation of fat metabolism in higher eukaryotes appears to be intimately associated with the activities of a family of recently identified nuclear receptors called 'peroxisome proliferator-activated receptors' (PPAR). In 1990, Issemann and Green[1] reported the cloning of a mouse steroid receptor superfamily member that they characterized as a peroxisome proliferator-responsive transcription factor. The steroid receptor family of genes consists of a group of ligand-activated DNA transcription factors that bind regulatory sequences upstream of their target gene(s) resulting in the activation or repression of specific gene transcription.[2–4] Subsequently it has been shown that PPAR is a small family of genes with reports of at least α, β and γ isoforms in mouse,[5,6] *Xenopus*,[7] rat[8] and humans.[9,10] An examination of PPAR regulatable promoters suggest this receptor family is intimately involved in fat metabolism including their breakdown,[11,12] storage[13] and synthesis.[14] The complexity of the PPAR activation pathway has been substantially enhanced by the demonstration that PPAR DNA binding is linked to heterodimerization with a member of the retinoid X (RXR) family of receptors,[15] and more recent PPAR transcription studies in yeast where it was demonstrated that PPAR activity is contingent upon both RXR and perhaps other mammalian cell-specific factor(s).[16] The RXR family of receptors appear to be a point of convergence for several members of the intracellular receptor superfamily of genes[17–19] and play a critical role in the transcriptional events associated with these receptors. All of the data accumulated to date clearly implicate PPAR-regulated events in the homeostasis of fats and suggest they may serve as viable targets for drug intervention and regulation of fat metabolism.

PPARα AS A MEDIATOR OF THE OXIDATIVE STRESS HYPOTHESIS

Obesity and fat metabolism have been correlated with a variety of cancers including: colorectal, prostate, cervical, ovarian, breast, esophageal, liver and pancreatic cancers.[20] In rodents, fat metabolism has been directly linked to chemically induced hepatocarcinogenesis.[21,22] Furthermore, this event appears to correlate with proliferation of liver peroxisomes. Peroxisomes are subcellular organelles that functionally compartmentalize cellular β-oxidation reactions. The oxidative enzymes found in peroxisomes are involved in a variety of metabolic pathways including: lipid metabolism, cholesterol metabolism, respiration, and gluconeogenesis. Of increasing interest is the large list of compounds that are capable of inducing the proliferation of peroxisomal structures in rodent livers. This list includes hypolipidemic drugs, environmental pollutants, analgesics, phthalates and dietary fats.[23] Rodent chemical hepatocarcinogenesis can also be closely correlated with the proliferation of liver peroxisomes,[24–26] and although these peroxisome proliferators appear to act as hepatocarcinogens, they do not show detectable mutagenic or genotoxic activity in suitable test systems.[27] Peroxisomes provide compartmentalization of a variety of oxidative events within the cell, and it has been proposed that stress of this system may be the mechanism by which carcinogenic events are induced by peroxisome proliferators.[25,28,29] A marriage of the events mediated by PPARα and the oxidative stress hypothesis (Figure 1) proposes that the stimulation of PPARα by peroxisome proliferators results in the production of enzymes for the peroxisomal-specific breakdown of fats. This event generates H_2O_2, which if in excess and unneutralized can create free radicals capable of mutating DNA. Oxidative stress-induced mutations to key cell cycle genes, proto-oncogenes, and genes regulating apoptotic events are ultimately proposed to both initiate and immortalize the cellular changes resulting in hepatocarcinogenesis. Defining PPAR-mediated peroxisomal events will be essential for understanding the mechanism(s) by which peroxisome proliferators induce hepatocarcinogenesis and may provide vital insights into how obesity and fat metabolism are linked to tumor formation. To this end it has recently been demonstrated that disruption of the gene for PPARα in a transgenic mouse model system specifically eliminates chemical induction of peroxisome proliferation in these mice,[30] and, although it has yet to be conclusively established, the suggestion is that these mice will be less susceptible to hepatocellular carcinomas. Even with this compelling evidence and the corroborative data suggesting PPARα can be directly implicated in the oxidative stress facets of this hypothesis, the specific biochemical mechanisms for tumor initiation and promotion are still unresolved.

PPARα AS A REGULATOR OF FAT CATABOLISM

Induction of peroxisome proliferation in rodent livers is accompanied by a stimulation of both peroxisomal and mitochondrial fatty acid catabolism.[31] PPARα appears to play a critical role in this event. As seen in Figure 2, when analyzed in an *in vitro* mammalian cell co-transfection assay, using a pRSVrPPARα expression plasmid and an acyl-CoA oxidase promoter region driven reporter plasmid, a variety of fatty acids can activate PPARα-mediated transcription. These include: saturated and unsaturated fatty acids as well as fatty acids whose catabolism is predicated by either β-oxidation (e.g. palmitic acid), α-oxidation (phytanic acid), or ω-hydroxylation (e.g. 12-hydroxydodecanoic acid).

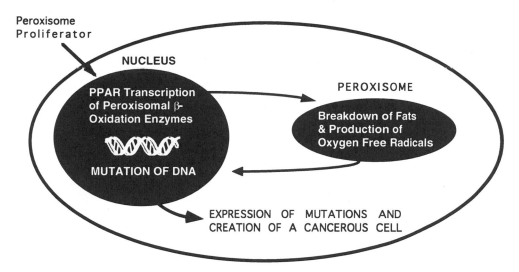

Figure 1. PPARα as a mediator of oxidative stress. PPARα integrates into the classical oxidative stress hypothesis through its role as a transcriptional regulator of peroxisome-specific β-oxidation enzymes.

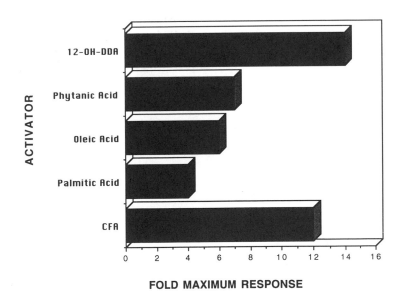

Figure 2. Fatty acids as activators of PPARα. The fatty acids 12-hydroxydodecanoic acid (12-OH-DDA), phytanic acid, oleic acid and palmitic acid, along with the classical peroxisome proliferator clofibric acid (CFA), were analyzed in a CV-1 cell in vitro co-transfection assay for their ability to activate PPARα-mediated transcription. The assay consisted of a pRSVrPPARα expression plasmid and an acyl-CoA oxidase promoter region driven luciferase reporter plasmid[16]. Values were normalized to a β-galactosidase expression plasmid included in all transfections, and fold maximum response was the concentration of activator giving the strongest response calculated with reference to the background generated in the absence of activator.

Furthermore, PPARα has been identified as a regulator of a variety of genes involved in the catabolism of fatty acids. These include the enzymes involved in peroxisomal-specific fat catabolism,[11,12] enzymes for mitochondrial fat catabolism,[32] enzymes for fat transport,[33] CYP450 enzymes involved in fatty acid ω-hydroxylation[34,35] and lipoprotein lipase,[6] an enzyme involved in the hydrolysis of triacylglycerols in lipoprotein particles. From these data it is apparent that PPARα plays a critical role in the breakdown of intracellular fats.

ISOPRENOIDS AS REGULATORS OF PPARα ACTIVITY

In addition to fatty acid catabolism, peroxisomes are centers for the synthesis of the complex lipids cholesterol and plasmalogens. Many of the intermediates in cholesterol synthesis have been identified in peroxisomes.[37] Furthermore, the initial step in cholesterol synthesis is mediated by the same enzyme (thiolase) responsible for the final step in peroxisome-specific fatty acid breakdown. In rats, this enzyme has been shown to contain a PPARα regulatable promoter and is substantially upregulated by peroxisome proliferators.[38–40] Cholesterol is an essential component of membrane structure and function, is a precursor to a variety of steroid based regulators of metabolic function, and is a precursor of protein prenylation events, a process linked to ras-mediated carcinogenesis.[41] Using inhibitors of the cholesterol synthesis (Figure 3A), this pathway was examined for its role in PPARα-mediated gene transcription (Figure 3B). The classical HMG-CoA reductase inhibitor Lovastatin, but not the squalene synthetase inhibitor Squalestatin, significantly repressed clofibric acid stimulated PPARα-mediated transcription.[42] Furthermore, this repression could be overcome with addition of mevalonate. These data support the suggestion that a cholesterol synthesis intermediate between mevalonate and farnesylpyrophosphate (FPP) can positively effect PPARα-mediated transcriptional events.

Several isoprenoid structures representing potential cholesterol pathway intermediates were analyzed for the ability to stimulate PPARα-mediated transcription in the in vi-

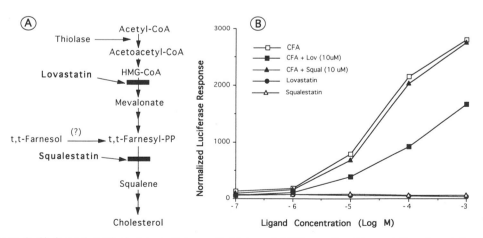

Figure 3. Cholesterol synthesis intermediates as facilitators of PPARα-mediated transcription. Lovastatin and Squalestatin, inhibitors of the cholesterol synthesis pathway (A), were analyzed in a CV-1 cell in vitro co-transfection assay for their ability to modulate CFA stimulated PPARα-mediated transcription (B). The assay consisted of a pRSVrPPARα expression plasmid and an acyl-CoA oxidase promoter region driven luciferase reporter plasmid.[16] Values were normalized to a β-galactosidase expression plasmid included in all transfections.

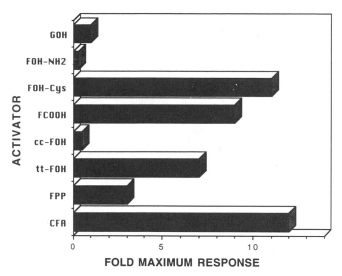

Figure 4. Isoprenoids as activators of PPARα. *Trans,trans*-farnesylpyrophosphate (tt-FPP), *trans,trans*-farnesol (tt-FOH), *cis,cis*-farnesol (cc-FOH), geranylgeranol (GOH), farnesol amine (FOH-NH2), farnesylcysteine (FOH-Cys) and farnesoic acid (FCOOH) were analyzed in a CV-1 cell in vitro co-transfection assay for their ability to stimulate PPARα-mediated transcription. The assay and fold maximum response were as described in figure 2.

tro co-transfection assay (Figure 4). These included the mammalian expressed *trans,trans*-farnesylpyrophosphate (tt-FPP), *trans,trans*-farnesol (tt-FOH), a plant derived *cis,cis*-farnesol (cc-FOH), the 20 carbon geranylgeranol (GOH), farnesol amine (FOH-NH2), farnesylcysteine (FOH-Cys) and farnesoic acid (FCOOH). Farnesol, FOH-Cys and FCOOH all demonstrated varying abilities to stimulate PPARα-mediated transcription. These data strongly implicate an isoprenoid intermediate of the cholesterol synthesis pathway as an activator of PPARα.

FARNESOL AS AN ACTIVATOR OF PPARα AND FXR

This PPAR/farnesol activation pathway interestingly converges with the pathway of another recently described steroid receptor orphan family member, farnesoid X-activated receptor: (FXR).[42,43] A comparison of the salient features of these two receptors suggest they may be coordinate regulators. Both are members of the steroid receptor superfamily, both are predominantly localized to the liver and kidney, both require heterodimerization with RXR and both can be activated by farnesol in an *in vitro* transcription assay. Although similar in many respects, a closer comparison of their in vitro activation parameters (Figure 5) suggests that these receptors also have major points of divergence. This is supported by the observations that: 1) peroxisome proliferators cannot activate FXR transcription, 2) the FXR activator juvenile hormone cannot activate PPARα transcription, 3) PPARα response elements do not work with FXR, and 4) the ecdysone response element used with FXR cannot mediate PPARα transcription. These data would suggest farnesol serves as a common intermediate for distinct endogenous activators of both PPARα and FXR. A potential scenario and ligand for PPAR would include conversion of farnesol to the fatty acid (farnesoic acid) or even one of the farnesol derived dicarboxylic acid forms.

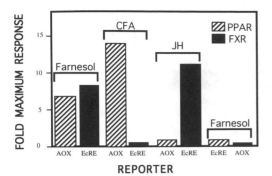

Figure 5. A comparison of activation parameters for PPARα and FXR. PPARα or FXR mammalian expression plasmids, and a designated luciferase reporter construct (AOX-RE or EcRE) were transfected into CV-1 cells and analyzed for (F-OH), CFA, and juvenile hormone (JH) stimulation of receptor activities. Activity is expressed as the average of triplicate luciferase responses normalized to their respective β-galactosidase rate.

This event has been recently shown to occur in rats fed large amounts of farnesol or a squalene synthetase inhibitor,[44] and is circumstantially supported by the assay data presented above. FXR on the other hand, might also be activated by some metabolic product of farnesol, or alternatively, may be activated by a product of the farnesylpyrophosphate pathway. Interestingly, juvenile hormone, a sesquiterpene derivative of FPP and regulator of insect morphogenesis, selectively activates transcription of FXR. Therefore, the data presented here argue effectively that the stimulation of PPARα-mediated fatty acid catabolism is ultimately going to provide the substrate (acetyl-CoA) for upregulating farnesol production and consequently the activities of FXR.

INTEGRATION OF FXR INTO THE OXIDATIVE STRESS HYPOTHESIS

From these data, FXR can be integrated into the PPARa-mediated oxidative stress hypothesis (Figure 6).[42] PPARα activation of oxidative catabolism of fatty acids is proposed to generate excess acetyl CoA, driving the overexpression of FPP, which subsequently feeds into the activation pathways for both PPARα and FXR. In the more speculative aspect of the hypothesis, FXR pathways are theorized to be linked to cell cycle or differentiation functions of the cell, providing the deregulatory phenomena associated with tumor promotion. This hypothesis does not eliminate the role of oxidative stress in tumorigenesis but rather creates a mechanism wherein the effects of oxidative stress might run rampant and/or be immortalized. The co-expression of these receptors in liver and kidney as well as their co-regulation by isoprenoid derivatives, strongly implicate them in some integrated cellular mechanism. Furthermore, FXR is specifically activated by a classical regulator of insect differentiation and development. Although rats and humans do not go through metamorphosis, it can be speculated that the function of juvenile hormone has some ancestral relationship with a mammalian isoprenoid-related transcriptional signaling pathway. A better understanding of FXR activation events may provide critical insights into a mechanism for dietary fat-associated carcinogenesis.

ACKNOWLEDGMENTS

The researchers and this research was supported in part through grants in aid from the National Institutes of Health (DK47132) and the American Cancer Society (CN124).

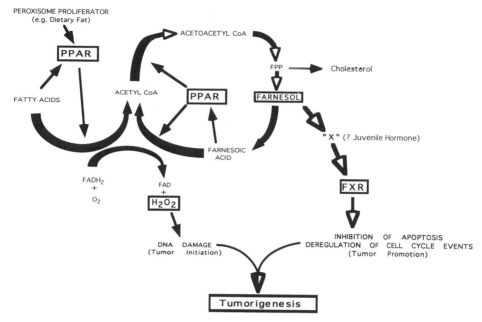

Figure 6. Integration of FXR into the oxidative stress hypothesis. Pathways for classical PPARα-mediated oxidative stress (solid arrowheads) and farnesol stimulated FXR transcription events (open arrowheads) are hypothesized to contribute cooperatively to events of tumor initiation and tumor promotion respectively.

REFERENCES

1. I. Issemann and S. Green. Activation of a member of the steroid hormone receptor superfamily by peroxisome proliferators. *Nature* 347:645–50 (1990).
2. R. M. Evans. The steroid and thyroid hormone receptor superfamily. *Science* 240:889–895 (1988).
3. M. Beato. Gene regulation by steroid hormones. *Cell* 56:335–344 (1989).
4. B. O'Malley. The steroid receptor superfamily: more excitement predicted for the future. *Molec. Endocrinol.* 4:363–369 (1990).
5. F. Chen, S. W. Law and B. W. O'Malley. Identification of two mPPAR related receptors and evidence for the existence of five subfamily members. *Biochem. Biophys. Res. Commun.* 196:671–677 (1993).
6. Y. Zhu, K. Alvares, Q. Huang, M. S. Rao and J. K. Reddy. Cloning of a new member of the peroxisome proliferator-activated gene family from mouse liver. *J. Biol. Chem.* 268:26817–26820 (1993).
7. C. Dreyer, G. Krey, H. Keller, F. Givel, G. Helftenbein and W. Wahli. Control of the peroxisomal β-oxidation pathway by a novel family of nuclear hormone receptors. *Cell* 68:879–887 (1992).
8. M. Göttlicher, E. Widmark, Q. Li and J.-Å. Gustafsson. Fatty acids activate a chimera of the clofibric acid-activated receptor and the glucocorticoid receptor. *Proc. Natl. Acad. Sci., USA* 89:4653–7 (1992).
9. T. Sher, H. F. Yi, O. W. McBride and F. J. Gonzalez. cDNA cloning, chromosomal mapping, and functional characterization of the human peroxisome proliferator activated receptor. *Biochemistry* 32:5598–604 (1993).
10. M. E. Greene, B. Blumberg, O. W. McBride, H. F. Yi, K. Kronquist, K. Kwan, L. Hsieh, G. Greene and S. D. Nimer. Isolation of the human peroxisome proliferator activated receptor gamma cDNA: expression in hematopoietic cells and chromosomal mapping. *Gene Expr.* 4:281–299 (1995).
11. J. D. Tugwood, I. Isseman, R. G. Anderson, K. R. Bundell, W. L. Mcpheat and S. Green. The mouse peroxisome proliferator activated receptor recognizes a response element in the 5'-flanking sequence of the rat acyl CoA oxidase gene. *EMBO J.* 11:433–439 (1992).
12. B. Zhang, S. L. Marcus, F. G. Sajjadi, K. Alvares, J. K. Reddy, S. Subramani, R. A. Rachubinski and J. P.-Capone. Identification of a peroxisome proliferator-responsive element upstream of the gene encoding rat peroxisomal enol-CoA hydratase/3-hydroxyacyl-CoA dehydrogenase. *Proc. Natl. Acad. Sci. U.S.A.* 89:7541–7545 (1992).

13. P. Tontonoz, E. Hu, R. A. Graves, A. I. Budavari and B. M. Spiegelman. mPPARg2: tissue-specific regulator of an adipocyte enhancer. *Genes Devel.* 8:1224–1234 (1994).

14. J. C. Rodriguez *et al.* Peroxisome proliferator-activated receptor mediates induction of the mitochondrial 3-hydroxy-3-methylglutaryl-CoA synthase gene by fatty acids. *J. Biol. Chem.* 269:18767–18772 (1994).

15. S. A. Kliewer, K. Umesono, D. J. Noonan, R. A. Heyman and R. M. Evans. Convergence of 9-cis retinoic acid and peroxisome proliferator signalling pathways through heterodimer formation of their receptors. *Nature* 358:771–774 (1992).

16. K. Henry, M. L. O'Brien, W. Clevenger, L. Jow and D. J. Noonan. Peroxisome proliferator-activated receptor response specificities as defined in yeast and mammalian cell transcription assays. *Toxicol. Appl. Pharmacol.* 132:317–324 (1995).

17. V. C. Yu, C. Delsert, B. Anderson, J. M. Holloway, O. V. Devary, A. M. Naar, S. Y. Kim, J. Boutin, C. K. Glass and M. G. Rosenfeld. A coregulator that enhances binding of retinoic acid, thyroid hormone, and vitamin D receptors to their cognate response elements. *Cell* 67:1251–1266 (1991).

18. X. Zhang, B. Hoffmann, P. B. V. Tran, G. Graupner and M. Pfahl. Retinoid X receptor is an auxilary protein for thyroid hormone and retinoic acid receptors. *Nature* 355:441–445 (1992).

19. S. A. Kliewer, K. Umesono, D. J. Mangelsdorf and R. M. Evans. Retinoid X receptor interacts with nuclear receptors in retinoic acid, thyroid hormone and vitamin D$_3$ signalling. *Nature* 355:446–9 (1992).

20. J. P. Deslypere. Obesity and cancer. *Metabolism* 44 (Suppl. 3):24–27 (1995).

21. H. P. Glauert and H. C. Pitot. Influence of dietary fat on the promotion of diethylnitrosamine-induced hepatocarcinogenesis in rats. *Proc. Soc. Exp. Biol. Med.* 181:498–506 (1986).

22. H. P. Glauert, L. T. Lay, W. S. Kennan and H. C. Pitot. Effect of dietary fat on the initiation of hepatocarcinogenesis by diethylnitrosamine or 2-acetylaminofluorene in rats. *Carcinogenesis* 12:991–995 (1991).

23. D. E. Moody, J. K. Reddy, B. G. Lake, J. A. Popp and D. H. Reese. Peroxisome proliferation and nongenotoxic carcinogenesis: commentary on a symposium. *Fundamental and Applied Toxicology* 16:233–48 (1991).

24. A. J. Cohen and P. Grasso. Review of the hepatic response to hypolipidaemic drugs in rodents and assessment of its toxicological significance to man. *Food and Cosmetic Toxicology* 19:585–605 (1981).

25. J. K. Reddy and N. D. Lalwani. Carcinogenesis by hepatic peroxisome proliferators: evaluation of risk of hypolipidemic drugs and industrial plastisizers to humans. *CRC Crit. Rev. Toxicol.* 12:1–58 (1983).

26. C. R. Elcombe. Species differences in carcinogenicity and peroxisome proliferation due to trichloroethylene: a biochemical human hazard assessment. *Archives of Toxicology. Supplement* 8:6–17 (1985).

27. P. Grasso, M. Sharratt and A. J. Cohen. Role of persistent, non-genotoxic tissue damage in rodent cancer and relevance to humans. *Annual Review of Pharmacology and Toxicology* 31:253–287 (1991).

28. M. S. Rao and J. K. Reddy. Peroxisome proliferation and hepatocarcinogenesis. *Carcinogenesis* 8:631–636 (1987).

29. J. K. Reddy and D. L. Azarnoff. Hypolipidaemic hepatic peroxisome proliferators form a novel class of chemical carcinogens. *Nature* 283:397–398 (1980).

30. S. S. Lee, T. Pineau, J. Drago, E. J. Lee, O. J.W., D. L. Kroetz, P. M. Fernandez-Salguero, H. Westphal and F. J. Gonzalez. Targeted disruption of the alpha isoform of the peroxisome proliferator-activated receptor gene in mice results in abolishment of the pleiotropic effects of peroxisome proliferators. *Mol Cell Biol* 15:3012–3022 (1995).

31. R. K. Berge, A. Aarsland, H. Kryvi, J. Bremer and N. Aarsaether. Alkylthio acetic acids (3-thia fatty acids)—a new group of non-beta-oxidizable peroxisome-inducing fatty acid analogues—II. Dose-response studies on hepatic peroxisomal- and mitochondrial changes and long-chain fatty acid metabolizing enzymes in rats. *Biochem. Pharmacol.* 38:3969–3979 (1989).

32. E. H. Hakkola, J. K. Hiltunen and H. I. Autio-Harmainen. Mitochondrial 2,4-dienoyl-CoA reductases in the rat: differential responses to clofibrate treatment. *J. Lipid Res.* 35:1820–1828 (1994).

33. I. Issemann, R. Prince, J. Tugwood and S. Green. A role for fatty acids and liver fatty acid binding protein in peroxisome proliferation? *Biochemical Society Transactions* 20:824–827 (1992).

34. A. S. Muerhoff, K. J. Griffen and E. R. Johnson. The peroxisome proliferator-activated receptor mediates the induction of cyp4A6, a cytochrome P450 fatty acid ω-hydroxylase, by clofibric acid. *J. Biol. Chem.* 267:19051–19053 (1992).

35. T. C. Aldridge, J. D. Tugwood and S. Green. Identification and characterization of DNA elements implicated in the regulation of CYP4A1 transcription. *Biochem. J.* 306:473–479 (1995).

36. J. B. Kim and B. M. Spiegelman. ADD1/SREBP1 promotes adipocyte differentiation and gene expression linked to fatty acid metabolism. *Genes Dev.* 10:1096–1107 (1996).

37. S. K. Krisans. The role of peroxisomes in cholesterol metabolism. *Am. J. Respir. Cell Molec. Biol.* 7:358–364 (1992).

38. P. B. Lazarow and C. De Duve. A fatty acyl-CoA oxidizing system in rat liver peroxisomes; enhancement by clofibrate, a hypolipidemic drug. *Proc. Natl. Acad. Sci., U.S.A.* 73:2043–6 (1976).

39. B. Chatterjee, W. F. Demyan, N. D. Lalwani, J. K. Reddy and A. K. Roy. Reversible alteration of hepatic messenger RNA species for peroxisomal and non-peroxisomal proteins induced by the hypolipidaemic drug Wy-14,643. *Biochemical Journal* 214:879–83 (1983).

40. T. Watanabe, N. D. Lalwani and J. K. Reddy. Specific changes in the protein composition of rat liver in response to the peroxisome proliferators ciprofibrate, Wy-14,643 and di-(2-ethylhexyl)phthalate. *Biochemical Journal* 227:767–75 (1985).

41. D. R. Lowry and B. M. Willumsen. Function and regulation of ras. *Annu. Rev. Biochem.* 62:851–891 (1993).

42. M. O'Brien, S. Rangwala, K. W. Henry, C. Weinberger, D. C. Crick, C. J. Waechter, D. R. Feller and D. J. Noonan. Convergence of three steroid receptor pathways in the mediation of nongenotoxic hepatocarcinogenesis. *Carcinogenesis* 17:185–190 (1996).

43. B. M. Forman, E. Goode, J. Chen, A. E. Oro, D. J. Bradley, T. Perlmann, D. J. Noonan, L. T. Burka, McMorris, T., W. W. Lamph, R. M. Evans and C. C. Weinberger. Identification of a nuclear receptor that is activated by farnesol metabolites. *Cell* 81:687–693 (1995).

44. J. D. Bergstrom, M. M. Kurtz, D. J. Rew, A. M. Amend, J. D. Karkas, R. G. Bostedor, V. S. Bansal, C. Dufrense, F. L. VanMiddlesworth, O. D. Hensens, J. M. Liesch, D. L. Zink, K. E. Wilson, J. Onishi, J. A. Milligan, G. Bills, L. Kaplan, M. N. Omstead, R. G. Jenkins, L. Huang, M. S. Meinz, L. Quinn, R. W. Burg, Y. L. Kong, S. Mochales, M. Mojena, I. Martin, F. Pelaez, M. T. Diez and A. W. Alberts. Zaragozic acids: A family of fungal metabolites that are picomolar competitive inhibitors of squalene synthetase. *Proc. Natl. Acad. Sci. U.S.A.* 90:80–84 (1993).

SHORT CHAIN FATTY ACID REGULATION OF INTESTINAL GENE EXPRESSION

John A. Barnard,[*] J. A. Delzell, and N. M. Bulus

Department of Pediatrics
Vanderbilt University School of Medicine
Nashville, Tennessee 37232–2576

INTRODUCTION

The molecular pathways involved in regulation of intestinal epithelial cell proliferation and differentiation have not been characterized to the extent that analogous pathways have been defined for many other cell types, especially those in the hematopoietic lineages. Much of the published work on intestinal cells has focused on regulation by polypeptide growth factors and extracellular matrix proteins while relatively less attention has been given to the contributions of luminal factors to growth and differentiation. Notwithstanding, luminal fluid in the colon contains a number of putative growth regulators. Foremost among these is the four carbon short chain fatty acid (SCFA) butyrate. Herein, we will review selected aspects of the cell physiology and biology of butyrate. Emphasis will be given to studies in epithelial systems, although a larger body of work has been conducted in cells of hematopoietic origin. We will also emphasize our own studies using the HT-29 colon adenocarcinoma cell line as a model for study of early cellular and molecular events associated with butyrate-mediated growth and differentiation.

BIOLOGY OF SHORT CHAIN FATTY ACIDS IN THE MAMMALIAN COLON

Large amounts of SCFA are generated in the mammalian colon as a result of bacterial fermentation of complex carbohydrates in dietary fiber (1,2). In fact, the most abundant intraluminal solutes in the colonic lumen are the three, four and five-carbon SCFAs, acetate, butyrate and propionate (3). Colonic concentrations of the SCFA have been esti-

---

* Address correspondence to: John Barnard, M.D., Associate Professor, Department of Pediatrics, Division of Gastroenterology and Nutrition, Vanderbilt University School of Medicine, Nashville, Tennessee 37232–2576. Phone: (615) 322–7449; Fax: (615) 343–8915.

Dietary Fat and Cancer, edited by AICR
Plenum Press, New York, 1997

mated in the range of 100 to 240 mM with the concentration of acetate being the highest and butyrate the lowest (2). Substantial variation in the concentration of SCFA has been found in the human intestine as a function of anatomic location. As expected, levels in the terminal ileum are significantly lower than concentrations in the cecum and more distally to it (4). Two transport pathways for uptake of SCFA into the colonic epithelial cell have been identified, passive transport and carrier-mediated apical anion exchange (5). Within the cell, SCFA are metabolized by β-oxidation and it has been estimated that as much as 80% of the energy requirements of the colonic epithelial cell are met by SCFA (6). Proportionately, butyrate contributes to energy metabolism to a greater degree than acetate and propionate. A small residual percentage of SCFA enter the portal circulation and are metabolized by the liver, often leaving concentrations in the peripheral blood below the level of detection (4).

SCFA have a wide variety of biological and clinical effects in the normal and diseased colon (2) and it has been hypothesized that the protective effects of dietary fiber on colon carcinogenesis in humans and animals are, in part, attributable to bacterial fermentation of fiber to SCFA (7–10). Most investigators have concluded that SCFA stimulate mucosal proliferation *in vivo* (11–13) although under specific circumstances, there may be a regional decrease in proliferation in ulcerative colitis (14). As discussed later in this review, much attention has been given to the potent effects of butyrate in stimulating cellular differentiation, but this has not been critically analyzed and conclusively shown in the gastrointestinal mucosa *in vivo*. Similarly, the effects of butyrate on apoptosis have recently received attention in cell culture models, but related observations have not been made *in vivo*. It has been reported that SCFA stimulate colonic blood flow, but this may occur because of the prominent effects of SCFA on energy metabolism in the colon epithelial cell. All these *in vivo* effects of SCFA on the colon have led clinical investigators to use SCFA in the treatment of a variety of colonic diseases. For example, Scheppach and coworkers found decreased levels of SCFA in colon effluents obtained from patients with ulcerative colitis and hypothesized that repletion of SCFA or butyrate would be efficacious in the management of patients with ulcerative colitis (14). Indeed, this proved to be the case and patients with ulcerative colitis as well as diversion colitis are significantly improved when SCFA or butyrate are provided in the form of an enema (14–17). Recent animal studies have demonstrated a protective effect of butyrate enemas against the development of azoxymethane-induced colon cancer in a rat model of inflammatory bowel disease, lending further support to the hypothesis that appropriate levels of fiber in the diet may be chemopreventive (8). Potential benefits of butyrate have also been realized in other areas of medicine. For example, butyrate has been used to raise the percent fetal hemoglobin levels in patients with hemoglobin SS disease, which may reduce the tendency for sickling and improve clinical outcomes (18).

MOLECULAR AND CELL BIOLOGY OF THE SHORT CHAIN FATTY ACID, BUTYRATE, IN CELLS OF GASTROINTESTINAL ORIGIN

Effects of Butyrate on Intestinal Cell Growth and Differentiation

Early studies showed that butyrate inhibits histone deacetylase activity, resulting in a nonspecific hyperacetylation of histones (19,20) and alteration of gene expression. Alterations in DNA methylation have also been reported (21). Notwithstanding these apparent nonspecific effects, the clear net result of butyrate treatment in a large variety of cell types *in vitro* is growth inhibition and induction of a more differentiated phenotype.

A variety of studies have shown that butyrate specifically regulates expression of growth-related genes, including p53 and thymidine kinase (22), c-*fos* and c-*sis* (23), pp60^c-^{src} and p56^{lck} (24), and activated N-*ras* (25). Several studies, including our own, have addressed the regulation of c-*myc* by butyrate. These studies are summarized below.

In cells of gastrointestinal origin, a large number of genes associated with differentiation are induced by butyrate. Some of these include lactase and sucrase (26,27), leucine aminopeptidase (28), mt3 mitochondrial genes (29), alkaline phosphatase (30,31), villin (32,33), the epidermal growth factor receptor (34) and carcinoembryonic antigen (35), among others. Recently, detailed analyses of the promoter regions of a variety of genes have identified highly specific butyrate-sensitive regions, suggesting heretofore unrecognized molecular specificity in the differentiation response to butyrate treatment. For example, Deng and coworkers (36) transiently transfected the human placental-like alkaline phosphatase (PLAP) promoter and a CAT reporter gene in colon carcinoma lines and identified a butyrate-responsive *cis* acting sequence. Souleimani and Asselin (37) identified a short sequence in the *fos* promoter of Caco-2 colon carcinoma cells that was responsible for butyrate induction of *fos* expression. This sequence contains an activating transcription factor-cyclic AMP response element (ATF-CRE). Gel mobility shift assays showed that butyrate induced an ATF-CRE binding activity (37). Similarly, butyrate-responsive *cis* regions have been identified in promoters of other genes of particular interest to the study of colon cancer biology, e.g. the *mdr*1 (multidrug resistance) gene (38). Butyrate-sensitive regions have also been found in genes relevant to growth and differentiation of myocytes (39) and hematopoietic cells (40).

The effect of butyrate on expression of the proto-oncogene c-*myc* deserves special discussion. c-*myc* is an immediate-early gene which encodes a sequence specific transcription factor localized to the nucleus (41). Expression of c-*myc* is required for G1 traverse and, in general, expression of c-*myc* is positively correlated with cellular proliferation and inversely correlated with cellular differentiation. In some instances, it appears that over expression of c-*myc* induces apoptosis (42). Over expression of the c-*myc* gene has been detected in approximately 70% of colorectal cancers (43) suggesting that aberrations in expression may contribute to colorectal carcinogenesis.

Modulation of c-*myc* expression is complex and regulatory mechanisms are not completely resolved. Mitogenic stimuli cause a marked transcriptional induction of c-*myc* RNA levels; however, the mRNA half-life is short, approximately 20–30 minutes in a wide variety of cells, including colorectal carcinoma cells. All the factors involved in the rapid destabilization of *myc* mRNA are not precisely defined. Inhibition of protein synthesis using cycloheximide results in a "superinduction" of *myc* expression in nearly all cell types with the exception of colorectal cancer cells (44–46 and Barnard, unpublished observations). Thus, regulation of *myc* RNA in colon carcinoma lines occurs by a novel mechanism which merits further study. Levels of c-*myc* mRNA (44–48) and protein (49) are decreased by butyrate treatment in a variety of epithelial cell lines. As emphasized below, our studies emphasize that down regulation of *myc* by butyrate occurs rapidly, within 30 minutes of exposure. There is general agreement, including our own results reported herein, that the rapid butyrate-mediated reduction in c-*myc* RNA levels is blocked by inhibition of protein synthesis (44–46). These results suggest that butyrate induces synthesis of a protein that has a negative effect on c-*myc* abundance. The mechanism by which this protein decreases *myc* mRNA abundance is far less clear. Heruth and coworkers (46) concluded that butyrate induces synthesis of a protein that decreases the rate of *myc* transcriptional elongation in SW837 colon carcinoma cells, while Souleimani and Asselin (47) concluded that regulation of c-*myc* by butyrate in Caco-2 cells occurs at a post-transcritpional level. Our own results reported below suggest that transcriptional mechanisms are operative.

Effects of Butyrate on Epithelial Cell Apoptosis

Several recent studies suggest that, in addition to effects on growth and differentiation, butyrate also induces apoptosis in certain cell types, including those of gastrointestinal origin. The precise conditions under which butyrate induces apoptosis and the mechanisms by which this occurs have not been defined, but according to one report, butyrate causes apoptosis in several human colon adenoma and carcinoma lines in a p53-independent manner (50). In a follow-up study, these same investigators found that propionate and acetate also induced apoptosis, but less well than butyrate (51). Heerdt and coworkers also detected apoptosis in butyrate treated HT-29 and SW620 colon adenocarcinoma cell lines (29). An interesting recent study using MCF-7 breast carcinoma cell lines showed that butyrate induced apoptosis is closely linked to down regulation of Bcl-2, a protein involved in suppression of entry into an apoptotic pathway (52). Additional investigation is required to dissect out the molecular pathways involved in butyrate-induced apoptosis. Conditions favoring entry into a differentiation pathway versus an apoptotic pathway must also be delineated.

SUMMARY OF STUDIES ON THE INDUCTION OF GROWTH INHIBITION AND DIFFERENTIATION BY BUTYRATE IN HT-29 CELLS

Butyrate Inhibits HT-29 Cell Growth

HT-29 colon adenocarcinoma cells differentiate along goblet cell and enterocytic lineages under a variety of experimental stimuli. We examined early molecular events associated with butyrate-induced growth arrest and differentiation along the enterocytic pathway (45). Butyrate inhibits HT-29 cell growth in a dose dependent manner. Cell count assays and thymidine incorporation assays detect inhibition at butyrate concentrations as low as 0.1 mM. Half-maximal inhibition occurs at a concentration between 1 and 5 mM. Similar concentrations of acetate and propionate do not inhibit HT-29 growth. The inhibitory effect of butyrate appears to be fully reversible, suggesting that growth inhibition is not due to cytotoxicity or apoptosis. Others have proposed that differentiating agents, including butyrate, may mediate their effects by induction of autocrine transforming growth factor (TGF) β activity (53,54), however, we found that HT-29 cells are resistant to growth inhibition by TGFβ1 and TGFβ2 (45).

Butyrate Inhibits Growth in Early G1

Early studies found that butyrate induces a G1 growth arrest (55). We designed experiments to more precisely determine the interval within G1 in which butyrate arrests growth, since such a delineation may provide clues to the molecular mechanism by which growth is arrested. Release of HT-29 cells from nocodazole arrest (metaphase arrest) results in S phase entry after a minimum G1 of approximately 20 hours. The kinetics of S phase entry after release from butyrate-arrest were quite similar, suggesting that butyrate arrest must occur within the first one or two hours of G1 traverse. This conclusion was confirmed by addition of butyrate to HT-29 cells after approximately 8 hours of G1 traverse. In this case, DNA synthesis is not affected, suggesting that the only targets for butyrate-mediated growth inhibition are early in G1. These data were confirmed by fluorescence-activated cell sorting (FACS) analysis (45).

Effect of Butyrate on Growth and Differentiation-Related Gene Expression in HT-29 Cells

Inasmuch as the aforementioned experiments identified the early G1 interval as a potential target for butyrate mediated growth inhibition, the effect of butyrate on immediate-early gene expression was examined. These findings were interpreted in the context of the appearance of a marker of differentiation, intestinal alkaline phosphatase (IAP). Treatment of rapidly growing HT-29 cells with 5 mM butyrate induces the appearance of alkaline phosphatase mRNA between hour 4 and 12 after treatment. The appearance of IAP mRNA is preceded by a marked reduction (~ 70%) in expression of the immediate-early gene c-*myc*. This is evident as early as 30 minutes after treatment and is maximal 4 hours after butyrate treatment. Expression of other immediate-early genes like c-*fos*, c-*jun*, *nup*/475, and *zif*/268 are not influenced by butyrate, suggesting specificity in the down regulation of c-*myc*. Interestingly, butyrate also induces expression of TGFβ1 mRNA with kinetics identical to the induction of IAP. Prior studies in our laboratory determined that TGFβ is a marker of differentiation in the intestinal epithelium, lending further support to the validity of the HT-29 cell line in studies of molecular events involved in differentiation. However, as pointed out earlier, HT-29 cells are resistant to the growth inhibitory effects of TGFβ, indicating that the differentiating effects of butyrate are not directly mediated by TGFβ.

The studies outlined above describe a valuable model for the study of the earliest molecular events associated with intestinal epithelial differentiation. Dissection of the molecular and cellular biological events occurring at these early time points will be critical to understanding both normal and dysregulated intestinal epithelial growth and differentiation. The studies described below represent a further effort to understand the mechanism by which butyrate down regulates c-*myc*.

Mechanism of Down Regulation of *myc* Expression by Butyrate

The mechanism by which butyrate down regulates c-*myc* in HT-29 cells was further examined. Nuclear run-on transcription assays indicate that butyrate decreases transcription of the c-*myc* gene (Barnard, unpublished data). Inhibition of RNA synthesis (transcription) by actinomycin followed by treatment with butyrate permitted determination of c-*myc* RNA half life. The half-life of c-*myc* in control and butyrate-treated cells was similar (20 minutes) suggesting that butyrate does not induce an activity that accelerated degradation of c-*myc* RNA (Barnard, unpublished data). Further experiments were designed to determine the importance of new protein synthesis in mediating the effect of butyrate on down regulation of c-*myc*. Inhibition of protein synthesis with cycloheximide markedly reduces the effect of butyrate on c-*myc* expression (Barnard, unpublished data). In contrast, the expected superinduction of the immediate early gene c-*fos* is observed after treatment with cycloheximide, even in the presence of butyrate. Additionally, treatment with cycloheximide alone does not result in the superinduction of c-*myc* as expected. This unique aspect of the regulation of *myc* expression in colonic cells lines was alluded to earlier. Collectively, these results, which are supported in part by others (46,47), suggest that butyrate regulates expression of the c-*myc* gene by induction of the synthesis of a new protein that decelerates *myc* transcription.

Current Studies

It is critical that the functional importance of butyrate-induced down regulation of *myc* in differentiation of HT-29 cells is determined. We are investigating this by stable transfection of a full length *myc*2 construct into HT-29 cells. The sensitivity of resulting clones to butyrate-induced differentiation is being analyzed. Experiments are also underway to isolate butyrate-inducible and butyrate-repressible genes by the technique of "differential display". Identification of such genes may further contribute to understanding the pathways by which butyrate induces growth arrest and differentiation in the colonic epithelium and contribute further to chemoprevention strategies in colorectal cancer.

REFERENCES

1. M. Bugaut, and M. Bentejac. Biological effects of short-chain fatty acids in nonruminant mammals. *Annu. Rev. Nutr.* 13:217 (1993).
2. K.H. Soergel. Colonic fermentation: Metabolic and clinical consequences. *Clin Investig* 72:742 (1994).
3. W.E.W. Roediger. Role of anaerobic bacteria in the metabolic welfare of the colonic mucosa in man. *Gut* 21.793–798 (1980).
4. J.H. Cummings, E.W. Pomare, W.J. Branch, C.P. E. Naylor, G.T. McFarlane. Short chain fatty acids in human large intestine, portal, hepatic and venous blood. *Gut* 28: 1221 (1987).
5. E. Titus, and G.A. Ahearn. Vertebrate gastrointestinal fermentation: transport mechanisms for volatile fatty acids. *American Physiological Society* (1992).
6. W.E.W. Roediger. Utilization of nutrients by isolated epithelial cells of the rat colon. *Gastroenterology.* 83:424 (1982).
7. M.J. Koruda. Dietary fiber and gastrointestinal disease. *Surgery, Gynecology & Obstetrics* 177:209 (1993).
8. G. D'Argenio, V. Cosenza, M. Delle Cave, P. Iovino, N. Della Valle, G. Lombardi, and G. Mazzacca. Butyrate enemas in experimental colitis and protection against large bowel cancer in a rat model. *Gastroenterology* 110:1727 (1996).
9. L.C. Boffa, J.R. Lupton, M.R. Mariani, M. Ceppi, H. L. Newmark, A. Scalmati, and M. Lipkin. Modulation of colonic epithelial cell proliferation, histone acetylation, and luminal short chain fatty acids by variation of dietary fiber (wheat bran) in rats. *Cancer Res.* 52:5906 (1992).
10. L.H. Augenlicht, A.Velcich, and B.G. Heerdt. Short-chain fatty acids and molecular and cellular mechanisms of colonic cell differentiation and transformation. *Advances Exp Med Biol* 375:137 (1995).
11. T. Sakata. Stimulatory effect of short-chain fatty acids on epithelial cell proliferation in the rat intestine: a possible explanation for trophic effects of fermentable fibre, gut microbes and luminal trophic factors. *J. Nutr.* 58:95 (1987).
12. S.A. Kripke, A.D. Fox, J.M. Berman, R.G. Settle, and J.L. Rombeau. Stimulation of intestinal mucosal growth with intracolonic infusion of short-chain fatty acids. *JPEN* 13:109 (1989).
13. W. Scheppach, P. Bartram, A. Richter, F. Richter, H. Liepold, G. Dusel, G. Hofstetter, J. Ruthlein, and H. Kasper. Effect of short-chain fatty acids on the human colonic mucosa *in vitro. JPEN* 16:43 (1992).
14. W. Scheppach, J. Sommer, T. Kirchner, G-M. Paganelli, P. Bartram, S. Christl, F. Richter, G. Dusel, and H. Kasper. Effect of butyrate enemas on the colonic mucosa in distal ulcerative colitis. *Gastroenterology* 103:51 (1992).
15. M.A.S. Chapman, M.F. Grahn, M.A. Boyle, M. Hutton, J. Rogers, and N.S. Williams. Butyrate oxidation is impaired in the colonic mucosa of sufferers of quiescent ulcerative colitis. *Gut* 35:73 (1994).
16. W. Frankel, J. Lew, B. Su, A. Bain, D. Klurfeld, E. Einhorn, R.P. MacDermott, and J. Rombeau. Butyrate increases colonocyte protein synthesis in ulcerative colitis. *J. Surgical Res.* 57:210 (1994).

17. A.H. Steinhart, A. Brzezinski, J.P. Baker. Treatment of refractory ulcerative proctosigmoiditis with butyrate enemas. *American J. Gastroenterology* 89:179 (1994).

18. S.P. Perrine, G.D. Ginder, D.V. Faller, G.H. Dover, T. Ikuta, H.E. Witkowska, S. Cai, E.P. Vichinsky, N.F. Olivieri. A short term trial of butyrate to stimulate fetal globin gene expression in the β globin gene disorders. *N Engl J Med* 328:81–86 (1993).

19. E.P.M. Candido, R. Reeves, J.R. Davie. Sodium butyrate inhibits histone acetylation in cultured cells. *Cell* 14: 105 (1978).

20. J.A. D'Anna, R.A. Tobey, and L.R. Gurley. Concentration-dependent effects of sodium butyrate in Chinese hamster cells: cell-cycle progression, inner-histone acetylation, histone H1 dephosphorylation, and induction of an H1-like protein. *Biochemistry* 19:2656 (1980).

21. D.E. Cosgrove and G.S. Cox. Effects of sodium butyrate and 5-azacytidine on DNA methylation in human tumor cell lines: variable response to drug treatment and withdrawal. *Biochimica et Biophysica Acta* 1087:80 (1990).

22. A. Toscani, D.R. Soprano, and K.J. Soprano. Molecular analysis of sodium butyrate-induced growth arrest. *Oncogene Res.* 3:223 (1988).

23. S-J. Tang, L-W. Ko. Y-H.W. Lee, and F-F. Wang. Induction of *fos* and *sis* proto-oncogenes and genes of the extracellular matrix proteins during butyrate induced glioma differentiation. *Biochimica et Biophysica Acta* 1048:59 (1990).

24. F.M. Foss, A. Veillette, O. Sartor, N. Rosen, and J.B. Bolen. Alterations in the expression of pp60^{c-src} and p56^{lck} associated with butyrate-induced differentiation of human colon carcinoma cells. *Oncogene Res.* 5:13 (1989).

25. J.H. Stodart, M.A. Lane, and R.M. Niles. Sodium butyrate suppresses the transforming activity of an activated N-*ras* oncogene in human colon carcinoma cells. *Experimental Cell Research* 184:16 (1989).

26. Y.S. Chung, I.S. Song, R.H. Erickson, M.H. Sleisenger and Y.S. Kim. Effect of growth and sodium butyrate on brush border membrane-associated hydrolases in human colorectal cancer cell lines. *Cancer Res.* 45:2976 (1985).

27. D. Tsao, A. Morita, A. Bella, Jr., P. Luu, and Y.S. Kim. Differential effects of sodium butyrate, dimethyl sulfoxide, and retinoic acid on membrane-associated antigen, enzymes, and glycoproteins of human rectal adenocarcinoma cells. *Cancer Res.* 42:1052 (1982).

28. R.H. Whitehead, G.P. Young, and P.S. Bhathal. Effects of short chain fatty acids on a new human colon carcinoma cell line (LIM1215). *Gut* 27:1457 (1986).

29. B.G. Heerdt, M.A. Houston, J.J. Rediske, and L.H. Augenlicht. Steady-state levels of mitochondrial messenger RNA species characterize a predominant pathway culminating in apoptosis and shedding of HT29 human colonic carcinoma cells. *Cell Growth & Differentiation* 7:101 (1996).

30. A. Morita, D. Tsao, and Y.S. Kim. Effect of sodium butyrate on alkaline phosphatase in HRT-18, a human rectal cancer cell line. *Cancer Res.* 42:4540 (1982).

31. F. Herz. Divergent effects of butyrate on the alkaline phosphatases of SW-620 cells. *Biochimica et Biophysica Acta* 1180:289 (1993).

32. R.A. Hodin, S. Meng, S. Archer, and R. Tang. Cellular growth state differentially regulates enterocyte gene expression in butyrate-treated HT-29 cells. *Cell Growth & Differentiation* 7:647 (1996).

33. I. Chantret, A. Barbat, E. Dussaulx, M.G. Brattain, and A. Zweibaum. Epithelial polarity, villin expression, and enterocytic differentiation of cultured human colon carcinoma cells: a survey of twenty cell lines. *Cancer Res.* 48:1936 (1988).

34. A. deFazio, Y-E. Chiew, C. Donoghue, C.S.L. Lee, and R.L. Sutherland. Effect of sodium butyrate on estrogen receptor and epidermal growth factor receptor gene expression in human breast cancer cell lines. *J. Biological Chemistry* 267:18008 (1992).

35. K. Saini, G. Steele, and P. Thomas. Induction of carcinoembryonic-antigen-gene expression in human colorectal carcinoma by sodium butyrate. *Biochem. J.* 272:541 (1990).

36. G. Deng, G. Liu, L. Hu, J.R. Gum, Jr., and Y.S. Kim. Transcriptional regulation of the human placenta-like alkaline phosphatase gene and mechanisms involved in its induction by sodium butyrate. *Cancer Res.* 52:3378 (1992).

37. A. Souleimani, and C. Asselin. Regulation of c-*fos* expression by sodium butyrate in the human colon carcinoma cell line Caco-2. *Biochemical and Biophysical Res. Comm.* 193:330 (1993).

38. C.S. Morrow, M. Nakagawa, M.E. Goldsmith, M.J. Madden, and K.H. Cowan. Reversible transcriptional activation of *mdr*1 by sodium butyrate treatment of human colon cancer cells. *J. Biological Chemistry* 269:10739 (1994).

39. L.A. Johnson, S.J. Tapscott, and H. Eisen. Sodium butyrate inhibits myogenesis by interfering with the transcriptional activation function of MyoD and myogenin. *Molecular and Cellular Biology* 12:5123 (1992).

40. J.G. Glauber, N.J. Wandersee, J.A. Little, and G.D. Ginder. 5'-flanking sequences mediate butyrate stimulation of embryonic globin gene expression in adult erythroid cells. *Molecular and Cellular Biology* 11:4690 (1991).

41. K.B. Marcu, S.A. Bossone, A.J. Patel. myc function and regulation. *Annu Rev Biochem* 61:809 (1992).

42. R. Bissonnette, F. Echeverri, A. Mahoubi, D.R. Green. Apoptotic cell death induced by c-*myc* is inhibited by *bcl*-2. *Nature* 359:552 (1992).

43. M.F. Melhem, A.I. Meisler, G.C. Finley, W.H. Bryce, M.O. Jones, I.I. Tribby, J.M. Pipas, and R.A. Koski. Distribution of cells expressing myc proteins in human colorectal epithelium, polyps, and malignant tumors. *Cancer Res.* 52:5853, (1992).

44. K.M. Herold, and P.G. Rothberg. Evidence for a labile intermediate in the butyrate induced reduction of the level of c-*myc* RNA in SW837 rectal carcinoma cells. *Oncogene* 3: 423, (1988).

45. J.A. Barnard, G. Warwick. Butyrate rapidly induces growth inhibition and differentiation in HT-29 cells. *Cell Growth and Differentiation* 4:495, (1993).

46. D.P. Heruth, G.W. Zirnstein, J.F. Bradley, and P.G. Rothberg. Sodium butyrate causes an increase in the block to transcriptional elongation in the c-*myc* gene in SW837 rectal carcinoma cells. *J. Biological Chemistry* 268:20466 (1993).

47. A. Souleimani, C. Asselin. Regulation of c-*myc* expression by sodium butyrate in the colon carcinoma cell line Caco-2. *FEBS Lett* 326:45, (1993).

48. G. Krupitza, S. Grill, H. Harant, W. Hulla, T. Szekeres, H. Huber, and C. Dirrich. Genes related to growth and invasiveness are repressed by sodium butyrate in ovarian carcinoma cells. *British J. Cancer* 73:433 (1996).

49. C.W. Taylor, Y.S. Kim, K.E. Childress-Fields, and L.C. Yeoman. Sensitivity of nuclear c-*myc* levels and induction to differentiation-inducing agents in human colon tumor cell lines. *Cancer Letters* 62:95 (1992).

50. A. Hague, A.M. Manning, K.A. Hanlon, L.I. Huschtscha, D. Hart, and C. Paraskeva. Sodium butyrate induces apoptosis in human colonic tumor cell lines in a p53-independent pathway: implications for the possible role of dietary fibre in the prevention of large-bowel cancer. *Int. J. Cancer* 55:498 (1993).

51. A. Hague, D.J.E. Elder, D.J. Hicks, C. Paraskeva. Apoptosis in colorectal tumor cells: Induction by the short chain fatty acids butyrate, propionate and acetate and by the bile salt deoxycholate. *Int J Cancer* 60:400, (1995).

52. M. Mandal, and R. Kumar. Bcl-2 expression regulates sodium butyrate-induced apoptosis in human MCF-7 breast cancer cells. *Cell Growth & Differentiation* 7:311 (1996).

53. L. Staiano-Coico, L. Khandke, J.F. Krane, S. Sharif, A.B. Gottlieb, J.G. Krueger, L. Heim, B. Rigas, and P.J. Higgins. TGF-α and TGF-β expression during sodium-N-butyrate-induced differentiation of human keratinocytes: evidence for subpopulation-specific up-regulation of TGF-β mRNA in suprabasal cells. *Experimental Cell Res.* 191:286 (1990).

54. P. Schroy, J. Rifkin, R.J. Coffey, S. Winawer, E. Friedman. Role of transforming growth factor β1 in induction of colon carcinoma differentiation by hexamethylene bisacetamide. *Cancer Res* 50:261, (1990).

55. E. Wintersberger, I. Mudrak, and U. Wintersberger. Butyrate inhibits mouse fibroblasts at a control point in the G1 phase. *J. Cellular Biochemistry* 21:239 (1983).

REGULATION OF GENE EXPRESSION IN ADIPOSE CELLS BY POLYUNSATURATED FATTY ACIDS

David A. Bernlohr,[*] Natalie Ribarik Coe, Melanie A. Simpson, and Ann Vogel Hertzel

Department of Biochemistry and Institute of Human Genetics
University of Minnesota
St. Paul, Minnesota 55108-1022

SUMMARY

In fat cells polyunsaturated fatty acids are both substrates for, and products of, triacylglycerol metabolism. Dietary fatty acids are efficiently incorporated into the triacylglycerol droplet under lipogenic conditions while rapidly mobilizing them during lipolytic stimulation. Hence, the flux and magnitude of the fatty acid pool in adipocytes is constantly changing in response to hormonal, metabolic and genetic determinants. Due to the rapidly changing flux of fatty acids, the majority of genes encoding enzymes and proteins of lipid metabolism are largely refractory to long-term regulatory control by fatty acids. Only at extremes of high or low lipid levels, or under pathophysiological conditions, do adipose genes respond by up- or down-regulating gene expression. Despite the lack of responsiveness to lipids in adipose tissue, a surprisingly large number of genes have been characterized recently as lipid responsive when assayed in heterologous systems. These observations suggest an endogenous negative element exists in the lipid signaling pathway in adipocytes.

The major intracellular lipid binding protein in adipose cells is the adipocyte lipid binding protein (ALBP), the product of the aP2 gene. This protein is 15 kDa, abundant and found exclusively in the cytoplasm of adipocytes. The protein binds fatty acids and related lipids in a 1:1 stoichiometry within a large water filled interior cavity. The lipid binding protein forms high affinity associations with polyunsaturated fatty acids such as arachidonic acid (K_d ~250 nM) but not with prostaglandins of the E, D or J series ($K_d > 4$ μM). The upstream region of the aP2 gene contains a peroxisome-proliferator activated re-

---

[*] Author to whom correspondence should be addressed: Dr. David A. Bernlohr, Department of Biochemistry, University of Minnesota, 1479 Gortner Avenue, St. Paul, Minnesota 55108–102. Phone: (612) 624–2712; FAX: (612) 625–5780; e-mail: david-b@biosci.cbs.umn.edu.

ceptor response element which associates with PPARs to regulate its expression. A positive autoregulatory circuit exists to upregulate lipid binding protein expression when polyunsaturated fatty acid levels are increased.

Analysis of adipose tissue from aP2 null animals generated by a targeted disruption revealed that the partial loss of ALBP expression in heterozygotes and complete lack of ALBP in the nulls was accompanied by a compensatory up-regulation of the keratinocyte lipid binding protein. However, the total amount of lipid binding protein in the nulls was less than 15 % that in the wild type littermates. No evidence was found for upregulation of other lipid binding proteins such as the heart FABP or liver FABP. In aP2 nulls, the fatty acid composition was unaltered but the mass of fatty acid per gram tissue more than doubled relative to wild type. In heterozygotes, the level of fatty acid was intermediate to that of wild-type and nulls, consistent with an intermediate level of lipid binding protein. These results indicate that the fatty acid pool level in adipocytes is inversely correlated with the amount of lipid binding protein. Since prostaglandin biosynthesis is dependent upon polyunsaturated fatty acid substrates, the intracellular lipid binding proteins control accessibility of substrates to the prostanoid pathway. Intracellular lipid binding proteins therefore are negative elements in polyunsaturated fatty acid control of gene expression.

INTRODUCTION

The regulation of gene expression in adipocytes by lipids is considered to be controlled in part by members of the peroxisome proliferator activated receptor (PPAR) family of transcription factors [1,2]. These multi-domain proteins are members of the steroid-thyroid supergene family and minimally possess a ligand binding domain, a zinc-finger type DNA binding domain and a transactivation domain [3]. Members of the PPAR family heterodimerize with the retinoid X receptor (RXR, 9-cis retinoic acid binding) family thereby activating or repressing target genes [4]. The PPAR/RXR heterodimer cooperatively activates transcription in conjunction with the CCAAT/enhancer binding protein [5]. A number of genes expressed in adipose cells, such as that for the adipocyte lipid binding protein (ALBP or aP2) and phosphoenolpyruvate carboxykinase (PEPCK), have binding sites for PPAR/RXR heterodimers [5,6]. Although polyunsaturated fatty acids do not associate directly with PPAR's with high affinity, products of arachidonic acid metabolism such as prostaglandin J_2 and its derivatives do form stable complexes with PPARs [7]. As such, expression constructs containing the upstream region of the aP2 and/or PEPCK gene fused to a reporter gene render such activity subject to polyunsaturated fatty acid or eicosanoid control when expressed in a heterologous cell system. Curiously, such genes are virtually nonresponsive to lipids in the adipocyte suggesting endogenous fat cell factors may play a role in regulating the expression of target genes.

Adipocytes synthesize and store large amounts of triacylglycerol when energy supplies are abundant and mobilize such lipid reserves when energy sources are depleted [8]. The cyclical conversion between lipogenesis and lipolysis is hormonally controlled through the concerted actions of insulin and β-adrenergic agonists, respectively. Therefore, the flux of fatty acids into and out of the adipocyte is constant and highly regulated.

Serving to solubilize fatty acids destined either for efflux or storage are the intracellular lipid binding proteins. The adipocyte member of the multigene family of intracellular lipid-binding proteins is termed the adipocyte lipid binding protein (ALBP or aP2 protein). The protein has been extensively characterized with regard to its ligand binding affinity and specificity as well as structural properties [9–11]. The X-ray structure of ALBP has been solved to 1.6 Å in the presence and absence of bound ligand. In general the lipid binding proteins form 10-stranded up-and down β-barrels. The up and down β-barrel is a common folding motif found frequently in proteins which bind and transport hydrophobic ligands. It is formed by an array of β strands arranged in an antiparallel manner with each strand hydrogen bonded to neighboring strands nearly always adjacent in the amino acid sequence. The arrangement is completed by forming hydrogen bonds between the first and last strands. The barrel motif so formed produces interior and exterior components. Interestingly, proteins belonging to this class of up and down β-barrels are found typically to be lipid binding proteins in which the interior surface forms a cavity or pit which serves as the ligand binding region. The 10-stranded β-barrels have a large, hydrophilic water-filled interior cavity which serves as the ligand binding domain. Hydrophobic lipids such as fatty acids bind within the cavity, totally sequestered from the external milieu.

Recently, Hotamisligil and colleagues have developed a line of transgenic mice containing a targeted disruption of the aP2 allele [12]. Such animals are fertile and developmentally normally. However, the aP2 null mice develop dietary obesity but not insulin resistance or diabetes, two conditions commonly associated with the obese phenotype. Obese aP2 null animals fail to express tumor necrosis factor-α, a molecule suggested to be causative in the development of insulin resistance. These results have implicated the intracellular lipid binding protein in linking obesity to insulin resistance via altered fatty acid metabolism. We hypothesized that the loss of the intracellular lipid binding protein would cause dramatic changes in the cellular fatty acid pool which in turn, would affect the expression of certain adipocyte genes. We report here that the targeted disruption of the aP2 gene results in the compensatory up-regulation of the keratinocyte lipid binding protein gene in adipose cell and the expansion of the fatty acid pool level.

MATERIAL AND METHODS

Materials

All fatty acids were purchased from Nu Chek Prep, Inc. (Elysian, MN). Prostaglandins were from Biomol (Plymouth Meeting, PA) and 1-anilinonaphthalene-8-sulfonate (1,8-ANS) was purchased from Molecular Probes, Inc. Nalidixic acid was purchased from Sigma. All molecular biology enzymes were purchased from Promega. Solvents and TLC silica G plates utilized in the fatty acid extraction, thin layer chromatography and GC-MS were purchased from Fisher Scientific. The DB-23 column (cyanopropylmethyl-polysiloxane coated) was from Chrom Tech (Apple Valley, MN). Collagenase was purchased from Worthington Biochemical (Freehold, NJ). Polyvinylidene fluoride membranes were purchased from Millipore (Medford, MA) and enhanced chemiluminescence (ECL) reagent was from Amersham. A plasmid containing the heart muscle FABP cDNA was subcloned into vector pSG5 and was a generous gift of Dr. J.H. Veerkamp. Plasmid pGEM11zf containing the liver FABP cDNA was obtained from Dr. Nathan Bass. The original line of C57/B6 transgenic mice were a generous gift of Bruce Spiegelman and Gokhan Hotamisligil, Dana-Farber Cancer Research Center, Harvard, Boston.

Lipid Binding Protein Purification

KLBP and ALBP were purified as previously described [13]. Briefly, *E. coli* cells containing either of the recombinant lipid binding proteins were pelleted by centrifugation, resuspended in lysis buffer (25 mM imidazole (pH 7.0), 50 mM NaCl, 5 mM EDTA, 1 mM β-mercaptoethanol, 1 mM PMSF, and 2μg/mL each of pepstatin, aprotinin, leupeptin) and sonicated. The insoluble cellular fraction was removed by centrifugation and the supernatant titrated with stirring with protamine sulfate (5% in lysis buffer) to a final concentration of 1%. The mixture was again centrifuged and the supernatant acidified to pH 5.0 with 2M sodium acetate, pH 5.0, and allowed to stir overnight at 4° C. After centrifugation at room temperature, the supernatant was concentrated and applied to a Sephadex G-75 column equilibrated in buffered saline (12.5 mM HEPES, 250 mM NaCl, pH 7.5). Fractions containing the lipid binding protein were identified by SDS-PAGE and immunochemical analysis, pooled, concentrated and loaded onto a Pharmacia Mono S FPLC column and eluted with a linear gradient from 0 to 0.3M NaCl in 50 mM sodium acetate, pH 5.2. Pure lipid binding proteins were identified by SDS-PAGE and immunoblotting dialyzed into sodium phosphate (50 mM, pH 7.4) and stored at -70° C.

Separation of Adipose Cells

All C57/B6 mice used for these studies were two to four month old males fed a standard laboratory chow diet (4% fat). Genotypes of wild type (aP2$^+$/aP2$^+$), heterozygous ALBP (aP2$^+$/aP2$^-$) and ALBP null (aP2$^-$/aP2$^-$) were determined by polymerase chain reaction amplification of the ALBP and neomycin loci [12]. The epididymal fat from C57/B6 mice was removed, weighed and the fat from three mice of each genotype was pooled. The fat pads were digested with collagenase buffer (3 mg/mL collagenase plus 20 mg/mL BSA in Krebs-Ringer solution) at 37°C with shaking until digestion was complete. The digested fat pads were centrifuged at 2000 rpm for ten minutes at room temperature. The floating fat cells were removed and washed with Krebs-Ringer solution by repeated centrifugation.

Extraction of Triglycerides and Fatty Acids from Fat Cells and Thin Layer Chromatography

Five mL of an extraction buffer containing 8 mL isopropanol, 2 mL heptane, 200 μL sulfuric acid and .005 g of butylated hydroxytoluene was added to each mL of washed fat cells and mixed vigorously. To this was added 4 mL of water and 6 mL of heptane. The organic phase was then evaporated under nitrogen. The oil remaining after solvent evaporation was applied directly to a prepared silica G thin layer chromatography (TLC) plate and developed in petroleum ether/ diethyl ether/ acetic acid (80/201; v/v/v). Triolein and oleic acid were run as standards to identify the regions corresponding to free fatty acids and triglycerides. The lipid regions were scraped and chloroform was added to the silica. Each sample was vortexed and centrifuged at 12,000 RPM for one minute at 4°C. The chloroform was evaporated under nitrogen and the fatty acids converted to their methyl esters. The triglyceride fractions were saponified to their corresponding fatty acids with .5M NaOH in methanol for 15 minutes at 80°C under nitrogen, then coverted to methyl esters. Samples were separated on a DB-23 column coated with cyanopropylmethylpolysiloxane from Chrom Tech (Apple Valley, MN). The temperature was increased from 70°C to 140°C at 20°C/min., 140°C to 220°C at 5°C/min., and then held at 220°C for 10 minutes. Mass spectroscopy of fatty acid methyl esters was performed using a Kratos MS 25 GC/MS system.

Determination of Total Free Fatty Acid and Triglyceride Levels

The level of free fatty acids in the adipose tissue of wild type, heterozygous and ALBP knockout mice was determined by utilizing a colorimetric nonesterified fatty acid kit (Wako, Richmond, VA.). The free fatty acid and triglyceride pools for each genotype were separated and isolated by thin layer chromatography as described above. Chloroform was added to the resin bound lipid fraction, vortexed, centrifuged, and the chloroform layer transferred to a second vial and evaporated under nitrogen. Extraction of lipid species from the silica was repeated twice for the free fatty acids fraction and four times for the triglyceride fractions. To the free fatty acids, 300 µl of methanol was added. The triglyceride fractions were saponified with .5M NaOH in methanol for 15 minutes at 80°C. The .5M NaOH solution was then evaporated under nitrogen and methanol added. Lipid fractions were then assayed for nonesterified fatty acids. The data represent the amount of nonesterified fatty acids (µmoles) per gram of fat tissue in the free fatty acid and triacylglycerol pools. The data represent the averages ± standard deviation of three experiments, each a pool of three mice for each genotype.

Western Analysis

Epididymal fat from male C57/B6 mice from each genotype (wild type (+/+), heterozygous (+/-), ALBP knockout (-/-)) was removed and homogenized in 50 mM phosphate buffer, pH 7.4, containing 2 ug/mL of aprotinin, leupeptin, and pepstatin as well as 1 mM PMSF and 5 mM EDTA. The fat pads were homogenized (Kinematica Polytron, Switzerland) and total protein concentration based on a bovine serum albumin standard was determined. Total fat protein for each genotype and pure KLBP and ALBP were separated by SDS-PAGE followed by protein transfer to polyvinylidene fluoride membranes. The membranes were then blocked with 3% nonfat milk and probed with either anti-KLBP or anti-ALBP polyclonal antibodies. The secondary antibody, goat anti-rabbit horseradish peroxidase conjugate, was added and the blots developed by enhanced chemiluminescence. Standard curves for ALBP and KLBP were determined from a plot of lipid binding protein (ng) versus densitometric signal. Anti-KLBP polyclonal antibodies were prepared by Dr. Soren Nielsen, this laboratory, from rabbits injected with glutathione S-transferase/KLBP fusion protein. Anti-ALBP antibodies were prepared in this laboratory from purified ALBP protein.

Northern Analysis

Total mRNA isolated from the epididymal fat pads of three male C57/B6 mice of each genotype (wild type (+/+), heterozygous (+/-), and ALBP knockout (-/-) was separated by agarose gel electrophoresis and the RNA transferred to nylon membranes by capillary action. The blot was hybridized with a ^{32}P-labeled KLBP cDNA probe. The blot was stripped and reprobed separately with a ^{32}P-labeled cDNA probe of ALBP, liver FABP, heart muscle FABP as well as a radiolabeled ribosomal 28S oligonucleotide. Densitometry was used to normalize the KLBP and ALBP mRNA levels by direct comparison with the corresponding ribosomal 28S value for each genotype.

Ligand Binding Studies

Ligand binding to KLBP was assessed using a fluorescence-based assay system utilizing the hydrophobic probe 1-anilinonaphthalene 8-sulfonic acid (1,8-ANS) [14]. 1,8-

ANS binds within the ligand binding cavity of the hydrophobic ligand binding proteins and can be displaced with fatty acids. The probe was dissolved in absolute ethanol and its concentration determined spectrophotometrically (ε_{372}= 8000 cm^{-1}, M^{-1}). Proteins were dialyzed into 50 mM sodium phosphate pH 7.4 and added in aliquots to ~500 nM 1,8-ANS in the same buffer (final ethanol 0.05 %, $^v/_v$). The samples were mixed for 1 minute under dim light and the fluorescence measured in a thermostated (37 °C) Perkin Elmer 650–10S fluorescence spectrophotometer. Relative fluorescence was plotted versus increasing protein concentration, and Scatchard analysis used to calculate the binding parameters. All values reported were calculated from two to five independent binding isotherms (mean ± standard deviation).

Competition Assays

Various lipids were assessed for their ability to displace 1,8-ANS bound to KLBP. KLBP (0.66 μM) in 50 mM sodium phosphate pH 7.4 was mixed with 500 nM 1,8-ANS at 37 °C and the fluorescence signal determined. Increasing concentrations of competitor lipid (each diluted from a 25 mM stock in absolute ethanol) were added to the KLBP/1,8-ANS complex, mixed for 30 seconds, and the fluorescence signal measured. Relative fluorescence as a function of increasing concentration of competitor lipid was determined and analyzed as described by Epps et al. [15]. The midpoint of the assay was defined as the point at which 50% of initial fluorescence had been lost. The I_{50} was used to calculate an apparent K_i using $K_i = [I_{50}]/ (1+ [L]/K_d)$, where K_i = apparent inhibitor constant (equivalent to the K_d), [L] = free concentration of 1,8-ANS and K_d = apparent dissociation constant of KLBP for 1,8-ANS.

RESULTS

Upregulation of KLBP Protein Levels in ALBP (aP2) Knockout Mice

The targeted disruption in the aP2 gene results in animals that are developmentally normal but exhibit no insulin resistance as a consequence of the development of metabolic obesity [12]. Hotamisligil and colleagues have reported that serum fatty acid and glucose levels are unaltered as a consequence of the disruption but that the expression of tumor necrosis factor-α is absent [12]. While it is not clear which molecules affect tumor necrosis factor-α gene expression, fatty acids are a reasonable possibility given the relationship between lipid metabolism and insulin resistance. Because of the centrality of the lipid binding proteins to fatty acid metabolism in fat cells, we examined the expression of the adipocyte lipid binding protein and other lipid binding proteins in aP2 heterozygotic and null mice. Epididymal fat pads were removed from wild type, heterozygous, and ALBP null mice and the expression of ALBP, KLBP, liver FABP and heart muscle FABP was determined by northern analysis. As shown in Figure 1, the expression of ALBP mRNA in heterozygotes was approximately one-half that of wild-type and undetectable in fat from null animals. Concomitant with the decrease in ALBP expression was an up-regulation in the expression of the keratinocyte lipid binding protein. Wild type mice express very low levels of KLBP mRNA, approximately 1 % that of ALBP. In heterozygotes, KLBP expression increased 7-fold while in adipose from aP2 null animals the KLBP expression was induced 40-fold relative to wild type. No evidence was found for upregulation of heart FABP or liver FABP. We have not systematically evaluated the expression of the

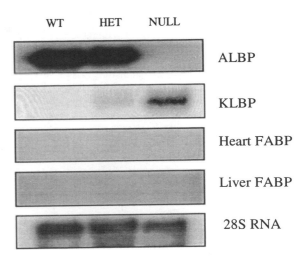

Figure 1. Expression of lipid binding protein mRNAs in adipose tissue from aP2$^+$/aP2$^+$, aP2$^+$/aP2$^-$ and aP2$^-$/aP2$^-$ mice. RNA isolated from adipose tissue from the indicated mouse genotype was separated by agarose gel electrophoresis, transferred to nylon membranes and probes with the indicated cDNA's corresponding to the various lipid binding proteins. An oligonucleotide corresponding to 28S RNA was used to evaluate the integrity of the RNA preparations and the loading amounts. The resultant blots were exposed to autoradiographic film at -70 °C.

other 20 members of the lipid binding protein multigene family for their upregulation in null animals.

To determine if the upregulation of KLBP mRNA in heterozygotic and null animals was accompanied by a corresponding increase in KLBP protein, immunochemical analysis was performed. Using antibodies specific for KLBP, immunoblotting experiments determined that there was a 10-fold increase in KLBP protein in heterozygotes and a 12-fold increase in aP2 null animals. However, since the level of expression of KLBP is far less than ALBP, the upregulation of KLBP does not fully compensate for the decrease in ALBP. These results are presented in Table 1. The total lipid binding protein level in aP2$^-$/aP$^-$ animals is less than 15 % that of wild type.

Because of the difference in expression levels of the two lipid binding proteins in the three genotypes, we systematically compared the binding activity of the two proteins. Table 2 presents the binding activities of ALBP and KLBP for a variety of lipophilic ligands. As seen, KLBP and ALBP have similar binding affinities for a variety of ligands. The highest affinity ligands were the long-chain fatty acids, particularly the polyunsaturated fatty acids where the apparent binding constants ranged from 200 to 500 nM. In contrast, neither protein exhibited any significant affinity for members of the eicosanoid family of lipids. Consequently, the intracellular lipid binding proteins do not compete with members of the PPAR family for ligands but they do associate with the precursor polyunsaturated fatty acids.

Due to the difference in expression of the two lipid binding proteins we considered that either the amount or distribution of fatty acids in fat cells from such animals may be

Table 1. Expression of lipid binding proteins and fatty acid pool levels in adipocytes from aP2$^+$/aP2$^+$, aP2$^+$/aP2$^-$, and aP2$^-$/aP2$^-$ mice

Genotype	ALBP*	KLBP*	Total LBP*	Total Fatty Acids*
aP2$^+$/aP2$^+$ (wt.)	40	0.4	40.4	812 ± 164
aP2$^+$/aP2$^-$ (het.)	26	4.5	30.5	1714 ± 313
aP2$^-$/aP2$^-$ (null)	0	5	5	1961 ± 150

*nmol/gm tissue. For lipid binding proteins determined by immunochemical titration, for fatty acid by indirect assay as described in Methods.

Table 2. Comparison of ligand binding properties of KLBP and ALBP
for various lipids using the 1,8-ANS displacement assay

Ligand	K_d (nM) KLBP	K_d (nM) ALBP
Decanoic Acid (C10:0)	> 8,000	2990
Myristic Acid (C14:0)	1409 ± 423	436 ± 94
Palmitic Acid (C16:0)	802 ± 164	336 ± 37
Oleic Acid (C18:1)	248 ± 12	185 ± 35
Linoleic Acid (C18:2)	313 ± 4	317 ± 2
Linoelaidic Acid (C18:2)	$220 \pm .5$	114 ± 8
Linolenic Acid (C18:3)	337 ± 29	476 ± 9
Linoelaidic Acid (C20:3)	146 ± 2	638 ± 26
Eicosatrienoic Acid (C20:3)	213 ± 37	502 ± 19
Arachidonic Acid (C20:4)	318 ± 14	245 ± 32
15-deoxy $\Delta^{12,14}$ PGJ$_2$	> 2μM	1620 ± 50
Prostaglandin J$_2$	> 4μM	> 4μM
Prostaglandin E$_2$	> 2μM	> 4μM
Prostaglandin H$_2$	> 4μM	> 4μM
Prostaglandin D$_2$	> 4μM	> 4μM

All values reported were determined at pH 7.4 at 37°C as described in Methods.

altered. We prepared the total lipid pool from adipocytes from the three genotypes and isolated the fatty acid fraction by preparative silica gel chromatography. Analysis of the fatty acid levels in the three samples revealed a surprising increase in the fatty acid pool from heterozygotes and wild type animals such that the fatty acid pool level increased 2.4-fold in null animals compared to wild type (Table 1). Consequently, while the total lipid binding protein level is decreasing, the fatty acid pool level is increasing. These results suggest that the cellular fatty acid levels are inversely correlated with the level of lipid binding proteins.

Given the marked expansion in the fatty acid pools between wild type, heterozygotes and null animals, we evaluated the fatty acid and corresponding triacylglycerol composition of each genotype by gas chromatography/mass spectrometry analysis. Molar percents of each fatty acid comprising the free fatty acid pool for wild type, heterozygous and ALBP knock out mice were determined and tabulated (Table 3). The esterified methyl esters representing the triglyceride pools of all three genotypes examined were principally myristic acid (C14:0), palmitic acid (C16:0), palmitelaidic acid (C16:1), oleic acid (C18:1) and linoleic acid (C18:2). The precise location of the double bonds in the unsaturated species was not determined. In all cases, C16:0 and C18:1 were the primary substituents. Trace amounts of 18:3 and 20:1 were also detected.

The free fatty acid composition of wild type, heterozygous and ALBP knockout mice was also determined by GC-MS analysis. The GC-MS spectrum of the free fatty acid pools for each genotype was dominated by a large peak corresponding to palmitic acid (C16:0). Smaller peaks representing methyl esters of oleic acid (C18:0), myristic acid (C14:0) and linoleic acid (C18:1) and arachidonic acid (20:4) were also identified. Again, as shown in Table 3, no significant differences in the composition of adipose triacylglycerol or fatty acid pools were detected. This is consistent with the very similar binding properties of the two lipid binding proteins (Table 2). The major finding concerning the fatty acid profiles in the wild type, heterozygotic and null animals is the expansion of the total fatty acid pool level without a significant change in the composition of such pools.

Table 3. Composition of fatty acid and triacylglycerol pools in adipocytes from aP2$^+$/aP2$^+$, aP2$^+$/aP2$^-$, and aP2$^-$/aP2$^-$ mice

Methyl Ester	Genotype		
	aP2$^\pm$/aP2$^\pm$	aP2$^\pm$/aP2$^-$	aP2$^-$/aP2$^-$
Fatty Acid Pool (mol%)*			
14:0	7.2	8.0	7.5
16:0	75.3	71.3	73.8
16:1	0.5	1.9	0.8
18:0	7.4	9.7	11.5
18:1	6.2	8.2	6.3
18:2	0.5	0.8	trace
Triacylglycerol Pool (mol %)			
14:0	3.4	2.9	3.2
14:1	0.3	0.3	0.3
16:0	33.8	30.4	30.0
16:1	9.4	12.0	11.8
18:0	2.4	1.8	1.8
18:1	34.6	36.2	35.8
18:2	14.4	15.1	15.6
18:3	0.6	0.6	0.5
20:0	0.1	0.1	0.1
20:1	0.4	0.5	0.5
20:2	0.2	0.1	0.2
20.4	0.5	0.2	0.2

*3.3% of fatty acid pool could not be assigned

DISCUSSION

The regulation of gene expression by fatty acids in adipocytes is considered to be dependent upon the action of the peroxisome proliferator activated receptor family of transcription factors [1–4]. The adipose members of the family, PPARα, PPARγ and PPARδ, can be activated by prostaglandin J$_2$ derivatives as well as several peroxisome proliferators or pharmacological agents such as the anti-diabetic thiazolidinediones [7]. PPAR's heterodimerize with RXRs to form transcriptionally competent factors [5,6]. Such heterodimers function synergistically with members of the CCAAT/enhancer binding protein family to regulate expression. Consequently, the regulation of prostaglandin biosynthesis and the availability of substrates for the prostanoid pathway is a key component of the lipid signaling system. The biosynthesis of prostaglandin J$_2$ utilizes the polyunsaturated fatty acid arachidonic acid as its starting point. Whereas the product of the prostaglandin J$_2$ pathway associates avidly with PPARs, it is not a high affinity ligand for intracellular lipid binding proteins. In contrast, lipid binding proteins bind polyunsaturated fatty acids such as arachidonic acid with high affinity but not the product prostaglandins. This implies that intracellular lipid binding proteins compete with the prostanoid pathway for arachidonic acid. Lipid binding proteins would, in effect, function as negative regulators of fatty acid-induced gene expression. Consistent with this is the observation that many fat cell genes regulated by polyunsaturated fatty acids are much more responsive to lipids when taken out of the context of an adipocyte and placed into a cell type devoid of lipid binding proteins.

To test the hypothesis that intracellular lipid binding proteins are negative regulators of gene expression, we have turned our attention to the adipocyte lipid binding protein (aP2) knock out mice developed by Hotamisligil and colleagues [12]. These mice harbor a targeted disruption in the aP2 gene which renders the loci inactive. Animals homozygous for such a disruption develop normally but fail to develop insulin resistance associated with diet-induced obesity [12]. $aP2^-/aP2^-$ mice failed to express the tumor necrosis factor-α but do express the insulin-stimulated glucose transporter, GLUT4, normally. Seeing as how a defect in the cellular lipid carrier resulted in a loss of tumor necrosis factor-α gene expression, models placing ALBP along the fatty acid signaling pathway have been developed. Consequently, we have analyzed the expression of genes within aP2 knock out mice and report that besides the change in expression of tumor necrosis factor-α, the keratinocyte lipid binding protein is significantly upregulated.

Analysis of the fatty acid composition of adipocytes from wild type, heterozygotes and aP2 null mice revealed a marked expansion of the fatty acid pool level accompanies the targeted disruption. A key finding is that the pool levels of fatty acids are inversely proportional to the lipid binding protein level. Total cellular fatty acid levels are increased more than 2-fold as a consequence of the disruption. No changes in the composition of the fatty acid pools were noted. As the pool levels have increased, the gene encoding the keratinocyte lipid binding protein became activated, resulting in the upregulation of KLBP expression at least 40-fold. There was not as large an increase in KLBP protein found in adipose cells (12-fold increase) possibly due to increased rate of KLBP degradation, however this point was not further examined. Although KLBP was significantly up-regulated, the level of KLBP reached less than 15 % that of ALBP in wild type cells. These observations suggest that the KLBP gene may contain a PPAR binding site and is regulated similarly to ALBP. This is reasonable for the liver FABP is known to be up-regulated by peroxisome proliferators in hepatocytes. Regulation by PPAR type transcription factors may be a common property of lipid binding protein genes.

The lack of responsiveness of wild type adipocytes to fatty acids is likely to be due to high levels of ALBP which binds avidly to polyunsaturated fatty acids and facilitates their metabolism [8,10]. In the presence of intracellular lipid binding proteins, the amount of polyunsaturated fatty acid available for eicosanoid biosynthesis would be minimal, making regulation of gene expression by prostaglandin J_2 derivatives unlikely. Under normal conditions where fatty acid pool levels in adipocytes rise due to dietary factors, the activated PPARs activate the ALBP gene, thereby increasing the amount of lipid binding protein which in turn reduces the fatty acid pool. The reduction in the fatty acid pool would result in decrease in prostaglandin J_2 metabolites and subsequent effects by PPARs. These observations are presented schematically in Figure 2. Under pathophysiological conditions such as the targeted disruption of the lipid binding protein gene, upregulation of ALBP is not possible and the cells partially compensate by upregulating the expression of the keratinocyte lipid binding protein gene. This gene is normally expressed at very low levels in adipose cells but is up-regulated significantly in the disruptants.

In sum, we find that intracellular lipid binding proteins are negative regulators of gene expression. The intracellular lipid binding proteins are a large multigene family of proteins found in a variety of eukaryotic cells. While the binding specificity of all family members has not been determined in detail, it is likely that a common feature is high affinity association with polyunsaturated fatty acids. Consequently, as a class, lipid binding proteins may be general negative regulators of gene expression. Clearly, pharmacological inactivation of lipid binding proteins would likely result in a cell type sensitized towards lipids as bioregulators. Also, polymorphisms in the lipid binding protein genes which re-

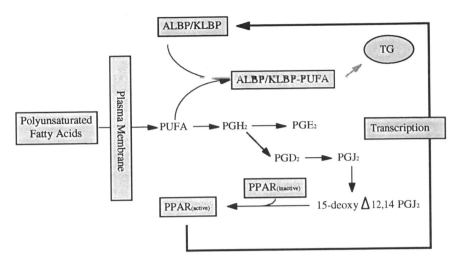

Figure 2. Schematic diagram representing the metabolism of polyunsaturated fatty acids leading to the activation of gene transcription.

sult in inactive proteins would also result in cell types with heightened responsiveness to polyunsaturated fatty acids. Such polymorphisms have been reported for the intestinal FABP [16]; it remains to be seen how widespread such mutations are in other lipid binding protein genes in the human population.

ABBREVIATIONS

LBP; lipid binding protein, KLBP; keratinocyte LBP, ALBP; adipocyte LBP, FABP; fatty acid binding protein, aP2; ALBP, PPAR; peroxisome-proliferator activated receptor, RXR; retinoid X receptor, 1,8-ANS; 1-anilinonaphthalene-8-sulfonate,

ACKNOWLEDGMENTS

The authors would like to acknowledge the members of the Bernlohr laboratory for their helpful discussions and to Ms. Anne Smith for development and maintenance of the transgenic mouse colony. In addition, the authors would like to thank Drs. Göhkan Hotamisligil and Bruce Spiegelman for making the aP2 null animals available to our laboratory.

Supported by funds from the American Institute for Cancer Research

REFERENCES

1. Tontonoz, P., Hu, E., Graves, R.A., Budavari, A.I., and Spiegelman, B.M. (1994) Genes Dev. 8; 1224–1234
2. Amri, E.Z., Bonino, F., Ailhaud, G., Abumrad, N.A., and Grimaldi, P.A. (1995) J. Biol. Chem. 270; 2367–2371
3. Isseman, I., and Green, S. (1990) Nature 347; 645–649

4. Tontonoz, P., Graves, R.A., Budavari, A.I., Erdjument-Bromage, H., Lui, M., Hu, E., Tempst, P., and Spiegelman, B.M. (1994) Nucleic Acids Res 22; 5628–5634
5. Tontonoz, P., Hu, E. and Spiegelman, B.M. (1994) Cell 79; 1147–1156
6. Tontonoz, P., Hu, E., Devine, J., Beale, E.G., and Spiegelman, B.M. (1995) Mol Cell Biol 15:;351–357
7. Yu, K., Bayona, W., Kallen, C.B., Harding, H.P., Ravera, C.P., McMahon, G., Brown, M. and Lazar, M.A. (1995) J. Biol. Chem. 23975–23983
8. Bernlohr, D.A. and Simpson, M.A. (1996) *Biochemistry of Lipids, Lipoproteins and Membranes* (Vance, D.E. and Vance, J., Eds) pp. 257–281 Elsevier, New York
9. Banaszak, L.J., Winter, N., Xu, Z., Bernlohr, D. A., Cowan, S., and Jones, T. A. (1994) *Advances in Protein Chemistry* (V.Schumaker, Ed) Vol 45, 89–151 Academic Press, San Diego
10. Lalonde, J., Levenson, M.A., Bernlohr, D.A., and Banaszak, L.J. (1994) J. Biol. Chem., 269, 25339–25347
11. Lalonde, J.M., Bernlohr, D.A., Banaszak, L.J. (1994) Up and down β-barrel proteins. FASEB J. 8; 1240–1247
12. Hotamisligil, G.S., Johnson, R.S., Distel, R.J., Ellis, R., Papaioannou, V.E., and Spiegelman, B.M. (1996) Science, in press
13. Kane, C. D., Ribarik-Coe, N., Krieg, P., and Bernlohr, D.A. (1996) Biochemistry 35; 2894–2900
14. Kane, C. D. and Bernlohr, D. A. (1996) Anal. Biochem. 233; 197–204
15. Epps, D.E., Raub, T.J. and Kédzy, F.J. (1995) Anal. Biochem. 227; 342–350
16. Baier, L.J., Sacchettini, J.C., Knowler, W.C., Eads, J., Paolisso, G., Tataranni, P.A., Mochizuki, H., Bennett, P.H., Bogardus, C. and Prochazka, M. (1995) J. Clin. Invest. 95; 1281–1287

REGULATION OF PEROXISOMAL FATTY ACYL-CoA OXIDASE IN THE YEAST
Saccharomyces cerevisiae

Gillian M. Small, Igor V. Karpichev, and Yi Luo

Department of Cell Biology and Anatomy
Mount Sinai Medical Center
New York, New York 10029

ABSTRACT

Peroxisomes are specialized organelles found in most eukaryote cells, where their major functions are in cellular respiration and fatty acid oxidation. Proliferation of this organelle, and induction of peroxisomal enzymes, is a phenomenon that occurs in diverse species, and is stimulated by a number of physiological and pharmacological stimuli. A large number of chemically diverse compounds, including hypolipidemic drugs and industrial plasticizers, have been shown to cause peroxisome proliferation and the induction of peroxisomal enzymes in rodents. Chronic exposure to these compounds produces hepatocellular carcinomas, however, the mechanism by which this tumorigenic event occurs is unknown. In the yeast *Saccharomyces cerevisiae* peroxisomes are induced when a fatty acid such as oleate is supplied as a carbon source in the growth medium. In addition, many peroxisomal enzymes are induced by growth on oleate; these include enzymes of the peroxisomal ß-oxidation cycle. This regulation occurs at the transcription level, and is controlled by specific *trans*-acting factors. The research in our laboratory has focused on the mechanisms involved in this regulation, and on the identification and characterization of the proteins involved. Our recent results, and current research directions are summarized.

INTRODUCTION

Peroxisomes are single membrane-bound organelles that play important roles in cellular respiration and lipid metabolism. Typically, peroxisomes contain enzymes involved in fatty acid oxidation, and catalase which decomposes the hydrogen peroxide generated from the oxidative reactions.[1] In some organisms other enzymes may be present, for example, mammalian peroxisomes contain enzymes involved in cholesterol and plasmalogen biosynthesis.[2] The functional importance of peroxisomes in mammalian metabolism is indicated by the existence of several human peroxisomal disorders (reviewed in 3). The most severe of these is the genetic disorder Zellweger syndrome which is characterized by the absence of morphologically distinguishable peroxisomes.[4]

Peroxisome Proliferation in Mammals

The size and number of peroxisomes varies between cells and organisms. In mammals peroxisomes are induced by a wide variety of physiological and pharmacological stimuli (for reviews see 5). In rodents, peroxisome proliferation is induced by feeding the animals a diet high in fat,[6,7] and by starvation.[8] Examples of chemicals that cause peroxisome proliferation include hypolipidemic drugs, certain phthalate ester plasticizers and several pesticides and herbicides.[9] Concomitant with an increase in peroxisome number is activation of a number of genes encoding peroxisomal proteins.

All peroxisome proliferators tested in long term studies have been found to induce liver tumors in rodents.[10] These compounds do not cause DNA damage directly, but do cause a substantial increase in liver size, and appear to have an effect on the growth of preneoplastic lesions and on the conversion of such lesions to tumors.[11] The mechanisms by which peroxisome proliferation leads to tumor formation are unknown. It has been postulated that the increased hydrogen peroxide produced from peroxisomal oxidative reactions would exceed the capacity for its decomposition by catalase, resulting in a gradual accumulation of oxidative damage.[12-14] Alternatively, it has been suggested that peroxisome proliferators may act as promoters of tumors, perhaps by having promoter activity in the generation of tumors but not foci.[15]

More recently a receptor-based mechanism for the peroxisome proliferation response was indicated following the identification of a peroxisome proliferator-activated receptor (PPAR).[16] PPAR consists of six functional domains which include DNA-binding and ligand-binding domains. These structural motifs place PPAR as a member of the steroid hormone receptor superfamily.[16] Three PPAR isoforms have since been isolated from the mouse; PPARα, PPARß and PPARγ.[17] Experiments with chimeric receptors containing the ligand binding domain of PPARα and the amino terminal and DNA-binding domains of either the estrogen or the glucocorticoid receptors, in the presence of peroxisome proliferators, suggested that PPAR modulates the ligand-induced activation of responsive genes. These include those genes encoding peroxisomal ß-oxidation enzymes.[16,18,19] This idea was validated by the discovery of a peroxisome proliferation response element (PPRE) in the 5' upstream regions of these genes. This element consists of an almost perfect direct repeat of the sequence TG A/T CCT, separated by one nucleotide.[20] PPARα heterodimerizes with the retinoic acid X receptor (RXR) to enhance the activation of the peroxisomal acyl-CoA oxidase gene.[21] Many of the details of this signaling pathway remain to be elucidated.

Peroxisome Proliferation in Yeast

Peroxisome proliferation also occurs in yeast. In *Saccharomyces cerevisiae* levels of peroxisomal ß-oxidation enzymes are regulated by the available carbon source. Expression of genes encoding these proteins is repressed when the yeast is grown in the presence of glucose, derepressed during growth in glycerol and activated when a fatty acid such as oleate is supplied for growth.[22] As with mammals, this regulation is controlled at the transcriptional level. Many genes encoding peroxisomal proteins in *S. cerevisiae* have been cloned, and this has allowed the identification and characterization of the upstream *cis* -

acting elements involved in the regulation. Major advances in the field of peroxisome bio-genesis have come from studies in yeast. *S. cerevisiae* is an ideal organism for such stud-ies because it is amenable to genetic manipulation as well as to molecular biological and biochemical techniques. Furthermore, many features of peroxisome biology are conserved from mammals to yeast.

REGULATION OF PEROXISOMAL ACYL-CoA OXIDASE

Identification of *cis*-Acting Elements in the Acyl-CoA Oxidase Gene

The focus of work from our laboratory has been on the regulation of *POX1*, the gene encoding peroxisomal acyl-CoA oxidase in *S. cerevisiae*.[23] This is the first and rate-limit-ing enzyme of the peroxisomal ß-oxidation cycle. We identified two glucose responsive elements in the *POX1* promoter that are involved in the repression of this gene. One se-quence consists of an inverted repeat and the other a direct repeat. Our experimental data suggests that the same protein binds to each element.[24] More recently we have charac-terized an upstream activating sequence (UAS) that is required for the oleate-specific acti-vation of *POX1*.[25] In a DNA band-shift experiment, using a labeled region of the POX1 promoter (-316 to -238) as a probe, we obtained a specific band shift with extracts from yeast cells grown in the presence of oleate, but not with extracts from cells grown in glu-cose or glycerol (Figure 1). Analysis of the DNA sequence in the region of the promoter that gave rise to this band-shift revealed the presence of two incomplete palindromes. The consensus sequence for these four repeats is (A/T) A (A/T)NNCCG(A/T)(A/T) (see Fig-ure 1).

To verify that the oleate responsive element is indeed an Upstream Activating Se-quence (UAS) we tested its ability to promote transcription in a heterologous promoter. For these experiments we used the promoter region of the *CYC1* gene from *S. cerevisiae*, lacking its own UAS sequence, and fused in-frame with the *lacZ* gene from *E. coli*. When a 56-bp fragment from the *POX1* promoter region that contained one of the two palin-dromes (-294 to -238) was inserted in front of the CYC/*lacZ* gene fusion, the ß-galactosi-dase activity produced from this construct was increased by 100-fold in oleate-grown cells (Table 1). An 80-mer encompassing both palindromes gave a 200-fold increase over back-ground under similar conditions (pNG1580, Table 1). These results confirm that this re-gion of *POX1* is indeed a UAS element. A similar UAS1 sequence (termed oleate response element) was identified in the upstream regions of other genes encoding enzymes of the peroxisomal ß-oxidation cycle, and is required for their activation.[26-28]

Purification of a Protein Required for Oleate-Induction of *POX1*

Having identified some of the DNA elements that serve as binding sites for specific transcription factors involved in *POX1* regulation, we next turned our attention towards identifying and characterizing the *trans*-acting proteins that bind to these elements. Using standard protein purification techniques followed by DNA-affinity column chromatogra-phy, we purified the UAS1-binding protein approximately 20,000-fold, as shown in Ta-ble 2.[29] Throughout the different stages of chromatography that we used to purify this protein we carried out DNA band-shift experiments to identify those fractions that con-tained activity. SDS-polyacrylamide gel electrophoresis of the fractions eluted from the affinity column revealed protein bands that consisted of a doublet, with a molecular mass

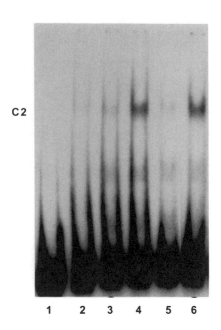

C 2

1   2   3   4   5   6

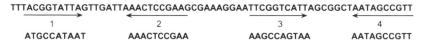

TTTACGGTATTAGTTGATTAAACTCCGAAGCGAAAGGAATTCGGTCATTAGCGGCTAATAGCCGTT

1   2   3   4

ATGCCATAAT   AAACTCCGAA   AAGCCAGTAA   AATAGCCGTT

Figure 1. A labeled DNA fragment including the UAS1 sequence was incubated with extracts from glucose- (lane 2), glycerol- (lane 3), or oleate-grown cells (lanes 4–6). Lanes 5 and 6 are as for lane 4 except that they were carried out in the presence of excess unlabelled UAS1 DNA (lane 5) or unrelated DNA (lane 6). The 65-nucleotide sequence within the DNA fragment, that includes the two inverted repeats of UAS1, is shown below. Reproduced from the Journal of Biological Chemistry 1994, 269: p24483, with copyright permission.

in the range of 110 - 120 kDa (Figure 2). The upper band of this doublet was subjected to tryptic digestion, and the peptides generated were resolved by reverse-phase high performance liquid chromatography. The peptides were separated and amino acid sequences were determined by Edman degradation. The microsequence analysis revealed that the purified protein was identical to a protein sequence in the GenBank sequence data base, described as a "putative 118.2-kDa transcriptional regulatory protein". We named this protein Oleate-Activated Factor 1 (Oaf1p).[29]

Table 1. ß-galactosidase activities of strains harboring *CYC1/lacZ* constructs in a pNG15 vector

pNG Construct	Glucose	Glycerol	Oleate
pNG15	0.375	0.623	0.611
pNG1556	0.269	8.93	61.1
pNG1580	0.211	15.62	130
pNG1599	0.738	0.967	33.63

Table 2. Purification of Oaf1p

	Specific Activity	Total Protein mg	Fold Purification	Recovery %
Lysate	16	2700	0	100
$(NH_4)_2SO_4$	24	1960	1.5	108
SP-Sepharose	610	60	38	84
DNA Cellulose	750	20	47	34
UAS1 Affinity	296,000	0.01	18,500	7

Using oligonuleotides homologous to the DNA sequence of the gene encoding this protein we obtained a PCR-amplified product, and confirmed that the resultant DNA corresponded to the *OAF1* gene. The effect of disrupting this gene in our yeast strain was determined in several ways. Firstly we found that the disrupted strain was unable to grow on plates containing oleic acid as the sole carbon source. It has previously been shown that yeast strains that lack functional peroxisomes are unable to grow under these conditions.[30,31] Thus these results confirmed that disrupting *OAF1* disabled peroxisomal function. Furthermore, when extracts from oleate-grown cells from the strain in which *OAF1* was disrupted were used in a DNA band shift assay, there was no specific DNA-protein complex detectable (Figure 3).

Electron microscope examination of the strain in which *OAF1* was disrupted revealed that the cells had a dramatically reduced number of peroxisomes when compared with the wild-type strain grown in the presence of oleate. In addition, peroxisomes present in the disrupted strain appeared smaller than in the wild-type cells, and could only be detected by immunogold-labeling of peroxisomal proteins (Figure 4). This finding suggests that the mechanisms involved in the induction of peroxisomal enzymes and in proliferation of the organelle are linked.

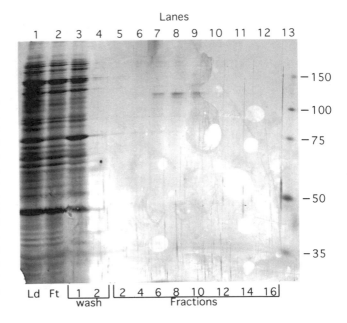

Figure 2. Silver-stained SDS-polyacrylamide gel containing fractions eluted from a UAS1-affinity column. Lane 1 contains 20 μl of the starting sample (concentrated fractions eluted from a calf thymus DNA column), lane 2 contains the flow-through, lanes 3 and 4 contain consecutive 0.25 M KCl washes, and lanes 4 through 13 contain alternate fractions eluted from the column. Sizes were approximated by using perfect protein™ markers (Novagen). Reproduced from the Journal of Biological Chemistry 1996, 271: p12071, with copyright permission.

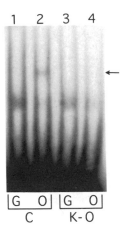

G O G O
C   K-O

Figure 3. DNA band shift assay with control (lanes 1 and 2) and *OAF1*-disruption (lanes 3 and 4) strains grown in glycerol (G) or oleate (O) medium. Labeled UAS1 DNA was used as a probe. Reproduced from the Journal of Biological Chemistry 1996, 271: p12073, with copyright permission.

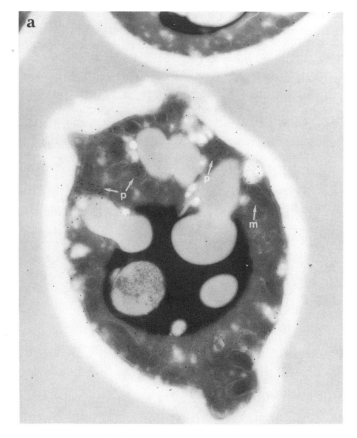

Figure 4. (a) Electron micrograph of a wild-type yeast cell grown in oleate medium and labeled immunocytochemically, using an antibody that recognizes peroxisomal proteins and Protein A conjugated to 15 nm gold particles. Arrows indicate labeled peroxisomes (p), and non-labeled mitochondria (m). (b) Electron micrograph from a strain in which *OAF1* is disrupted, labeled as described in (a). Arrows mark a peroxisome (p) and a lipid droplet (L). Magnification x40,000.

CURRENT AND FUTURE RESEARCH

In order to gain a full understanding of the mechanisms involved in peroxisome pro-liferation and enzyme induction it is necessary to characterize all of the factors in the pathway leading to oleate induction. With this goal we initiated a genetic approach to iso-late yeast strains that are mutated in their ability to undergo *POX1* activation, and thus acyl-CoA oxidase induction. Using a screening strategy designed to select for loss of *POX1* activation we obtained a number of mutants in which peroxisomal ß-oxidation en-zymes are not induced when the cells are grown in the presence of oleate. One of these mutants appears to exhibit a similar phenotype to that of the strain in which *OAF1* is dis-rupted. We have cloned the mutated gene in this strain, and will briefly describe its char-acteristics here; details of this mutant will be communicated elsewhere. The protein encoded by this gene has high homology to Oaf1p, each protein has a zinc cluster motif common to yeast transcription factors.[32] In addition they both contain leucine zipper mo-tifs, which are known to mediate dimer formation.[33] Both proteins are required for the for-mation of a specific UAS-protein complex, and for the induction of acyl-CoA oxidase and other peroxisomal ß-oxidation enzymes. The data described above have led us to postulate a model for the regulation of *POX1*. Glucose repression is mediated through two upstream repressing sequences. In the presence of oleate the two transcription factors Oaf1p and

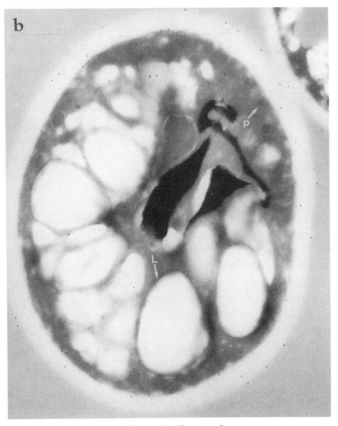

Figure 4. *(Continued)*

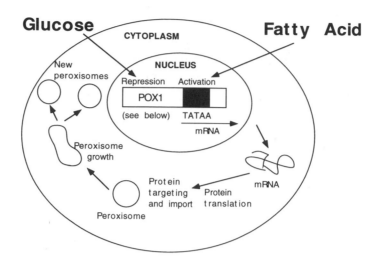

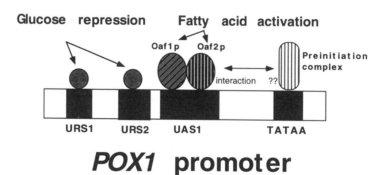

POX1 promoter

Figure 5. Cartoon depicting our model for the regulation of peroxisomal acyl-CoA oxidase. *POX1*, encoding acyl-CoA oxidase, undergoes glucose repression and oleate induction. The translated protein is targeted to and imported inside peroxisomes, which then grow and divide to form new organelles. The lower part of the figure shows how Oaf1p and Oaf2p may dimerize with each other, and bind to UAS1 to mediate induction of *POX1*.

Oaf2p bind to UAS1 in the form of a heterodimer (Figure 5). This complex is activated by some manner, as yet unknown, and this activation leads to the transcription of *POX1*.

This model resembles the mechanism by which PPAR and RXR regulate mammalian peroxisomal enzymes. In the mammalian system each receptor alone is able to stimulate gene expression, but the two receptors together act synergistically. In yeast it appears that both Oaf1p and Oaf2p are required for gene activation. Further studies in our laboratory are being directed towards identifying other factors involved in this signaling pathway and to characterize the molecular mechanisms involved.

ACKNOWLEDGMENTS

We would like to thank Vladimir Protopopov for performing the electron microscopy. This work was supported by grants AHA-95008910 and AHA-92001690 from the American Heart Association.

REFERENCES

1. C. de Duve and P. Baudhuin. Peroxisomes (microbodies and related particles). *Physiol.Rev.* 46:323–357, (1966).
2. H. van den Bosch, R.B.H. Schutgens, R.J.A. Wanders and J.M. Tager. Biochemistry of peroxisomes. *Annu.Rev.Biochem.* 61:157–197, (1992).
3. Lazarow, P.B. and Moser, H.W. Disorders of peroxisomal biogenesis. In: "The Metabolic Basis of Inherited Diseases", edited by Scriver, C.R., Beaudet, A.L., Sly, W.S. and Valle, D. New York: McGraw-Hill Co., p. 1479–1509.(1989).
4. S. Goldfischer, C.L. Moore, A.B. Johnson, A.J. Spiro, M.P. Valsamis, H.K. Wisniewski, R.H. Ritch, W.T. Norton, I. Rapin and L.M. Gartner. Peroxisomal and mitochondrial defects in the cerebro-hepato-renal syndrome. *Science* 182:62–64, (1973).
5. E.A. Lock, A.M. Mitchell and C.R. Elcombe. Biochemical mechanisms of induction of hepatic peroxisome proliferation. *Annu.Rev.Pharmacol.Toxicol.* 29:145–163, (1989).
6. C.E. Neat, M.S. Thomassen and H. Osmundsen. Induction of peroxisomal B-oxidation in rat liver by high fat diets. *Biochem.J.* 186:369–371, (1980).
7. C.E. Neat, M.S. Thomassen and H. Osmundsen. Effects of high-fat diets on hepatic fatty acid oxidation in the rat. Isolation of rat liver peroxisomes by vertical-rotor, iso-osmotic, Percoll gradient. *Biochem.J.* 196:149–159, (1981).
8. H. Ishii, S. Horie and T. Suga. Physiological role of peroxisomal β-oxidation in liver of fasted rats. *J.Biochem.* 87:1855–1858, (1980).
9. J.K. Reddy and N.D. Lalwani. Carcinogenesis by hepatic peroxisome proliferators: Evaluation of the risk of hypolipidemic drugs and industrial plasticizers to humans. *Crit.Rev.Toxicol.* 12:1–58, (1983).
10. M.S. Rao and J.K. Reddy. Peroxisome proliferation and hepatocarcinogenesis. *Carcinogenesis* 8:631–636, (1987).
11. Popp, J.A. and Cattley, R.C. Peroxisome proliferators as initiators and promoters of rodent hepatocarcinogenesis. In: "Peroxisomes: Biology and importance in toxicology and medicine", edited by Gibson, G. and Lake, B. London: Taylor and Francis, p. 653–665 (1993).
12. J.K. Reddy, N.D. Lalwani, M.K. Reddy and S.A. Qureshi. Excessive accumulation of autofluorescent lipofuscin in the liver during hepatocarcinogenesis by methyl clofenapate and other hypolipidemic peroxisome proliferators. *Cancer Res.* 42:259–266, (1982).
13. J.K. Reddy and M.S. Rao. Oxidative DNA damage caused by persistent peroxisome proliferation: its role in hepatocarcinogenesis. *Mutation Research* 214:63–68, (1989).
14. M.S. Rao and J.K. Reddy. An overview of peroxisome proliferator-induced hepatocarcinogenesis. *Environ.Health Perspect.* 93:205–209, (1991).
15. R.C. Cattley and J.A. Popp. Differences between the promoting activities of the peroxisome proliferator WY-14,643 and phenobarbital in rat liver. *Cancer Res.* 49:3246–3251, (1989).
16. I. Issemann and S. Green. Activation of a member of the steroid hormone receptor superfamily by peroxisome proliferators. *Nature* 347:645–650, (1990).
17. S.A. Kliewer, B.M. Forman, B. Blumberg, E.S. Ong, U. Borgmeyer, D.J. Mangelsdorf, K. Umesong and R.M. Evans. Differential expression and activation of a family of murine peroxisome proliferator-activated receptors. *Proc.Natl.Acad.Sci.USA* 91:7355–7359, (1996).
18. C. Dreyer, G. Krey, H. Keller, F. Givel, G. Helftenbein and W. Wahli. Control of the peroxisomal β-oxidation pathway by a novel family of nuclear hormone receptors. *Cell* 68:879–887, (1992).
19. M. Gottlicher, E. Widmark, Q. Li and J-A. Gustafsson. Fatty acids activate a chimera of the clofibrate acid-activated receptor and the glucocorticoid receptor. *Proc.Natl.Acad.Sci. USA* 89:4653–4657, (1992).
20. J.D. Tugwood, I. Issemann, R.G. Anderson, K.R. Bundell, W.L. McPheat and S. Green. The mouse peroxisome proliferator activated receptor recognizes a response element in the 5' flanking sequence of the rat acyl CoA oxidase gene. *EMBO J.* 11:433–439, (1992).
21. S.A. Kliewer, K. Umesono, D.J. Noonan, R.A. Heyman and R.M. Evans. Convergence of 9-cis retinoic acid and peroxisome proliferator signalling pathways through heterodimer formation of their receptors. *Nature* 358:771–774, (1992).
22. M. Veenhuis, M. Mateblowski, W.H. Kunau and W. Harder. Proliferation of microbodies in *Saccharomyces cerevisiae*. *Yeast* 3:77–84, (1987).
23. A. Dmochowska, D. Dignard, R. Maleszka and D.Y. Thomas. Structure and transcriptional control of the *Saccharomyces cerevisiae POX1* gene encoding acyl-coenzyme A oxidase. *Gene* 88:247–252, (1990).
24. T.W. Wang, A.S. Lewin and G.M. Small. A negative regulating element controlling transcription of the gene encoding acyl-CoA oxidase in *Saccharomyces cerevisiae*. *Nucleic Acids Res.* 20:3495–3500, (1992).

25. T. Wang, Y. Luo and G.M. Small. The *POX1* gene encoding peroxisomal acyl-CoA oxidase in *Saccharomyces cerevisiae* is under the control of multiple regulatory elements. *J.Biol.Chem.* 269:24480–24485, (1994).

26. A.W.C. Einerhand, T.M. Voorn-Brouwer, R. Erdmann, W-H. Kunau and H.F. Tabak. Regulation of transcription of the gene coding for peroxisomal 3-oxoacyl-CoA thiolase of *Saccharomyces cerevisiae*. *Eur.J.Biochem.* 200:113–122, (1991).

27. A.W.C Einerhand, W.T. Kos, B. Distel and H.F. Tabak. Characterization of a transcriptional control element involved in proliferation of peroxisomes in yeast in response to oleate. *Eur.J.Biochem.* 314:323–331, (1993).

28. W. Kos, A.J. Kal, S. van Wilpe and H.F. Tabak. Expression of genes encoding peroxisomal proteins in *Saccharomyces cerevisiae* is regulated by different circuits of transcriptional control. *Biochim.Biophys.Acta* 1264:79–86, (1995).

29. Y. Luo, I.V. Karpichev, R.A. Kohanski and G.M. Small. Purification, identification and properties of a *Saccharomyces cerevisiae* oleate-activated upstream activating sequence-binding protein that is involved in the activation of *POX1*. *J.Biol.Chem.* 271:12068–12075, (1996).

30. J.W. Zhang, Y. Han and P.B. Lazarow. Novel peroxisome clustering mutants and peroxisome biogenesis mutants of *Saccharomyces cerevisiae*. *J.Cell Biol.* 123:1133–1147, (1993).

31. R. Erdmann, M. Veenhuis, D. Mertens and W-H. Kunau. Isolation of peroxisome-deficient mutants of *Saccharomyces cerevisiae*. *Proc.Natl.Acad.Sci. USA* 86:5419–5423, (1989).

32. R.J. Reece and M. Ptashne. Determinants of binding-site specificity among yeast C6 zinc cluster proteins. *Science* 261:909–911, (1993).

33. W.H. Landschulz, P.F. Johnson and S.L. McKnight. The leucine zipper: a hypothetical structure common to a new class of DNA binding proteins. *Science* 240:1759–1764, (1988).

DIETARY FAT, GENES, AND HUMAN HEALTH[*]

Donald B. Jump,[1†] Steven D. Clarke,[2] Annette Thelen,[1] Marya Liimatta,[1] Bing Ren,[1] and and Maria V. Badin[1]

[1]Departments of Physiology and Biochemistry
Michigan State University
East Lansing, Michigan 48824-1101
[2]Division of Nutritional Sciences
Department of Human Ecology
University of Texas-Austin
Austin, Texas 78712-1907

INTRODUCTION

Dietary fat is an essential macronutrient in the diet of all animals. It provides a source of energy and hydrophobic components for biomolecule synthesis. Fatty acids also are used for the synthesis of signaling molecules like steroids and prostanoids as well as being covalently linked to specific proteins.[1,2] In addition to these well established roles, recent studies indicate that fatty acids have pronounced effects on gene expression leading to changes in metabolism, cell growth and differentiation.[3–22] While many of these effects are beneficial to human health, dietary fat appears to become a problem when humans or animals ingest high fat diets and/or diets that are disproportionately enriched in saturated or polyunsaturated fatty acids. Numerous epidemiologic, clinical and animal studies have studied the link of dietary fat to the onset and progression of chronic diseases like breast, colon and prostate cancer, coronary heart disease, insulin resistance, hypertension and obesity.[23–40] How these diets contribute to disease is unclear. In our view, the recent advances defining the cellular and molecular basis of dietary fat action are likely to provide important clues to explain how fats alter cell function and lead to chronic disease. In this presentation, we will first briefly discuss the diverse effects of fatty acids on cell function and then focus on dietary fat regulation of gene transcription.

* The research was supported by a grant (to DBJ) from the National Institutes of Health (DK43220) and a predoctoral fellowship (to MVB) from the National Science Foundation.
† To whom all correspondence should be addressed.

Dietary Fat Effects on Cell Function

How does an essential macronutrient like dietary fat affect cell function? Fat administration both in vivo and to cultured cells has been reported to induce many changes in function. For example, increased cellular fatty acid levels (as fatty CoA derivatives) stimulate fatty acid oxidation (both mitochondrial and peroxisomal), inhibit glucose uptake and glycolysis and promote gluconeogenesis. This may be the basis for fatty acid-induced insulin resistance in muscle and liver.[27–31,41–43] Changes in membrane phospholipid composition in cells treated with various fatty acids have been correlated with changes in cellular growth rates and hormone signaling from the plasma membrane.[24,25,44–47] For example, a decrease in muscle membrane polyunsaturated (PUFA) phospholipid composition is associated with a decline in insulin action. The type of fatty acid is also important. This is illustrated by the differential rates of synthesis of prostanoids from n-6 and n-3 PUFA; n-3 PUFA are poor substrates for prostaglandin synthases.[48–50] Thus, the balance of n-6 and n-3 in the diet may affect phospholipid pools that contribute to local prostanoid synthesis and action. The highly unsaturated fatty acids like 20:5, n-3 and 22:6, n-3 can be oxidized to form lipid peroxides.[51] Oxidized lipids might induce oxidant stress and promote changes in gene expression through changes in protein redox state and transcription factor function.[52,53] This may be a contributing factor in atherosclerosis.[52] Peroxisomal β-oxidation of PUFA generates hydrogen peroxide which may lead to oxidant stress and changes in gene expression.[18,54,55] Finally, both in vivo and cell culture studies show that specific fatty acids have pronounced effects on gene expression leading to changes in cell metabolism, cell growth and differentiation.[3–22] The molecular basis for the effects on gene expression are not fully understood. However, this brief summary illustrates that dietary fat can affect cell function through multiple pathways. Substantial changes in either dietary fat composition or quantity might distort normal physiological function and shift cells into different metabolic, differentiation or growth patterns. Such changes over the long term may lead to an adaptive response in which cells display abnormal patterns of growth or metabolism and lead to chronic disease.

PUFA Effects on Hepatic Lipid Synthesis and Metabolism

One area where dietary fat has been found to have pronounced effects on gene expression is in the regulation of hepatic lipid synthesis and oxidation.[11–16] In the rat, hepatic lipogenesis (synthesis of palmitate) is induced by high carbohydrate-fat free diets, thyroid hormone (T_3) and insulin and repressed by hormones that elevate cellular cAMP levels or by starvation or diabetes. In humans, de novo lipogenesis is operative and is induced with high carbohydrate-low fat (<10–20% calories) diets.[56] Elevation of the dietary fat to >20% calories suppresses de novo lipogenesis in humans. In the rat, suppression of de novo lipogenesis is dependent on the type of fat ingested.[11] While polyunsaturated fat, like corn, safflower, or menhaden containing 60, 75 and 43% PUFA, respective, inhibit lipogenesis, diets containing predominantly saturated (tallow or lard) or monounsaturated fat (olive oil) have little or no inhibitory effect on lipogenesis. Inhibition of hepatic lipogenesis is characterized by a decline in both lipogenic and some glycolytic enzymes. The principal target of control of these enzymes is at the level of gene transcription, making dietary fat an important modulator of cell function at the genomic level.[3–9]

Structure function studies have shown that when compared to 18:2,n-2, the long chain highly unsaturated PUFA like eicosapentaenoic acid (20:5,n-3) and docosahexaenoic acid (22:6, n-6) are particularly potent inhibitors of hepatic lipogenesis. This appears to be due to a requirement of Δ6-desaturase activity to convert 18:2, n-6 and 18:3, n-3 to more unsaturated fatty acids. Inhibition of lipogenic gene expression requires at least 18 carbons with 2 conjugated double bonds located at the 9 and 12 positions. The inhibitory effect on lipogenesis and inhibition of transcription of specific genes is rapid (occurs within hours of switching animals onto a menhaden oil diet) and persists so long as PUFA remains in the diet. Moreover, PUFA acts directly on cultured hepatocytes indicating the lack of a requirement for extrahepatic hormones or fatty acid metabolism.[11]

In addition to effects on lipogenesis, the highly unsaturated fatty acids found in fish oil, like 20:5,n-3 and 22:6,n-3 are potent suppressors of hepatic triglyceride synthesis and VLDL assembly and secretion.[37–40] Ingestion of these fatty acids lowers serum VLDL-triglycerides. Elevation of both VLDL and cholesterol is positively correlated with coronary heart disease.[33] The decline in VLDL-triglyceride may be due to an elevation in both mitochondrial and peroxisomal β-oxidation.[57–61] Recent human studies suggest that eicosapentaenoic acid (20:5, n-3) might be a potent activator of both mitochondrial and peroxisomal β-oxidation, while docosahexaeonic acid (22:6, n-3) activates peroxisomal β-oxidation.[61] The elevation of peroxisomal β-oxidation by hypolipemic agents, e.g. gemfibrozil, and fatty acids requires the induction of several enzymes involved in peroxisomal fatty acid oxidation, including acyl CoA oxidase (AOX) and the bifunctional enzyme.[54] Hypolipemic agents activate transcription of the AOX gene through a transcription factor termed peroxisome proliferator activated receptor (PPAR).[13,14,16] PPARs are steroid-like receptors that bind DNA at direct repeats of TGACCT with a 1-nucleotide spacer (DR+1). Binding to DNA requires a second steroid-like receptor, designated retinoid X receptor (RXR). Together, PPAR-RXR bind as heterodimers to DR+1 type motifs. In the cases of AOX, binding of activated PPAR-RXR to DR+1 motifs upstream from the AOX gene stimulates transcription leading to a rise in AOX activity and peroxisomal β-oxidation.

PUFA EFFECTS ON HEPATIC GENE EXPRESSION

The effect of PUFA on lipogenesis and β-oxidation is due to a change in the hepatic level of specific enzymes (proteins) and their corresponding mRNAs. For example, mRNAs encoding L-type pyruvate kinase (L-PK), fatty acid synthase (FAS), pyruvate dehydrogenase, glucose 6-phosphate dehydrogenase, acetyl CoA carboxylase, stearoyl CoA desaturase 1 (SCD-1), malic enzyme and the S14 protein (S14) are suppressed in livers of rats fed PUFA.[3–12,62] In contrast, hepatic mRNAs encoding phosphoenolpyruvate carboxykinase, β-actin are marginally induced (2-fold) while the peroxisomal enzyme (AOX) is induced 4 to 5-fold. mRNAs encoding tyrosine aminotransferase, thyroid hormone receptor β1 and c/EBPα remain unaffected by these treatments[7]. Treatment of cultured primary hepatocytes with specific fatty acids, like 18:2 n-6 18:3 (n-3 and n-6), 20:4 n-6 and 20:5 n-3, suppress the mRNAs encoding the lipogenic enzymes and induce AOX mRNA.[7,62]

PUFA-mediated suppression of hepatic levels of mRNAs encoding fatty acid synthase (FAS), stearoyl CoA synthase-1 (SCD-1), L-type pyruvate kinase (L-PK) and the S14 protein (S14) is due to an inhibition of gene transcription.[6–9] For both FAS and S14, this inhibition is rapid and occurs within hours of PUFA ingestion. The molecular basis for this inhibition has been examined in two genes, the L-PK and the S14 genes.[6,7,9]

PUFA-Mediated Suppression of L-Pyruvate Kinase Gene Transcription

L-PK is a glycolytic enzyme that plays a central role in hepatic glucose metabolism.[63] The activity of this hepatic enzyme is subject to complex regulation. Acute control

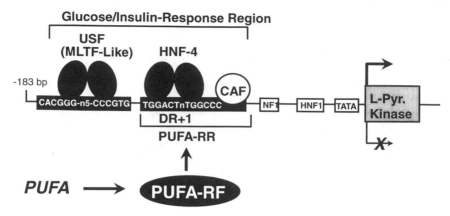

Figure 1. PUFA targets the HNF-4 region of the L-pyruvate kinase promoter. The L-pyruvate kinase (L-PK) promoter has a TATA box (TATA), and binding sites for HNF1 and NF1. L-PK gene transcription is induced by the combined action of glucose and insulin. The target for this control is designated as the Glucose/Insulin Response Region. This region contains binding sites for USF, a c-myc-related factor, hepatic nuclear factor-4 (HNF-4) and a carbohydrate ancillary factor (CAF). The DNA motif for USF is similar to the hexamer repeat (CACGTG) with a 5 nucleotide spacer. The DNA motif binding HNF-4 is similar to a direct repeat (TGGACT) with a one nucleotide spacer, i.e. (DR+1). The target for PUFA control, i.e. the PUFA-regulatory region (PUFA-RR), is the HNF4/CAF portion of the glucose/insulin response region. The factor mediating PUFA action is designated PUFA-regulatory factor (PUFA-RF).

is mediated through allosteric and covalent modification effectors, while chronic control is achieved by regulating the cellular level of $mRNA_{LPK}$ through transcriptional mechanisms. L-PK gene transcription is induced by insulin and dietary glucose and suppressed by hormones that elevate hepatic cAMP levels and dietary PUFA.[63-66]

The L-PK promoter has been cloned and extensively characterized (Figure 1).[63-66] The principal cis-regulatory elements are found within 200 bp of the 5' end of the gene. The proximal elements contain binding sites for a preinitiation complex (TATA-box), and transcription factor, HNF1. The targets for glucose/insulin control are located between -183 and -97 bp and bind 4 transcription factors: a) members of the c-myc transcription factor family, USF (upstream stimulatory factor) or major late transcription factor-like (MLTF-like) bind between -178 and -144 bp; b) HNF4 and an uncharacterized factor called carbohydrate ancillary factor (CAF) bind between -145 and -125 bp, c) NF1 binds between -115 and 97 bp. The USF factors represent the principal target for glucose/insulin control. Factors binding the HNF4/CAF region appear to play a key ancillary role for glucose/insulin control within the contexts of the L-PK promoter.[63-66]

Transfection of primary hepatocytes with L-PKCAT (CAT:chloramphenicol acetyltransferase) fusion genes coupled with deletion and linker scanning analysis has shown that cis-regulatory targets for PUFA inhibition of L-PK gene transcription are located within the region binding HNF-4/CAF.[9] This target of control suggests that because PUFA-regulatory factors (PUFA-RF) do not co-localize with the key elements involved in insulin/glucose control, PUFA may not directly interfere with insulin/glucose activation of L-PK. This is significant because receptors for insulin and glucagon are located within the plasma membrane. PUFA are known to alter plasma membrane phospholipid composition and potentially alter signaling from the plasma membrane.[44-47] This observation suggest that PUFA action on L-PK probably does not involve such interference. Instead, PUFA targets a key ancillary factor that is required for insulin/glucose-mediated activation of L-PK gene transcription.

Efforts to define further the PUFA control of L-PK have shown that inhibitors of A-kinase, C-kinase (staurosporin and H7) or phosphatases (okadaic acid) do not affect the PUFA control of this gene. Moreover, recent reports have suggested that genes harboring HNF-4 binding sites might be targets for the peroxisome proliferator activated receptors (PPAR). PPAR binding to HNF-4 sites might inhibit gene transcription.[67,68] Gel shift studies using in vitro synthesized PPAR and RXR have failed to show binding to the L-PK promoter. Taken together, these studies argue against PUFA control of L-PK by affecting phosphorylation-mediated signaling pathways or by activating the PPAR pathway. The mechanism of PUFA control of L-PK remains under investigation.

PUFA-Mediated Suppression of S14 Gene Transcription

The S14 gene encodes a 17 kd, 4.9 pI protein of unknown function. However, recent studies suggest that the S14 protein might be involved in the control of lipogenic enzymes, like malic enzyme and fatty acid synthase.[11,12] Despite this limited understanding of its function, the S14 protein has proven to be a very useful model to define the molecular basis of hormone and nutrient control of hepatic lipogenic gene transcription. S14 gene transcription is regulated in parallel with other lipogenic enzymes. It is rapidly induced by T_3 and insulin and inhibited by cAMP (starvation and diabetes). Moreover, the PUFA control of S14 shows essentially the same pattern as seen for FAS.[3-7] These features have made the S14 gene a suitable model to define the mechanism of PUFA regulation of gene transcription.

The S14 gene is flanked by 3 upstream cis-regulatory regions that span a 3 kb region (Figure 2).[11,12] The proximal promoter region extending to -300 bp contains elements involved in both constitutive and tissue specific initiation of gene transcription. These ele-

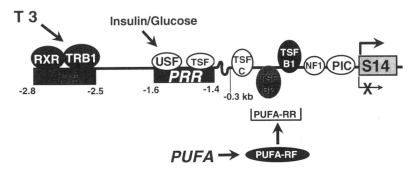

Figure 2. PUFA targets the proximal promoter region of the S14 gene. The key cis-regulatory elements controlling transcription of the S14 gene in rat liver are located within 2.8 kb of the transcription start site. The proximal promoter region extending to -300 bp contains a modified TATA-box at -27 bp that binds a preinitiation complex (PIC), an NF-1 binding site at -48 to -63 bp and binding sites for tissue-specific factors (TSF-B1, TSF-B2 and TSF-C) at -88 to -244 bp, respectively. This region functions to confer hepatic-specific initiation of gene transcription. Two enhancers are located upstream: a pluripotent response region (PRR) and a thyroid hormone response region (TRR). The PRR is located between -1.6 and -1.4 kb and contains cis-acting elements regulated by glucose, insulin and glucocorticoids. The insulin/glucose response element binds upstream stimulatory factors (USF), a factor similar to the one involved in L-PK response to glucose/insulin. The TRR is located between -2.8 and -2.5 kb and contains 3 thyroid hormone response elements (TRE). Each TRE binds thyroid hormone receptor β1 and retinoid X receptor (RXR) as a heterodimer. In rat liver, T_3 (through its nuclear T_3 receptor, TRβ1) is the principal activator of S14 gene transcription. The target for PUFA control in the S14 gene, i.e. the PUFA-regulatory region (PUFA-RR), is the region binding TSF-B1 and TSF-B2 in the proximal promoter region.

ments bind known transcription factors (preinitiation complex and NF1) and unknown tissue-specific factors (TSF-B1 and TSF-B2). Two enhancers are located upstream. One enhancer located between -1.4 and -1.6 kb is a target for additional tissue-specific factors, insulin and dietary carbohydrate control. In adipocytes, this region contains functional response elements for both glucocorticoid and retinoic acid regulation. Because of the multiple regulatory elements within this region, it is designated a pluripotent response region, PRR. A second enhancer is located between -2.5 and -2.8 kb and is known as a thyroid hormone response region (TRR). This region contains 3 thyroid hormone response elements (TRE), each binding thyroid hormone receptors (TRβ1) and RXR as heterodimers. In liver, T_3 binding TRβ1 functions as the principal activator of S14 gene transcription.

T_3 treatment of primary hepatocytes transfected with S14CAT fusion genes leads to a 50-fold induction of CAT activity.[6] Treatment of cells 18:1, n-9 (250 μM) has a marginal effect on T_3 regulation of this gene. However, treatment with 18:2 (n-6), 18:3 (n-3), 18:3 (n-6), 20:4 (n-6) or 20:5 (n-3) leads to a 60 to 80% suppression of CAT activity. This suppressive effect of PUFA is due to suppression of both basal and T_3-stimulated S14CAT activity. In contrast, transfection of cells with RSVCAT (RSV, Rous Sarcoma Virus promoter) was unaffected by these fatty acid treatments indicating the promoter-specificity of the PUFA effect on transcription.

Deletion analysis showed that the principal target for the PUFA control of S14 was the proximal promoter region. The PUFA-regulatory region (PUFA-RR) mapped to sequences between -80 and -220 bp upstream from the S14 gene, a region binding at least 2 tissue-specific factors, TSF-B1 and TSF-B2. This finding was interesting because neither of the two hormone-regulated enhancers located farther upstream was targeted by PUFA-regulatory factor (PUFA-RF). Additional deletion and linker scanning studies suggest that factors binding the -220/-80 bp region are required to regulate both the basal level and T_3-mediated transactivation of S14 gene transcription.

T_3-receptors interact with preinitiation complexes either directly or through co-activators and co-repressors.[69] In the S14 gene, T_3-receptors bind at 3 sites 2.5 to 2.8 kb from the preinitiation complex. T_3 receptor-preinitiation interaction might require factors binding the -220/-80 bp region for stability. We speculate that PUFA-RF interferes with this interaction and leads to inhibition of S14 gene transcription. Further definition of this system requires the isolation and identification of PUFA-RF and the tissue-specific factors binding between -80 and -220 bp.

PUFA Regulation of Peroxisome Proliferator Activated Receptor (PPAR)

Because PPAR are regulated by fatty acids, we speculated that PPAR, might be PUFA-RF.[10,70] The strong peroxisome proliferator, WY14,643, suppressed $mRNA_{S14}$ and $mRNA_{FAS}$ and inhibited S14CAT in primary hepatocytes. In contrast to expectations, deletion studies indicated that the target for both WY14,643 and co-transfected PPAR control was localized to the S14 TRR and not the S14 PUFA-RE. The lack of co-localization of PUFA and WY14,643/PPAR control indicated that PPAR cannot be viewed as the principal mediator of PUFA suppression of S14 gene transcription. Thus, there appear to be at least two mechanisms for fatty acid regulation of gene transcription in the liver: one is PPAR-dependent and one is PPAR-independent.

Additional studies to define the mechanism of PPAR suppression of S14 gene transcription showed that while PPAR did not directly bind the S14 TRR it affected the binding of other transcription factors. Both PPAR and T_3 receptors require RXR for DNA

binding. PPAR interacts with RXR to inhibit TRβ1 binding to the S14 TRR.[70] This observation suggested that in primary hepatocytes RXR might be limiting. To test this possibility, RXR levels were elevated in hepatocytes by co-transfection. These studies showed
that increasing RXR expression essentially abrogated the PPAR-mediated inhibition of
S14 gene transcription. Thus, PPAR can be viewed as an inhibitor of S14 gene transcription and it does so by sequestering RXR.[70]

SUMMARY AND CONCLUSIONS

These studies show that a macronutrient like dietary fat plays an important role in gene
expression. In the cases presented here, dietary fat regulates gene expression leading to
changes in carbohydrate and lipid metabolism. The interesting outcome of these studies is the
finding that the molecular targets for dietary fat action did not converge with the principal targets for hormonal regulation of gene transcription, like hormone receptors. Instead, PUFA-RF
targets elements that play key ancillary roles in gene transcription. This is important because
it shows how PUFA can interfere with hormone regulation of a specific gene without having
generalized effect on overall hormonal control, i.e. PUFA effects are promoter-specific. How
PUFA-RF interferes with gene transcription will require the isolation and characterization of
PUFA-RF along with the tissue-specific factors targeted by PUFA-RF.

A different story emerges when fatty acids activate PPAR. Based on the studies presented here and elsewhere, long chain-highly unsaturated fatty acids (like 20:5,n-3 and
22:6, n-3) or high levels of fat activate PPAR.[15,57-61] PPAR directly activates genes like
AOX, but also inhibits transcription of genes like S14, FAS, apolipoprotein CIII, transferrin.[10,67,68] For S14, the mechanism of inhibition involves sequestration of RXR, a critical
factor for T_3 receptor binding to DNA. Thus, PPAR can have generalized effects on T_3 action or on other nuclear receptors, like vit. D (VDR) and retinoic acid (RAR) receptors,
that require RXR for action. For apolipoprotein CIII and transferrin, PPAR/RXR heterodimers compete for HNF-4 binding sites (DR+1).[67,68] In addition to HNF-4, COUP-TF,
ARP-1 and RXR all bind the DR+1 type motif.[69] These factors are important for tissue-
specific regulation of gene transcription. PPAR can potentially interfere with the transcription of multiple genes through disruption of nuclear receptor signaling leading to
changes in phenotype. Clearly, more studies are required to assess the role PPAR plays in
the fatty acid regulation of gene transcription and its contribution to chronic disease.

Finally, it is clear that dietary fat has the potential to affect gene expression through
multiple pathways. Depending on the gene examined, PUFA might augment or abrogate
gene transcription which leads to specific phenotypic changes altering metabolism, differentiation or cell growth. These effects can be beneficial to the organism, such as the n-3
PUFA-mediated suppression of serum triglycerides[37-40] or detrimental, like the saturated
and n-6 PUFA-mediated promotion of insulin resistance.[23-31] How such effects contribute
to the onset or progression of specific neoplasia is unclear. However, studies in metabolism might provide important clues for this connection.

REFERENCES

1. D.J. McGarry, Lipid metabolism I: Utilization and storage of energy in lipid form, in: Textbook of Biochemistry with Clinical Correlations Devlin, T.M. ed. 3rd ed. p. 387–420 Wiley-Liss, Inc.New York
   (1992).
2. J.I. Gordon, R.H. Duronio, D.A. Rudnick, S.P.Adams, and G.W. Gokel, Protein N- myristoylation. *J. Biol.
   Chem.* 266: 8647 (1991).

3. S.D. Clarke, M.K. Armstrong and D.B. Jump, Nutritional control of rat liver fatty acid synthase and S14 mRNA abundance. *J. Nutr.* 120: 218 (1990).
4. S.D. Clarke, M.K.Armstrong, and D.B. Jump, Dietary polyunsaturated fats uniquely suppress rat liver fatty acid synthase and S14 mRNA content, *J. Nutr.* 120:225 (1990).
5. W.L.Blake and S.D. Clarke, Suppression of hepatic fatty acid synthase and S14 gene transcripiton by dietary polyunsaturated fatty acids, *J. Nutr.* 120:1727 (1990).
6. D.B. Jump, S.D. Clarke, O.A. MacDougald and A. Thelen, Polyunsaturated fatty acids inhibit S14 gene transcription in rat liver and cultured hepatocytes. *Proc. Natl. Acad. Sci. USA* 90: 8454 (1993).
7. D.B. Jump, S.D. Clarke, A. Thelen and M. Liimatta, Coordinate regulation of glycolytic and lipogenic gene expression by polyunsaturated fatty acids. *J. Lipid Res.* 35:1076 (1994).
8. K.T. Landchultz, D.B. Jump, O.A. MacDougald and M.D. Lane, Transcriptional control of the stearoyl-CoA desaturase-1 gene by polyunsaturated fatty acids. *Biochem. Biophys. Res. Comm.* 200: 763 (1994).
9. M. Liimatta, H.C. Towle, S.D. Clarke, and D.B. Jump, Dietary polyunsaturated fatty acids interfere with the insulin/glucose activation of L-type pyruvate kinase, *Molecular Endocrinology* 8:1147 (1994).
10. D.B. Jump, B. Ren, S.D. Clarke and A. Thelen, A., Effects of fatty acids on hepatic gene expression, *Prostaglandins, Leukot. Essent. Fatty Acids* 52:107 (1995).
11. S.D. Clarke and D.B. Jump, Dietary polyunsaturated fatty acids regulaion of gene transcription. *Annu. Rev. Nutr.* 14: 83 (1994).
12. D.B. Jump, S.D. Clarke, A. Thelen, M. Liimatta, B.Ren and M. Badin. Dietary polyunsaturated fatty acid regulation of gene transcription. *Progress in Lipid Research* (In Press)(1996).
13. I. Issemann and S. Green Activation of a member of the steroid hormone receptor superfamily by peroxisome proliferators. *Nature* 347:645 (1990).
14. M. Gottlicher, E. Widmark, Q. Li and J.A. Gustafsson, Fatty acids activate a chimera of the clofibric acid-activated receptor and the glucocorticoid receptor. *Proc. Natl. Acad. Sci. USA* 89: 4653 (1992).
15. J.K. Reddy and N.D. Lalwani, Carcinogenesis by hepatic peroxisome proliferator: evaluation of the risk of hypolipidemic drugs and industrial plasticizers to humans. *CRC Crit. Rev. Toxicol.* 12: 1 (1983).
16. S. Green, Receptor-mediated mechanism of peroxisome proliferators. *Biochem. Pharmacol.* 43: 393 (1992).
17. P. Tontonoz, E. Hu, R.A. Graves, A.I. Budavari and B.M. Spiegelman, mPPARγ2: tissue-specific regulator of an adipocyte enhancer, *Genes Dev.* 8: 1224 (1994).
18. E-Z. Amri, B. Bertrand, G. Ailhaud and P. Grimaldi, Regulation of adipose cell diffeentiation. I. Fatty acids are inducers of aP2 gene expression. *J. Lipid Res.* 32: 1449 (1991).
19. E-Z. Amri, G. Ailhaud, and P. Grimaldi, Regulation of adipose cell differentiation. II. Kinetics of induction of the aP2 gene by fatty acids and modulation by dexamethasone. *J. Lipid Res.* 32: 1457 (1991).
20. E-Z. Amri, G. Ailhaud and P. Grimaldi, Fatty acids as signal transducing molecules: involvement in the differentiation of preadipose to adipose cells. *J. Lipid Res.* 35: 930 (1994)
21. E-Z. Amri, F. Bonino, G. Ailhaud, N.A. Abumrad, and P. Grimaldi, Cloning of a protein that mediates transcription effects of fatty acids in preadipocytes. *J. Biol. Chem.* 270: 2367 (1995).
22. B.M. Forman, P. Tontonoz, J. Chen, R.P. Brun, B.M. Spiegelman and R.M. Evans, 15- Deoxy-$\Delta^{12,14}$-prostaglandin J$_2$ is a ligand for the adipocyte determination factor, PPARγ, *Cell* 83: 803 (1995).
23. E. Ferrannini, E.J. Barrett, S. Bevilacqua, R.A. DeFronzo, Effect of fatty acids on glucose production and utilization in man. *J. Clin. Invest.* 72: 1737 (1983).
24. L.H. Storlien, E.W. Kraegen, D.J. Chisholm, G.L. Ford, D.G. Bruce, and W.S. Pascoe, Fish oil prevents insulin resistance induced by high fat feeding in rats. *Science* 237: 885 (1987).
25. L.H. Storlien, D.E. James, K.M. Burleigh, D.J. Chisholm, E.W. Kraegen, Fat feeding causes widespread in vivo insulin resistance, decreased energy expenditure and obesity in rats. *Am. J. Physiol.* 251: E576 (1986).
26. J.D. McGarry, What if Minkowski has been ageusic? An alternative angle on diabetes. *Science* 258: 766 (1992).
27. G.M. Reaven, Role of insulin resistance in human disease. *Diabetes* 37: 1595 (1988).
28. T.H. Malasanos and P.W. Stacpoole, Biological effects of ω-3 fatty acids in diabetes mellitus. *Diabetes Care* 14: 1160 (1991).
29. L.H. Storlien, A.B. Jenkins, D.J. Chisholm, W.S. Pascoe, S. Khouri and E.W. Kraegen, Influence of diatary fat composition on development of insulin resistance in rats: Relationship to muscle triglyceride and ω-3 fatty acids in muscle phospholipid. *Diabetes* 40: 280 (1991).
30. M. Borkman, L.H. Storlien, D.A. Pan, A.B. Jenkins, D.J. Chisholm and L.V. Campbell, The relation between insulin sensitivity and the fatty-acid composition of skeletal muscle phospholipids, *N.Engl. J. Med.* 328:238 (1993).
31. D.A. Pan, A.J. Hulbert and L.H. Storlien. Dietary fats, membrane phospholipids and obesity. *J. Nutr.* 124: 1555 (1994)

32. H. Hallaq, T.W. Smith and A. Leaf, Modulation of dihydropyridine-sensitive calcium channels in heart cells by fish oil fatty acids. *Proc. Natl. Acad. Sci. USA.* 89: 1760 (1992)

33. H.N. Hodis, W.J. Mack, S.P. Azen, P. Alaupovic, J.M. Pogoda, L. LaBree, L.C. Hemphill, D.M. Kramsch and D.H. Blankenhorn, Triglyceride- and cholesterol-rich lipoproteins have a differential effect on mild/moderate and severe lesion progression as assessed by quantitative coronary angiography in a controlled trial of lovastatin. *Circulation* 90: 42 (1994).

34. W.T. Cave, Dietary n-3 PUFA effects on animal tumorigenesis. *FASEB J.* 5:2160 (1991).

35. C.W. Welsch, Dietary fat, calories and experimental mammary gland tumorigenesis. *Cancer Res.* 52: S2040 (1992).

36. W.C. Willett, Diet and Health: What should we eat? *Science* 264: 532 (1994).

37. W.S. Harris, Fish oils and plasma lipid and lipoprotein metabolism in humans: a critical review. *J. Lipid Res.* 30: 785.

38. B.E. Phillipson, D.W. Rothrock, W.E. Connor, W.S. Harris, D.R. Illingworth, Reduction of plasma lipids, lipoproteins and apolipoproteis by dietary fish oils in patients with hypertriglyceridemia. *N. Engl. J. Med.* 312:1210 (1985).

39. D. Kromhout, E.B. Bosschieter and C.L. Coulander, The inverse relation between fish consumption and 20-years mortality from coronary heart disease. *N. Engl. J. Med.* 312: 1205 (1985).

40. P.J. Nestel, W.E. Connor, M.F. Reardon, S. Connor, S. Wong and R. Boston, Suppression by diets rich in fish oil of very low density lipoprotein production in man. *J. Clin. Invest.* 74: 82 (1984).

41. A.F. Goodridge, Regulation of the activity of acetyl coenzyme A carboxylase by palmitoyl coenzyme A and citrate. *J. Biol. Chem.* 247: 6946 (1972).

42. P.F. Randle, P.B. Garland, C.N. Hales and E.A. Newsholme The glucose-fatty acid cycle. Its role in insulin sensitivity and metabolic disturbances of diabetes mellitus. *Lancet* 1:785 (1963).

43. E.W. Kraegen, P.W. Clarke, A.B. Jenkins, E.A. Daley, D.J. Chisholm and L.H. Storlien, Development of muscle insulin resistance after liver insulin resistance in high-fat- fed rats. *Diabetes* 40: 1397 (1991).

44. M.T. Clandinin, S. Cheema, C.J. Field, M.L. Garg, J. Vendatraman and T.R. Clandinin, Dietary fat: exogenous determination of membrane structure and cell function. *FASEB J.* 5:2761 (1991)

45. A.A. Spector and M.A. York, Membrane lipid composition and cellular function. *J. Lipid Res.* 26: 1015 (1985).

46. C.J. Field, E.A. Ryan, A.B.R. Thomson and M.T. Clandinin., Diet fat composition alters membrane phospholipid composition, insulin binding and glucose metabolism in adipocytes from control and diabetic animals. *J. Biol. Chem.* 265: 11143 (1990).

47. E.M. Dax, J.S. Partilla, M.A. Pilleyro and R.I. Gregerman, Altered glucagon and catecholamine hormone sensitive adenylyl cyclase responsiveness in rat liver membranes induced by manipulation of dietary fatty acid intake. *Endocrinology* 127: 2236 (1990).

48. O. Laneuville, D.K. Breuer, N. Xu, Z.H. Huang, D.A. Gage, J.T. Watson, M. Lagarde, D.L. DeWitt and W.L. Smith, Fatty acid substrate specificities of human prostaglandin-endoperoxide H synthase-1 and -2, *J. Biol. Chem.* 270: 19330 (1995).

49. J.E. Kinsella, K.S. Broughton and J.W. Whelan, Dietary unsaturated fatty acids: interaction and possible needs in relation to eicosanoid synthesis. *J. Nutr. Biochem.* 1:123- 141 (1990).

50. K.S. Broughton, J. Whelan, I. Hardardottir and J.E. Kinsella, Effect of increasing the dietary (n-3) to n-6) polyunsaturated fatty acid ration on murine liver and peritoneal cell fatty acids and eicosanoid formation. *J. Nutr.* 121: 155 (1991).

51. Halliwell, B. and Gutterridge, J.M.C. Free Radicals in Biology and Medicine, Clarendon Press, Oxford. pp.139 (1985).

52. B. Henning, J.H. Diana, M. Toborek, C.J. McClain, Influence of nutrients and cytokines on endothelial cell metabolism. *J. Amer. Coll. Nutr.* 31: 224 (1994).

53. C.K. Sen and L. Packer, Antioxidant and redox regulation of gene transcription. *FASEB J.* 10: 709 (1996).

54. J.K. Reddy and G.P. Mannaerts, Peroxisomal lipid metabolism. *Annu. Rev. Nutr.* 14: 343 (1994).

55. A.B. Lock, A.M. Mitchell and E.R. Elcombe, Biochemical mechanism of induction of hepatic peroxisome proliferation, *Annu. Rev. Pharmacol. Toxicol.* 29:145 (1989).

56. M.K. Hellerstein, J-M. Schwarz and R.A. Neese, Regulation of hepatic de novo lipogenesis in humans. *Annu Rev. Nutr.* 16: 523 (1996).

57. C.E. Neat, M.S. Thomassen and H. Osmundsen, Induction of peroxisomal β-oxidation of rat liver by high-fat diets. *Biochem. J.* 186: 369 (1980).

58. M.S. Thomassen, E.N. Christiansen and K.R. Norum, Characterizaion of the stimulatory effect of high-fat diets on peroxisomal β-oxidation in rat liver. *Biochem. J.* 206:195 (1982).

59. T. Flatmark, A. Nilsson, J. Kvannes, T.S. Eikhom, M.H. Fukami, H. Kryvi and E.N. Christiansen, On the mechanism of induction of enzyme systems for peroxisomal β- oxidation of fatty acids in rat liver by diets rich in partially hydrogenated fish oil. *Biochim. Biophys. Acta* 962: 122 (1988).

60. A. Aarsland, M. Lundquist, B. Borretsen and R.K.Berge, On the effect of peroxisomal β- oxidation and carnitine palmitoyltransferase activity by eicosapentaenoic acid in liver and heart from rats. *Lipids* 25: 546 (1990).

61. G.S. Rambjor, A.I. Walen, S.L. Windsor and W.S. Harris, Eicosapentaenoic acid is primarily respnsible for hypotriglyceridemic effect of fish oil in humans. *Lipids* 31: S45 (1996)

62. B. Ren, A. Thelen and D. Jump Unpublished observation (1996).

63. S. Vaulont and A. Kahn, Transcriptional control of metabolic regulation by carbohydrates. *FASEB J.* 8: 28 (1994).

64. M.O. Bergot, M-J.M. Diaz-Guerra, N. Puzenat, M. Raymondjean and A.Kahn, Cis- regulation of the L-type pyruvate kinase gene promoter by glucose, insulin and cAMP. *Nucleic Acids Res.* 20: 1871 (1992).

65. Z. Liu, K.S.Thompson and H.C. Towle, Carbohydrate regulation of the rat L-type pyruvate kinase gene requires two nuclear factors: LF-A1 and a member of the c-myc family, *J. Biol. Chem.* 268: 12787 (1993).

66. H-M. Shih, Z. Liu and H.C. Towle, Two GACGTG motifs with proper spacing dictate the carbohydrate regulation of hepatic gene transcription. *J. Biol. Chem.* 270: 21991 (1995).

67. R. Hertz, J. Bishara-Shieban and J. Bar-Tana, Mode of action of peroxisome proliferators as hypolipidemic drugs, *J. Biol. Chem.* 270:13470 (1995).

68. R.Hertz, M. Seckbach, M.M. Zakin and J. Bar-Tana, Transcription suppression of the transferrin gene by hypolipidemic peroxisome proliferators. *J. Biol. Chem.* 271: 218 (1996).

69. D.J. Mangelsdorf and R.M. Evans, The RXR heterodimers and orphan receptors. *Cell* 83: 841 (1995).

70. B. Ren, A. Thelen, and D.B. Jump, Peroxisome proliferator-activated receptor-α inhibits hepatic S14 gene transcription. *J. Biol. Chem.* 271 :17167 (1996)

SESSION V

Future Directions and Implications of Research on Dietary Fat and Genetics

Diane F. Birt

Eppley Institute for Research in Cancer
University of Nebraska Medical Center
Omaha, Nebraska 68198–6805

DISCUSSION

The discussion was opened by Dr. Diane Birt with a summary of results from an epidemiological study and an animal investigation which support the interaction of genetics and nutrition in cancer development. Sellers et al.[1] assessed the influence of body fat distribution and family history of breast and ovarian cancer on the risk of postmenopausal breast cancer. Linkage studies of the breast cancer susceptibility gene BRCA-1 on chromosome 17p demonstrate a high association with families with a dual history of premenopausal breast and ovarian cancer. However, breast cancer family history alone is only weakly associated with a specific genetic mutation. The results from Sellers et al.[1] suggested that a waist to hip ratio of >0.906 was associated with a higher risk of breast cancer amongst women with a family history of breast and ovarian cancer (relative risk = 4.8) than in women with a family history of breast cancer alone (relative risk = 2.1) or than women with no family history of breast or ovarian cancer (relative risk = 1.1). Although there were few patients with a family history of both breast and ovarian cancer, these results supported a strong relationship between obesity, or as was emphasized during the discussion, body fat distribution and breast cancer in women with genetic predisposition.

The study by Lu et al.[2] of breast and colon cancer assessed the influence of dietary fat on mammary carcinogenesis in 1-methyl-1-nitrosourea treated Sprague-Dawley rats. These investigators determined that the excess in mammary carcinomas observed in the rats fed high fat diet (24.6%) was entirely due to increases in the yield of carcinomas with wild type ras at codon 12. They measured the GGA->GAA mutation in codon 12 of c-Ha-ras and demonstrated that the number of tumors with this mutation was not different in rats fed low or high fat diet. These studies used a different rat strain and carcinogen than the studies of Dr. Archer presented at this conference; however both approaches indicated that dietary fat may have a preferential influence on increasing the number of tumors with particular mutations.

The second phase of the discussion provided time for several of the session moderators and participants from the conference to comment on future directions: Dr Kenneth Carroll, University of Western Ontario, London, ON, stressed the importance of the role of environment in the causation of human cancer. Clearly, evidence for a role of specific mutations in tumor suppressor genes or ongogenes in human cancer is growing. However, environmental factors continue to be the most important determinant of human cancer risk. Dr. Carroll also spoke in favor of public education campaigns for increasing the public's participation in cancer prevention through healthy lifestyles.

Dr. Laurence Kolonel, University of Hawaii at Manoa, Honolulu, HI, reiterated the need for human studies which demonstrate a role of dietary fat in the modulation of human cancer risk. He suggested that studies of populations with particular cancer susceptibilities may be helpful in seeking human data in support of a role of dietary fat in cancer.

Dr. Steve Zeisel, University of North Carolina at Chapel Hill, indicated that it is clear that lipids are involved in human cancer susceptibility since lipids play such an important role in cellular processes which are critical in transformation and the regulation of proliferation and differentiation pathways. He suggested that people who are not responsive to dietary fat may have masked the impact on the responders in the earlier epidemiological investigations where dietary fat was not observed as a factor in cancer risk. He stressed the need for integrative studies in the area of nutrition and cancer.

Dr. Robert Clarke, Georgetown University Medical Center, Washington, DC, discussed the need to focus studies on dietary fat and cancer genetics on the time in life when exposure is important and on the mix of dietary fat to which humans are exposed. He commented on the need for better animal models of human cancer which will mimic the endocrinology of the human disease process.

Dr. Leonard Augenlicht, Albert Einstein Cancer Center, Bronx, NY, indicated the need for epidemiological studies which assess cancer genetics using molecular biology techniques. He commented that since at the cellular level cancer is a rare event, it is difficult to study the impact of dietary fat on the development of cancer in vivo. We cannot identify the initiated or early transformed cell in vivo and hence we cannot assess differential lipid metabolism at these early stages when effects are most important in influencing the progression of the cell to a tumor. Furthermore, subtle genetic variation in the population which determines the relative uptake and utilization of lipids must be understood and incorporated into epidemiological studies in order to be able to determine the relationship of fat consumption to eventual differences in cancer incidence.

The final phase of the discussion was opened to the audience. There was considerable interest in the observation that the distribution of body fat, and thus body shape, has been associated with breast and colon cancer risk in humans. It is possible that elevated risk of these cancers in individuals with higher abdominal fat is related to genetic factors which control fat distribution. For example, genetics may influence hormonal or enzymatic regulation of lipid deposition and/or mobilization. Although we believe that body shape is genetically determined, the particular genes involved have not been identified. This could certainly be a fertile area for future research endeavors.

A related topic of discussion was the observation that body fat from separate anatomic locations within an individual is mobilized at different rates and under different conditions. This observation suggests differential hormonal or enzymatic regulation of lipid metabolism in tissues within an individual. This was suggested by the group as a fertile model for future research on factors which regulate lipid metabolism and signalling.

The discussion was closed by Dr. Ritva Butrum thanking all participants for a stimulating and lively meeting.

REFERENCES

1. Sellers TA, Gapstur SM, Potter JD, Kushi LH, Bostick RM, Folsom AR. Association of body fat distribution and family histories of breast and ovarian cancer with risk of postmenopausal breast cancer. Am J Epidemiol 1993;138:799–803.
2. Lu J, Jiang C, Fontaine S, Thompson HJ. *ras* may mediate mammary cancer promotion by high fat. Nutr Cancer 1995;23:283–290.

ABSTRACTS

Poster Abstract # 1

Proliferating nuclear cell antigen (PCNA) in the colonic epithelium as affected by the type and level of dietary lipids and a high calcium diet . RP Bird, P Kumarathasan and L Lafave Department of Foods and Nutrition, University of Manitoba, Winnipeg, MB R3T 2N2

Consumption of dietary lipids varying in fatty acid composition has been implicated in the etiology and prevention of a variety of chronic illnesses including colon cancer. It is well established that fatty acid composition of a high fat diet is an important determining factor in exerting a tumor enhancing or an inhibitory effect. A variety of mechanism(s) has been proposed to explain how different lipids may modulate tumor growth. It is proposed that a tumor promoting diet would increase and a tumor inhibitory diet would decrease colonic cell proliferation. In the present study we assessed the number of cycling cells by quantifying the cells exhibiting PCNA among the colons of rats fed 5% or 23.4% corn oil, canola oil, olive oil and beef tallow diets. A group of rats were fed high fat diets along with 2% calcium (4X the normal level). All diets were based on AIN-76A diet modified to contain high fat diets. Male Sprague Dawley rats (90-100g) were fed the test diets for four weeks. The PCNA labelling index (PCNA-LI) were assessed in the distal and mid colon.The PCNA-LI were lower in the distal colon than their counterparts in the mid colon. All groups fed 5% fat had lower PCNA-LI than those fed high fat diets (p< 0.05). The PCNA-LI among the low fat groups ranged from, distal colon 5-9 and mid colon 6-11. The PCNA-LI among the high fat groups ranged from distal colon 14-33 and mid colon 10-30. Among the high fat groups the lowest PCNA-LI were noted for the corn oil (distal 14, mid 10) and highest in the beef tallow group (distal 33, mid 30). Additional calcium in the diet further increased the PCNA-LI among the corn oil, canola oil and olive oil groups and decreased it in the beef tallow group. The highest increment in the PCNA-LI was noted for the high calcium-corn oil group when compared to its normal calcium counterpart (37,35 vs 14,10). This study demonstrated that A) a high corn oil and a high beef tallow tumor promotary diets did not alter the proliferative responses in the colon in a similar manner;B) all high fat diets increased the proliferative status of the colonic epithelium; C) distal colon responded differently to dietary lipids than mid colon and D) fatty acid composition of dietary lipids was an important variable in modulating the PCNA-LI in the colon, however, only in the high fat environment and was influenced by a high calcium diet. Particulate and soluble protein kinase-C activity did not correlate with the PCNA- LI in all groups. (Supported by the American Inst. Cancer Research and the Canadian Cancer Research Society Inc.)

Poster Abstract # 2

FOOD INTAKE REGULATORY MECHANISMS IN CANCER ANOREXIA: INVOLVEMENT OF
VMN NEUROTRANSMITTERS DOPAMINE (DA) AND SEROTONIN (5HT)

Vladimir Blaha, MD,* Zhong-jin Yang, MD, Michael Meguid, MD, PhD, Alessandro Laviano, MD,
Zdenek Zadak, MD,* and Filippo Rossi-Fanelli, MD. Surgical Metabolism & Nutrition Laboratory,
Department of Surgery, University Hospital, Syracuse, NY; and *Department of Metabolic Care and
Gerontology, Charles University, Hradec Kralove, Czech Republic.

Hypothalamic neurotransmitters have been candidates as anorectic factors in cancer. Among them,
serotonin is recognized as promoting satiety, acting mainly in paraventricular (PVN) and ventromedial
(VMN) nuclei. Although an increase in dopamine release has been postulated in tumor anorexia,
anorexigenic effects of IL-1α were not mediated through dopamine activity in the lateral hypothalamic
area (LHA). Based on recent reports that low dopamine in VMN is associated with post-prandial
satiety, we investigated whether the mechanism of cancer anorexia is mediated by combined high
serotonin and low dopamine in VMN. **METHODS**: A microdialysis catheter was surgically placed via
stereotactic technique into VMN of methylcholanthrene sarcoma-bearing (TB) rats and non-TB (NTB)
controls under general anesthesia. When TB rats manifested anorexia, in vivo VMN serotonin and
dopamine were measured using HPLC at baseline, in response to food, and after feeding. Thereafter,
TB rats had tumor removed and VMN microdialysis performed 7 days later. Intergroup data compared
using Student's t-test; sig: $p < 0.05$. **RESULTS**: At baseline, VMN-5HT in TB vs overnight fasted NTB
rats was increased and VMN-DA was reduced. When food was offered, intake in TB vs NTB rats was
significantly lower. During eating, VMN-5HT rose significantly in TB vs NTB rats and remained
elevated after cessation of eating while VMN-DA was significantly lower vs NTB, remaining lower after
eating. After tumor removal, food intake increased but did not normalize. VMN-5HT normalized but
DA remained depressed at baseline, during, and post-eating. **CONCLUSION**: In TB rats, anorexia is
associated with low VMN-DA and high VMN-5HT. After tumor removal, VMN-5HT normalized while
VMN-DA remained low, accounting for non-normalization of food intake. Thus, we hypothesize that
a dynamic interaction of VMN dopamine and serotonin are involved in cancer anorexia.

Poster Abstract # 3

INTERLEUKIN-1 AND SEROTONIN IN THE HYPOTHALAMIC VENTROMEDIAL NUCLEUS (VMN) CONTRIBUTE TO ANOREXIA IN THE TUMOR BEARING RAT.

Michael M. Meguid, MD, PhD, Alessandro Laviano, MD, Zhong-Jin Yang, MD, and Vladimir Blaha, MD. Surgical Metabolism and Nutrition Laboratory, Department of Surgery, University Hospital, SUNY Health Science Center at Syracuse, NY.

Under normal conditions, brain interleukin-1 (IL-1) and serotonin (5-HT) induce satiety, while during tumor growth, increased VMN activity contributes cancer anorexia. Since IL-1 and 5-HT act via the VMN, we hypothesized that intra-VMN IL-1 and 5-HT are key in mediating cancer anorexia. To test this hypothesis, the effects on food intake of stereotactically located intra-VMN microinjections of the IL-1 receptor antagonist (IL-1ra; expt.1) or the 5-HT antagonist mianserin (MIA, expt.2) were examined in anorectic TB (ATB) rats inoculated with methylcholanthrene sarcoma cells. **METHODS**: ATB Fischer rats (~300g) were studied; controls were non-TB (NTB). Rats had chow and water throughout the study, and food intake was measured daily. Anorexia occurred when TB rats' daily food intake was 1 g less than the mean of NTB rats for 3 consecutive days. Expt.#1: when TB rats became anorectic, they and their NTB controls were injected bilaterally in the VMN with IL-1ra (25 ng in 250 nl of vehicle; TB-IL-1ra, n=6; NTB-IL-1ra, n=7) or the vehicle (TB-V, n=6; NTB-V, n=7). Expt.#2: when anorexia developed, TB and NTB rats received bilaterally in the VMN MIA (200 mMol in 200 nl of normal saline; TB-MIA, n=6; NTB-MIA, n=7) or normal saline (TB-NS, n=6; NTB-NS, n=6). Data were analyzed using ANOVA. **RESULTS**: Expt #1: Food intake improved significantly only in TB-IL-1ra rats, while in TB-V rats food intake progressively declined. Expt #2: Intra-VMN microinjection of MIA improved significantly food intake in TB-MIA rats, while no effect was observed in TB-NS or NTB control rats. **CONCLUSION**: Data show that intra-VMN IL-1 and 5-HT are involved in the pathogenesis of cancer anorexia, by possibly acting synergically.

Carotene: Pro- and Anti-Oxidant Reaction Mechanisms and Interactions with Vitamins E and C

R. Edge, D. J. McGarvey and T. G. Truscott

Chemistry Department, Keele University, Staffs. ST5 5BG, UK

Dietary carotenoids and vitamins E and C are well known anti-oxidants. However, β-carotene also acts as a pro-oxidant at high oxygen pressure [1]. In anti-oxidant processes a damaging oxy-radical ($RO_2^{\cdot}$) reacts with an anti-oxidant such as vitamin E (EOH): $RO_2^{\cdot} + EOH \rightarrow RO_2H + EO^{\cdot}$ via electron transfer. For the protection to be maintained the vitamin E must be regenerated from $EO^{\cdot}$. The roles of the anti-oxidants, including the carotenoids, depend on their redox potentials. Using pulsed techniques we have shown that the following scheme can arise:

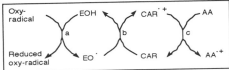

where β-carotene repairs vitamin E and the carotene radical itself is repaired by ascorbic acid (AA). Consistent with this mechanism we observed lymphoid cell protection by carotenoids [2] which was very enhanced by the addition of vitamins E and C. Smokers have low levels of ascorbic acid so that $CAR^{\cdot+}$ may not be repaired without vitamin C supplementation and this could be related to some of the disappointing carotene anti-cancer trials [3].

The pro-oxidant behaviour of carotenoids could also be of importance and we [4] have speculated a molecular mechanism for the switch from anti- to pro-oxidant behaviour as the $[O_2]$ is increased. The key step is that carbon centred radicals ($R^{\cdot}$) react with carotene to give $CAR^{\cdot+}$ leading to an anti-oxidant effect, but react with O_2 leading to pro-oxidant cycling processes via reactions such as:

$$RO_2^{\cdot} + CAR \rightarrow R^{\cdot} + CARO_2 \rightarrow \text{lipid peroxidation}$$

$$R^{\cdot} + O_2 \rightarrow RO_2^{\cdot}$$

Acknowledgements: We thank AICR for support and Drs. Land and Böhm for collaboration.

1.   G. W. Burton and K. U. Ingold, *Science*, **224**, 569-573 (1984).

2.   F. Böhm, J. H. Tinkler and T. G. Truscott, *Nature Medicine*, **1**, 98-99 (1995).

3.   C. H. Hennekens *et al.*, *New Eng. J. Med.*, **334**, 1145-1149 (1996).

4.   T. G. Truscott, *J. Photochem. Photobiol. B: Biology*, In Press (1996)

Poster Abstract # 5

Inhibitory Effect of Curcumin on Tumor Initiation and Promotion Stages in Mouse Skin Tumorigenesis.

P Limtrakul[1], A Apisariyakul[2], O Namwong[2] and FW. Dunn[1].

[1]Department of Biochemistry and [2]Department of pharmacology, Faculty of Medicine, Chiang Mai University, Chiang Mai, Thailand.

Curcumin, the yellow pigment, that is obtained from rhizomes of the plant curcuma longa Linn, is commonly used as a spice and food coloring agent normally eaten by humans. The purpose of the present investigation is to study the effect of curcumin on 7,12-dimethylbenz-(a)anthracene (DMBA)-initiated and 12,0-tetradecanoylphorbol-13-acetate (TPA) promoted skin tumor formation in male Swiss Albino mice. In a two-stage skin tumorigenesis model, single topical application of 390 nmol DMBA to the backs of mice followed a week later by promotion with 4 n mol TPA twice weekly for 15 weeks resulted in the formation of 7.5 skin tumors per mouse, and 90% of the mice had tumors. In a parallel group of mice, which the animals were treated with 10 or 30 µmol curcumin 5 min before the topical application of 390 nmol DMBA, the number of tumors per mouse was decreased by 36 or 57% respectively. The percentage of tumor-bearing mice was decreased by 27 or 17% and the relative tumor volume per mouse was decreased by 69 or 87% respectively. In an additional study, topical application of 10 µmol curcumin at 5 min or one day prior to each application of 4 n mol TPA twice weekly for 15 weeks decreased the number of tumors per mouse by 91 or 50% and decreased the percentage of tumor-bearing mice by 82 or 17%, respectively. The relative tumor volume per mouse was decreased by 97 or 68%. These results indicate that curcumin treatment significantly inhibit both tumor initiation (first stage) and promotion (second stage) in the two-stage mouse skin tumorigenesis. Furthermore, the inhibitory effect of curcumin on the tumor initiation stage was less effective than the promotion step in mouse skin tumorigenesis. To further elucidate the mechanism of antineoplastic action of curcumin, its effect on cellular oncogenes is now under investigation.

Poster Abstract # 6

The Effects at 2 Years of a Low-Fat High-Carbohydrate Diet on Radiological Features of the Breast: Results from a Randomized Trial. NF Boyd, C Greenberg, G Lockwood, L Little, L Martin, J Byng, M Yaffe, D Tritchler for the Canadian Diet and Breast Cancer Prevention Study Group. The Division of Epidemiology and Statistics, Ontario Cancer Institute, Division of Preventive Oncology, Ontario Cancer Treatment and Research Foundation, and the Reichman Research Institute, Sunnybrook Health Sciences Centre, Toronto, Canada.

Extensive areas of radiologically dense breast tissue seen on mammography are associated with an increased risk of breast cancer. The purpose of the present study was to determine whether the adoption of low-fat high-carbohydrate diet for two years would reduce breast density. Women with radiological densities in more than 50% of the breast area on mammography were recruited and randomly allocated to an intervention group taught to reduce intake of dietary fat (mean 21% of calories) and increase complex carbohydrate (mean 60% of calories) or to a control group (mean fat 32% and carbohydrate 50% of calories). Mammograms from 817 subjects taken at baseline and compared with those taken 2 years after randomization using a quantitative image analysis system, without knowledge of the dietary group of the subjects or of the sequence in which pairs of images had been taken.

After 2 years, the total area of the breast was reduced by an average of 246.2 mm^2 (2.5%) in the intervention group, and an average of 9.2 mm^2 (<0.1%) in controls (p=0.01). The area of density was reduced by 376.6 mm^2 (6%) in the intervention group and by an average of 153.1 mm^2 (2.5%) in controls (p=0.02). Weight loss was associated with a reduction in breast area and the effect of the intervention on breast area was no longer statistically significant after weight change was taken into account. Greater age at entry to the trial, weight loss, and becoming postmenopausal were all associated with statistically significant reductions in the area of density on the mammogram at two years. The effect of the intervention on area of density remained statistically significant after controlling for all of these influences. These results show that after 2 years a low-fat high-carbohydrate diet reduced the area of mammographic density, a radiographic feature of the breast that is a risk factor for breast cancer. Longer observation of a larger number of subjects will be required to determine whether these effects are associated with changes in risk of breast cancer.

Poster Abstract # 7

INDUCTION OF HETEROCYCLIC AMINE METABOLISM IN RATS BY DIETS CONTAINING WELL-BROWNED BEEF AND A HIGH FAT CONTENT

B.C. Pence, C.-L. Shen, M. Landers, and L.P. Walsh
Dept. of Pathology, Texas Tech University Health Sciences Center, Lubbock, TX

Epidemiological studies have linked the consumption of red meat, the consumption of highly browned meats containing high levels of heterocyclic amines (HCA), and the inducibility of cytochrome P4501A2 (CYP1A2) and N-acetyltransferase (NAT2) which metabolize HCA, to increased risk of colorectal cancer or polyps. The present study was undertaken to examine the effects of feeding beef containing low and high levels of HCA, in the context of a low or high beef tallow diet, on the inducibility of CYP1A2 and NAT1 and NAT2 in liver and colon of Sprague-Dawley rats. Very lean beef was cooked in a variety of ways at different temperatures and the levels of the HCA (DiMeIQx, MeIQx, and PhIP) were measured by HPLC. Cooking beef to 65° only to remove water and pink color in a stainless steel electric skillet, produced almost no HCA during cooking, where as heavy browning of beef patties in an iron skillet to an internal temperature of 85° resulted in a mean concentration of 2.10 ng DiMeIQx, 12.12 ng MeIQx, and 71.39 ng PhIP per g of cooked beef patty. Beef cooked in these two methods was then mixed with an AIN-76A diet base providing all other nutrients isocalorically, and beef tallow was added to either 5 or 20% by weight. These diets were fed for 6 weeks and the animals' livers and colons were excised for measurement of CYP1A2, NAT1, and NAT2 activities. Diets containing high HCA and 20% tallow induced the highest CYP1A2 activity in liver with an apparent synergistic effect between HCA and fat. Similarly for NAT1 and NAT2 in liver, fat level seemed to be a stronger inducer than HCA, although not statistically significant. Colon enzymes were not responsive to dietary induction. These results indicate that the fat content of the diet is as important as the HCA content in the induction of the enzymes responsible for metabolism of cooked meat carcinogens. (Supported by the American Institute for Cancer Research)

Poster Abstract # 8

Effects of Palm Oil Tocotrienols, Flavonoids and Tamoxifen on Proliferation of MDA-MB-435 and MCF-7 Human Breast Cancer Cells

Guthrie, N.[a], Gapor, A.[b], Chambers, A.F.[c] and Carroll K.K.[a]   Departments of Biochemistry[a] & Oncology[c], University of Western Ontario, London, ON, Canada, N6A 5C1 and Palm Oil Research Institute of Malaysia[b], Kuala Lumpur, Malaysia.

Previous studies have provided evidence for anti-cancer properties of tocotrienols from palm oil (Nesaretnam et al., Nutr. Res. 12:63-75, 1992) and of flavonoids from various sources (Middleton & Kandaswami, Food Technol. 48(11):115-119, 1994). Experiments in our laboratory have shown that palm oil tocotrienols inhibit proliferation of MDA-MB-435 estrogen receptor-negative (ER-) as well as MCF-7 estrogen receptor-positive (ER+) human breast cancer cells at IC_{50}s of 30-180 μg/mL in ER- cells & 2-125 μg/mL in ER+ cells, whereas α-tocopherol was ineffective. These cells were also inhibited by flavonoids, including genistein, hesperitin, naringenin and quercetin, at IC_{50}s of 18-140 μg/mL in ER- cells and 2-18 μg/mL in ER+ cells.

In the present studies, we tested apigenin (from various plants) and two citrus flavonoids, tangeretin and nobiletin (from tangerines), alone and in 1:1 combinations with α-tocopherol or the individual tocotrienols, with or without tamoxifen, in both MDA-MB-435 and MCF-7 cells. Apigenin inhibited both cells lines at similar concentrations (IC_{50}s-2.4 & 3 μg/mL for ER- cells and ER+ cells respectively). Tangeretin and nobiletin inhibited these cells more effectively than other flavonoids tested to date, with an IC_{50} of 0.5 μg/mL for both compounds in the ER- cells and IC_{50}s of 0.4 & 0.8 μg/mL respectively in the ER+ cells. When the flavonoids were tested in 1:1 combinations with tocotrienols, tangeretin and γ-tocotrienol gave the lowest IC_{50}s (0.05 μg/mL in ER- cells and 0.02 μg/mL in ER+ cells). The addition of tamoxifen gave a lower IC_{50} of 0.01 μg/mL in ER- cells but did not alter the IC_{50} in ER+ cells. Tamoxifen also decreased the IC_{50} for nobiletin and δ-tocotrienol from 0.8 to 0.001 μg/mL in ER+ cells. These results suggest that tocotrienols and citrus flavonoids merit further investigation as possible agents in the treatment of breast cancer. (Supported by the Palm Oil Research & Development Board of Malaysia).

Poster Abstract # 9

Immunoprevention of Fat-Modulated Responses to UV-Induced Tumors

H. S. Black, G. Okotie-Eboh, and J. Gerguis

Veterans Affairs Medical Center and Baylor College of Medicine, Houston, TX

Previous studies demonstrated that high levels of dietary fat exacerbate UV-carcinogenic expression. Further, this dietary effect occurs primarily at the post-initiation, or promotion, stage of carcinogenesis, a stage thought to be modulated immunologically. Indeed, we have shown that level of dietary fat has a marked effect upon delayed type hypersensitivity; upon I-J$^+$ T-cell populations; and upon rejection of transplanted tumors. The latter two effects occurred only after a tumor initiating dose of UV radiation had been administered. Thus, level of dietary fat modulates immunoresponsiveness in UV-irradiated animals and may account for the exacerbation of carcinogenic expression by high dietary fat. Here, we explored the preventive potential of fat-modulated immune responses. HRA.HRII-c/+/Skh hairless mice were fed isocaloric diets containing high (12%, w/w) levels of corn oil; irradiated 5 days/wk (1.0 J/cm^2/day) for 11 wk with filtered FS-40 sunlamps; immediately afterwhich some animals were fed diets containing low (0.75%) corn oil. A soluble cell-free fraction (equivalent to 2.5 x 10^6 cells / 0.1 ml; from cultured T-14 cells derived from UV-induced squamous cell carcinomas) was subcutaneously injected at axillae and inguen of one group of animals fed the high-fat diet. Animals were treated weekly for three wk post-UV. At the fourth wk, all animals receiving the high-fat diet were challenged with 1.7 x 10^6 T-14 cells injected subcutaneously at both flanks. At day 21 post-challenge, tumor volumes in low-fat and immunized, high-fat groups were 0 vs 593 mm^3 for the high-fat group (P<0.007). Subsequently, pre-immunization, i.e., during the first 3 wk of UV, was found to be most effective for tumor rejection and that such treatment increased (P<0.03) the latent period of UV-induced primary tumors. These data suggest that fat-modulated carcinogenesis can, itself, be regulated immunologically. Supported by AICR grant 92B01.

Poster Abstract # 10

Effects of Folate Deficiency on the Development of Leukemia in Friend Virus-infected Mice.

Mark J. Koury and Donald W. Horne, Departments of Internal Medicine and Biochemistry, Vanderbilt University and Nashville Veterans Administration Medical Centers, Nashville, Tennessee.

Nutritional deficiency of folate causes megaloblastic anemia, in which blood cell production is decreased because hematopoietic cells undergo programmed cell death (apoptosis) in response to DNA damage. Megaloblastic anemia can be reversed by a folate-rich diet or supplemental folic acid, but cells that survive with damaged DNA can undergo malignant transformation into leukemia. In some previously reported studies, folate deficiency retarded tumor growth while in others it increased the incidence of tumor development. We examined the effects of folate deficiency on leukemia development in mice infected with Friend leukemia virus. Mice were fed an amino acid-based, folate-free diet for one or two months around the time of infection before being switched to the control diet (the folate-free diet with 2 mg folic acid per kg). Control mice were fed the control diet throughout. After a transient erythroblastosis, the mice entered a latent phase during which they appeared hematologically normal. The latent phase lasted from two months to greater than nine months. The latent phase ended when the mice developed overt leukemia with massive spleen and liver involvement. In the short-term folate deficiency delayed leukemogenesis but in the long-term it resulted in accelerated leukemogenesis. Thus, our experiments provide an explanation for the seemingly paradoxical previous reports of folate deficiency increasing or decreasing malignant tumor development. We believe that the delayed onset of leukemia results when cells with genetic alterations and proliferative potentials to become leukemic clones are killed by apoptosis in animals that are on the folate-free diet. After these mice have been switched to the control diet, these potentially leukemic cells develop, like their counterparts in control mice, into leukemic clones. However, the hematopoietic cells that can become leukemic in the mice that experienced folate deficiency are either greater in number or have more leukemic potential than their counterparts in control mice.

Poster Abstract # 11

Helicobacter pylori infection and Gastric Cancer in Mexico.

Lizbeth López-Carrillo, Dr. P.H.[1], Cielo Fernández-Ortega, M. Sc.[1], Guillermo Robles, Ph.D.[2]

Ramón A. Rascón, M. Sc.[1]

[1] Mexico Institute of Public Health, [2] Mexico Institute of Nutrition.

The International Agency for Research on Cancer (IARC) of the WHO has recognized a cause-effect relationship between Helicobacter pylori (Hp) infection and stomach cancer of such magnitude that the presence of this infection increases the risk of cancer approximately four times. Stomach cancer is currently the second cause of mortality due to malignant neoplasms in Mexico City

The association between Hp infection, dietary factors and stomach cancer incidence is being evaluated through an ongoing clinical-based case-control study, in three geographic areas of Mexico.

These preliminary results, based on 109 gastric cancer patients and 177 hospital controls in Mexico City, show that Hp infection was present in 87.2% of the cases, compared with 82.5% of the controls. The odds ratio of having stomach cancer if infected with Hp was 1.44 $IC_{95\%}$ 0.7-2.8. In addition, it was calculated that with eradication of Hp infection in the general population, stomach cancer incidence would decrease by at least 26.6%.

An improvement of the actual sanitary conditions along with the development of an effective vaccine for Hp infection and the existence of increasingly effective treatments to eradicate the bacteria are the necessary next step for populational prevention and control of gastric cancer. The role of the dietary factors in the presence of Hp infection requires to be evaluated.

Poster Abstract # 12

DIETARY REGULATION OF GENE EXPRESSION AND MAMMARY TUMOR DEVELOPMENT IN MICE. Nurul H. Sarkar and Jinwen Chen. Medical College of Georgia, Augusta, GA. 30912, USA.

Our previous studies have shown that a low calorie (LC; 10 kcal/day/mouse) diet reduces the incidence of mouse mammary tumor virus (MMTV)-induced mammary tumors in mice, MMTV expression , and prolactin secretion. Since the amplification of a variety of oncogenes have been implicated in human breast cancer, we have investigated whether a LC diet would reduce the expression of an oncogene(s) in an appropriate animal model, and, if so, study the underlying mechanism and demonstrate that a LC diet also reduces the incidence of tumors in that model. We fed two groups of 4-5 week old transgenic mice carrying the oncogene *int*-1 with a LC diet and a high calorie (HC; 16 kcal/day/mouse) diet. Groups of tumor-free mice fed LC and HC diets were sacrificed at different time points. Their fourth pair of mammary glands were isolated and RNA extracted. Groups of animals were also kept for tumor development. Our results show that 1) the level of *int*-1 RNA in the mammary glands is significantly reduced in those mice that were fed with a LC diet for a period of 3 months or more as compared to those mice that were fed a HC diet, and 2) the incidence of mammary tumors in the LC group is 50% lower than the HC diet fed mice. The kinetics of tumor appearance in the two groups of mice were also found to be different: a high percentage of LC diet-fed mice remained tumor-free for a much longer period of time than those mice fed a HC diet. The effects of a LC diet on hormones and growth factors that are known to play roles in mammary tumorigenesis in both experimental animals and humans, are being investigated.

Poster Abstract # 13

Indole-3-carbinol abrogates the estrogen enhancement

of the expression of oncogenes of papillomaviruses

Fang Yuan and <u>Karen Auborn,</u> Department of Otolaryngology, Long Island Jewish Medical Center, The

Long Island Campus of Albert Einstein College of Medicine, New Hyde Park, NY 11040.

Indole-3-carbinol (I3C) is an active chemopreventive compound present in cruciferous vegetables (cabbage, broccoli, cauliflower and brussel sprouts). Its anticancer properties in breast tissue are attributed to induction of 2-hydroxylation of estradiol, and 2-hydroxy. Most cervical cancers have a papillomavirus (PV) component and a hormonal component. PV oncoproteins inactivate the tumor suppressers, RB and p53, and immortalize cells. Sex hormones augment the PV transformation of cells. The potential exists that I3C can abrogate the hormone enhancement. Our studies show that PV increases 16α hydroxylation of estradiol. Estradiol and 16α-hydroxyestone promote proliferation, including anchorage independent growth, of the PV infected cells. I3C is efficacious in preventing benign PV tumors in mice and humans. In current studies, we show that estradiol increases expression of the viral oncogenes in CaSki cell line derived from a cervical cancer. Increase in viral RNA occur as early as four hours after the addition of estradiol. I3C or 2-hydroxyestrone abrogate the enhanced expression of viral oncogenes. Like estradiol, progesterone increases expression of the viral oncogenes but this increase is not detected until 7-10 days. The progesterone effect depends on the progesterone receptor because it is abrogated by RU486 and competition with nuclease resistent oligomers containing the progesterone receptor responsive DNA binding motif. The interplay of the sex hormones and induction of 2-hydroxylation of estradiol by I3C are being futher investigated in oncogene expression.

This work was supported by Grant #95A111 from The American Institute for Cancer Research.

Poster Abstract # 14

Curcumin inhibits the activation of NFκB, and production of TNFα, IL-1β and nitric oxide.

Marion Man-Ying Chan[1,2], Hsing-I Huang[2] and Dunne Fong[2]

[1] Department of Biomedical Science, Pennsylvania College of Podiatric Medicine, 8th at Race Street, Philadelphia, PA

[2] Department of Biological Sciences, Rutgers, The State University of New Jersey, Piscataway, NJ

Curcumin, contained in the rhizome of the plant *Curcuma longa* Linn, is a naturally occurring phytochemical that is used widely in Asia for the treatment of inflammation. It has also been shown to prevent colon and skin cancers in rodents. Chronic inflammation may lead to the development of cancer and some anti-inflammatory agents are cancer preventive. Therefore, to better understand the mechanism of its action, our laboratory have studied the effect of curcumin on a sequence of cellular events that potentiate inflammation.

We have found that, at 5 μM, curcumin inhibits the activation of nuclear factor kappa B (NFκB), a free radical-activated transcription factor that participates in the gene expression of many inflammation mediators, including tumor necrosis factor-α (TNFα) and inducible nitric oxide synthase (iNOS). Correspondingly, we found that curcumin reduces TNFα production by 57%. TNFα induces the production of interleukin-1β (IL-1β), another pro-inflammatory factor with similar biological function. Curcumin reduces its production by 69%.

IL-1β and TNFα induce the production of nitric oxide synthase. The product of this enzyme, nitric oxide, is a free radical with tumorigenic and damaging potentials. It may react rapidly with superoxides to form potent N-nitrosating agents such as peroxynitirite and nitrite. At 1 - 10 μM, curcumin inhibits nitric oxide activity by 37 - 46%, as indicated by production of nitrite, a stable metabolite of nitric oxide.

TNFα has many biological functions. One of which, with respect to tumorigenesis, is lysis of tumor cells. We have found that curcumin, at 50 µM, reduces the lytic activity of 100-500 ng/mL of recombinant TNF by 30-40%.

Base on these observations, we can speculate that curcumin may prevent tumorigenesis by suppressing inflammation, thus reducing the associated free radical formation and cellular proliferation. It is action, most likely, is not through removal of transformed cells.

Poster Abstract # 15

**MODULATION OF *RAS* EXPRESSION IN HUMAN MALIGNANT CELLS BY
DIETARY SUPPLEMENTS.** R.J. Hohl and K. Lewis-Tibesar., University of Iowa College of
Medicine, Iowa City, IA.

RAS mutations that promote excessive cell proliferation are present in approximately 30 % of
human malignancies. The function of the growth-promoting RAS protein is dependent upon its
attachment to farnesyl pyrophosphate which is an intermediate in the cholesterol biosynthetic
pathway. Farnesylation is important for RAS to be localized to the inner surface of the cell
membrane and this localization is critical for RAS to exert growth-promoting activities. The n-3
polyunsaturated fatty acids have been shown to inhibit the growth of a variety of malignant cell
types. We have investigated the hypothesis that dietary supplements, such as the n-3
polyunsaturated fatty acids, have growth-inhibitory effects that are due to decreased RAS activity
as a consequence of either decreased RAS farnesylation or interference with RAS membrane
localization. Human-derived lymphoid leukemia cells, RPMI-8402, were incubated in tissue
culture with up to 40 micromolar docosahexanoic or linoleic acid for 24 to 72 hours. Cell
viability, assessed as trypan blue exclusion, was greater than 80% for all conditions. New DNA
synthesis, assessed as radiothymidine incorporation into precipitable DNA, markedly decreased to
30 - 50% control values with docosahexanoic, but not linoleic, fatty acid supplementation. RAS
levels, measured in Western analysis, did not change with supplementation of cells with either
fatty acid. RAS localization to cytosolic and membrane fractions was also not altered by fatty
acid supplementation. These findings contrast with control incubations that showed perillyl
alcohol, a naturally occurring monoterpene, to decrease RAS levels. Furthermore, lovastatin, an
inhibitor of hydroxymethylglutaryl coenzyme A (HMG CoA) reductase and therefore farnesyl
pyrophosphate synthesis, decreased RAS farnesylation and resulted in the accumulation of RAS in
the cytosolic compartment. We conclude that the diminution in cell proliferation induced by
supplementation with n-3 polyunsaturated fatty acids does not appear to be a consequence of
induced alterations in *RAS* expression. Additional studies have shown that fatty acid
supplementation does not interfere with perillyl alcohol's ability to downregulate *RAS* expression.
These latter findings raise the possibility that combined fatty acid and perillyl alcohol
supplementation may be an effective strategy to limit proliferation of malignant cells.

Poster Abstract # 16

Lipid Peroxidation and DNA Damage

Harold C. Box, Sek-Wen Hui, Alexander E. Maccubbin, Thomas M. Nicotera

Biophysics Department, Roswell Park Cancer Institute, Buffalo, NY 14263

Antioxidants in the diet, such as vitamin E, may mediate lipid peroxidation and the consequences thereof. Two mechanisms are often postulated whereby lipid peroxidation can lead to DNA damage with implications for cancer: (1) Lipid peroxidation generates reactive aldehydes which react with DNA. Malondialdehyde is considered the most likely aldehyde to damage DNA. It forms an adduct with guanine. (2) Metabolic processes involving lipid peroxidation products may generate superoxide which, in the presence of metal catalysts, generates hydroxyl radicals. The 8-hydroxyguanine (7,8-dihydro-8-oxoguanine) lesion is a principal free radical-induced DNA lesion. Its presence in cellular DNA is often taken as an indication of oxidative stress.

We have been adapting ^{32}P-postlabeling assays for sensitive detection of the malondialdehyde and the 8-hydroxyguanine lesions. The basic plan is to assay for these damages at the dimer level. The damaged DNA can be hydrolyzed enzymatically such that the lesions of interest are present in the hydrolysate predominantly in the form of modified dinucleoside monophosphates. The advantages we hope to realize from this approach to DNA damage analysis are the following: (1) The dinucleoside monophosphates can be selectively labeled using T4 polynucleotide kinase and radioactive ATP. The bulk of the DNA digest is in the form of nucleosides which are not acceptable substrates for the T4 kinase and consequently are not phosphorylated. (2) After postlabeling, but prior to TLC, it is a considerable advantage to remove radioactivity due to inorganic phosphate or unused ATP from the DNA digest. This is readily accomplished using solid phase extraction columns (e.g. SEP-PAK C18 cartridges). Complete removal of this radioactive debris can be accomplished because the DNA lesions are manipulated at the dimer level; removal of the unwanted material could be effected if the lesion were manipulated at the monomer level. (3) The literature shows that the TLC assay conventionally used in ^{32}P-postlabeling assays will not distinguish from the inevitable background of the labeled entities. The background arises because of the extreme sensitivity of the technique. The last step in the assay permits excellent discrimination against background through the use of cold carrier bearing the lesion of interest. Because most radiation-induced base modifications have a negligible absorption in the ultraviolet, they are difficult to detect and make poor carriers. In contrast, the carrier in our assay is readily detected because of the residual absorption of the unmodified nucleoside of the dimer. The lesion is quantitated by measurement of radioactivity co-chromatographed with the carrier. We are able to generate the requisite carriers because of our past experience with the identification and characterization of oxidative damages in DNA oligomers. Progress to date in developing these assays will be reported.

Poster Abstract # 17

Inhibitory Effect of Inositol Hexaphosphate on
Phorbol Ester-induced Ornithine Decarboxylase Activation

Kwangok P. Nickel and Martha A. Belury

Department of Foods and Nutrition, Purdue University, West Lafayette, IN 47907

High-fiber diets have been shown to have beneficial effects on preventing tumorigenesis. Inositol hexaphosphate ($InsP_6$ or phytic acid) which is a fiber-associated component of cereals and legumes has been demonstrated to inhibit cell proliferation and enhance cell differentiation, indicating the potential of chemopreventive roles. We investigated the effect of $InsP_6$ on 12-O-tetradecanoylphorbol-13-acetate (TPA)-induced ornithine decarboxylase (ODC) activity, an essential event in tumor promotion, both *in vitro* using an established murine epidermal cell line (HEL-30) and *in vivo* using female SENCAR mouse skin. When several doses of $InsP_6$ ranging from 0.05% to 0.5% were added to serum- and inositol-free media of HEL-30 cells with TPA, induction of ODC was significantly reduced ($p<0.05$). We found that ^{3}H-$InsP_6$ was incorporated into HEL-30 cells in a time-dependent manner when the uptake was expressed as percent of the control (no ^{3}H-$InsP_6$), although less than 1% of dosed radioactivity was found in the cells after 36 hr incubation. In addition, topical application of 1% $InsP_6$ in 200 µl methanol on SENCAR mouse skin prior to TPA treatment significantly inhibited ODC activity ($p<0.05$). These results suggest that $InsP_6$, a chelator of Ca^{2+} and/or precursor of inositol triphosphate (IP_3), is involved in the inhibition of TPA-induced tumor promotion.

Poster Abstract # 18

Dietary Conjugated Linoleic Acid Induces Peroxisome Proliferation and Ornithine Decarboxylase Activity in Mouse Liver

Moya-Camarena SY, Vanden Heuvel JP, Liu KL, and Belury MA.
Department of Foods and Nutrition, Purdue University, West Lafayette, IN 47907

Peroxisome proliferators are considered non-genotoxic hepatocarcinogens in rodents. Several studies have found that the dienoic fatty acid derivative conjugated linoleic acid (CLA) acts as an anticarcinogen in rodent mammary gland, colon, forestomach and skin. Because polyunsaturated fatty acids such as CLA have structural features similar to peroxisome proliferators, we hypothesize that CLA could act in a similar manner to this group of chemicals. In this study we determined the effect of increasing levels of dietary CLA (0.0, 0.5, 1.0, 1.5% by weight) on the induction of peroxisome proliferation in murine liver. Female SENCAR mice were fed experimental diets for six weeks, livers were excised and total RNA and protein were extracted. Quantitative reverse-transcriptase polymerase chain reaction (qRT-PCR) was used to determine levels of acyl-CoA (ACO) mRNA expression. In addition, the expression of several proteins including ACO, as well as other fatty acid metabolizing enzymes (fatty acid binding protein, FABP and palmitoyl-CoA synthase, PCS) were analyzed by Western blots. Finally, the effect of dietary CLA on ornithine decarboxylase (ODC) activity, a hallmark event associated with cell proliferation and tumor promotion in rodent liver, was studied. ACO mRNA accumulation and ACO, FABP and PCS protein levels were significantly enhanced ($p<0.05$) in livers of mice fed 1.0 and 1.5% CLA diets compared with mRNA and protein levels in the 0.0% CLA diet group. In addition, hepatic ODC activity was increased ($p<0.05$) in mice fed 1.0 and 1.5 % CLA compared to activity in the 0.0% CLA diet group. These data suggest that CLA displays the typical peroxisome proliferator response, i.e. induction of ACO, FABP and PCS. Also, since dietary CLA induces ODC activity, this fatty acid derivative may modulate hepatocarcinogenesis in a manner similar to other peroxisome proliferators.

Poster Abstract # 19

Antagonistic interaction of unsaturated fatty acids and selenium with protein kinase C at its cysteine-rich regions

Rayudu Gopalakrishna, Usha Gundimeda, and Zhen-hai Chen

Department of Cell and Neurobiology, School of Medicine

University of Southern California, Los Angeles, CA 90033

Cis-unsaturated fatty acids at higher concentrations can increase tumorigenesis process. Selenite as well as certain other forms of selenium have been shown to decrease carcinogenesis. An antagonistic interaction might be occurring between these two types of permissive and protective dietary factors to influence various stages of multistage carcinogenesis. Protein kinase C (PKC) is implicated in both tumor promotion and progression related events such as invasion and metastasis. Therefore, it is of our interest to understand whether there is any antagonistic interaction between these nutrients with PKC as well as to understand the unique structural aspects of this kinase that make it a specific target for this interaction. Selenite reversibly inactivated PKC by oxidizing the cysteine-rich regions present within the catalytic domain (at least 4 cysteine residues present in C3 and C4 regions). The reversibility of this modification can be achieved with thiol agents in the test tube, while in the cells it depended on the availability of the cellular reducing equivalents. *Cis*-unsaturated fatty acids (arachidonic acid, and linoleic acid) and branched chain fatty acids can activate PKC by binding to the regulatory domain. We have observed that they also can inactivate PKC if the cofactors are not protecting the catalytic domain. Prior modification of PKC with selenite, selenocystine, or selenodiglutathione made this enzyme refractory to regulation by unsaturated fatty acids. This suggested that the cysteine-rich regions present within the catalytic domain may be required to mediate the effects of unsaturated fatty acids, and selenium blocks these effects by oxidizing these cysteine-rich regions. Selenite-mediated regulation of PKC was well correlated with the inhibition of induction of ornithine decorboxylase, a cell transforamtion related marker, induced by these unsaturated fatty acids. Taken together these results suggested that the unique structural aspects of PKC is well suited for the interaction of both tumor promoting agents and chemopreventive agents. This enzyme may be a good candidate for understanding the interaction between the permissive and protective factors present within the diet that can influence the carcinogenesis process.

Supported by American Institute for Cancer Research grant 93B43.

Poster Abstract # 20

OXIDATIVE DNA DAMAGE LEVELS IN MAMMARY GLAND OF RATS FED VARYING AMOUNTS OF CORN OIL. Z. Djuric[1], S.M. Lewis[2], M.H. Lu[2], D.A. Luongo[1], L. Barno-Winarski[1], L. Naegeli[1] and R.W. Hart[2]. [1]Karmanos Cancer Institute, Wayne State University, Detroit, MI; [2]National Center for Toxicological Research, Jefferson, AR.

Dietary intake of corn oil has been shown in many studies to influence mammary gland tumorigenesis in rats. One mechanism behind this association is the effect dietary fat can have on endogenous levels of oxidative DNA damage. We have examined the levels of 5-hydroxymethyl-2'-deoxyuridine, an oxidized thymine residue, in the DNA of mammary gland, liver and blood from rats fed varying levels of corn oil. After a 2 week equilibration period using a 5% corn oil diet, Fischer 344 rats were fed 3, 5, 10, 15 or 20% corn oil for 20 weeks. In liver and blood, the levels of DNA damage were not significantly affected by dietary fat level, although DNA damage did increase in blood with increasing dietary fat. The levels of 5-hydroxymethyl-2'-deoxyuridine in DNA from mammary gland were significantly lower in rats fed 3% versus 10% fat. The levels of this oxidized base were not significantly increased in rats fed higher amounts of fat. This is unlike our previous results after 2 weeks of feeding where DNA damage levels were maximal using 5% corn oil. These results indicate that there are time-dependent effects on oxidative stress by dietary fat. In addition, there was also a plateau phenomenon observed. Although protection against DNA damage was achieved with the low-fat diets, increased fat intake above 10% oil (22% of calories from fat) did not further appreciably increase oxidative DNA damage levels in rat mammary gland.

Supported by grant number NC93B61 from the American Institute for Cancer Research.

Poster Abstract # 21

CHRONIC WEIGHT-CYCLING INCREASES OXIDATIVE DNA DAMAGE LEVELS IN MAMMARY GLAND OF FEMALE RATS FED A HIGH-FAT DIET. V. Uhley[1], M. Pellizzon[2], A. Buison[2], F. Guo[2], Z. Djuric[1], and K-L.C. Jen[2]. [1]Karmanos Cancer Institute, Wayne State University, Detroit, MI; [2]Nutrition and Food Science, Wayne State University, Detroit, MI.

Increased fat and caloric intakes have been associated with increased mammary tumor incidence. Oxidative DNA damage levels have also been shown to be modulated by these factors and have been implicated in the process of tumor promotion. We examined the effects of a high-fat (HF) diet (40% by weight) fed in varying feeding regimens on levels of 5-hydroxymethyl-2'-deoxyuridine, an oxidized thymine residue, in DNA from mammary gland. Aging female Sprague Dawley rats (11 months old) fed the HF diet were randomly divided into 3 groups: ad libitum fed (AL), weight-cycling (WC), or energy restricted (ER). WC was induced by repeated ad-libitum/restricted (50%) feeding of the HF diet. The ER group was fed a restricted amount of HF diet in order to maintain baseline weight. All groups were sacrificed after 28 weeks. At sacrifice the WC rats had similar body weights as the ER group, but had significantly greater levels of 5-hydroxymethyl-2'-deoxyuridine than the ER rats ($p < 0.01$). WC rats also had higher levels of this oxidized base than the rats in the AL group, but this achieved only borderline significance. This is significant since the WC group had significantly lower body weights than the AL fed group. There was no significant difference in DNA damage levels between the AL fed and ER groups, although the levels were relatively lower in ER rats. In humans body weight has been associated with cancer risk. These results indicate that a history of WC, even when body weight is reduced, can increase levels of 5-hydroxymethyl-2'-deoxyuridine in mammary gland DNA, a potential biomarker of cancer risk, while constant control of calories for maintenance of body weight may be more beneficial.

Poster Abstract #22

Regulation of neurofibromin and p120 GTPase activating protein (GAP) by fatty acids

Joung H. Lee[1], Sophia Bryant[2], Jyoti A. Harwalkar[1], Dennis W. Stacey[3], Vidyodhaya Sundaram[3], Richard Jove[2], and Mladen Golubic[1]

[1]Department of Neurosurgery, Cleveland Clinic Foundation, 9500 Euclid Ave./NC2-150, Cleveland, OH 44195, E-mail:golubim@cesmtp.ccf.org
[2]Moffitt Cancer Center, 12902 Magnolia Drive, Tampa, FL 33612
[3]Department of Molecular Biology, Cleveland Clinic Foundation, Cleveland, OH 44195

Introduction: Neurofibromin is a protein product of neurofibromatosis type 1 tumor suppressor gene which is frequently mutated in various types of nervous system tumors. Neurofibromin and p120 GAP down-regulate the activity of cellular p21 ras proteins. How the activity of these two proteins is controlled is not yet clear. In this study, we analyzed the effects of eight nutritionally relevant fatty acids (FAs) on GTPase stimulatory activity of neurofibromin and p120 GAP. The FAs tested were: saturated stearic acid (SA, 18:0), monounsaturated oleic acid (OA, 18:1), three ω-6 FAs, linoleic acid (LA, 18:2), dihomo-γ-linolenic acid (DGLA, 20:3) and arachidonic acid (AA, 20:4), and three members of ω-3 FAs, α-linolenic acid (LNA, 18:3), eicosapentaenoic acid (EPA, 20:5), and docosahexaenoic acid (DHA, 22:6). **Methods**: Analysis was performed by p21 ras immunoprecipitation GTPase assay. The full-length p120 GAP expressed in insect Sf9 cells and immunoaffinity purified full-length neurofibromin were used. Inhibition by FAs was analyzed at seven different concentrations, ranging from 0.5 μg/ml to 50 μg/ml. **Results:** p120 GAP was weakly, if at all, inhibited by SA and OA. These FAs in concentrations 100-fold higher were necessary in order to inhibit p120 GAP compared to neurofibromin. Neurofibromin was also more strongly inhibited by other FAs tested. The inhibition of neurofibromin was 3-fold stronger by ω-6 LA and AA (IC_{50} of 12.2 μM and 6.6 μM, respectively) than by ω-3 LNA and EPA (IC_{50} of 39.8 μM and 17.5 μM, respectively). **Discussion and Conclusions**: Alterations in content of some FAs studied were documented earlier in several types of human cancer and central nervous system tumors. Here we show that tumor promoting effects of some dietary FAs could be, at least in part, mediated through inhibition of GTPase stimulatory activity of neurofibromin upon p21 ras. The lowest inhibitory concentrations were in micromolar range and thus could be physiologically important.

Supported by the American Institute for Cancer Research grant # 94B63.

Poster Abstract # 23

REVERSION OF CONSTITUTIVE *pim-1* EXPRESSION TO GROWTH FACTOR-DEPENDENCY IN RAT Nb2 LYMPHOMA AND HUMAN B-CLL CELLS BY BUTYRATE.

Arthur R. Buckley, Donna J. Buckley, Joshua S. Krumenacker, Matthew A. Leff, John A. Laurie, Peter W. Gout, Gary de Jong, and Nancy S. Magnuson. University of North Dakota School of Medicine and Health Sciences and the Grand Forks Clinic, Grand Forks, ND, the B.C. Cancer Agency, Vancouver, B.C., and Washington State University, Pullman, WA.

The pre-T lymphoma, Nb2-SFJCD1, is growth factor-independent for proliferation and highly metastatic when implanted into Nb rats. *In vitro* treatment with the dietary differentiating agent, sodium butyrate (NaBT), growth-arrests Nb2-SFJCD1 cells and transiently reverts them to growth factor (prolactin, PRL)-dependency. In the present study, the relationship between NaBT-induced growth-arrest and cell cycle progression was examined. Additionally, the effect of PRL on expression of the immediate-early protooncogene, *pim-1* in Nb2-SFJCD1 and primary B-CLL cultures was evaluated. Pretreatment with 2 mM NaBT for 72 h growth-arrested Nb2-SFJCD1 cultures in the G_1 phase of cell cycle; PRL stimulated a concentration-dependent resumption of proliferation. In other experiments, NaBT significantly reduced steady state levels of *pim-1* mRNA in Nb2-SFJCD1 and B-CLL cultures. Stimulation with PRL induced a rapid and concentration-dependent biphasic accumulation of the *pim-1* transcript in Nb2-SFJCD1 cells and a similar response in B-CLL. Results from stability studies suggest that increased *pim-1* in this Nb2 subline most likely does not reflect augmented mRNA stability. These results indicate that NaBT-induced accumulation of Nb2-SFJCD1 cultures in the G_1 phase of cell cycle and markedly attenuated constitutive *pim-1* expression. Mitogenic stimulation with PRL reinitiated cell cycle progression characterized by the rapid expression of *pim-1*. We suggest that NaBT may reverse the consequences of malignant progression from growth factor-dependency to autonomy in certain hematopoietic tumors. Supported in part by the AICR (95B089).

Poster Abstract # 24

An Abnormal Receptor for a Vitamin A Derivative in Promyelocytic Leukemia

Robert L. Redner, Elizabeth A. Rush, and Sheri L. Pollock. University of Pittsburgh Medical Center and the University of Pittsburgh Cancer Institute, Pittsburgh PA 15213

Acute Promyelocytic Leukemia is characterized by maturational arrest of malignant bone marrow cells. The Vitamin A derivative retinoic acid induces differentiation of these cells and induces clinical remissions. We have reported a patient with APL who manifested t(5;17). This translocation gives rise to fusion of the genes encoding nucleophosmin (NPM) and the retinoic acid receptor alpha (RAR), a retinoic acid-dependent nuclear transcription factor. Two NPM-RAR fusion proteins, derived form alternative splicing of the transcript, are expressed in the leukemic cells. To determine whether the NPM-RAR fusion protein underlies the maturation arrest in this patient we investigated whether ectopic NPM-RAR expression would inhibit myeloid differentiation. We chose to use U937 cells, a monocytic cell line capable of differentiating into mature monocytes upon stimulation with Vitamin D3 and TGFß. An NPM-RAR expression vector with a Neomycin selectable marker, or control plasmid, was electroporated into U937 cells. Subclones were selected by limiting dilution in G418. Clonality was determined by Southern analysis of integration sites, and expression of the NPM-RAR protein was verified by immunoblotting. To assess the ability of the subclones to differentiate, cells were incubated in medium containing Vitamin D3 and TGFß. After 5 days cells were analyzed by flow cytometry for acquisition of the CD14 differentiation marker. The level of CD14 positivity rose to over 70% in parental and vector control U937 clones. By contrast, the NPM-RAR expressing U937 clones showed 20% or less CD14 positivity after exposure to the differentiating agents. These results suggest that NPM-RAR interferes with the differentiation of U937 cells. This model system will prove valuable in our future studies of the mechanism of NPM-RAR action.

Poster Abstract # 25

Effects of docosahexaenoic acid feeding on tumor and host arabinosylcytosine sensitivity of Fisher rats.
Atkinson, T., Murray, L., Berry, D., Ruthig, D. and Meckling-Gill, K.A. Department of Human Biology and Nutritional Sciences, University of Guelph, Guelph, ON, Canada, N1G 2W1

Weanling male Fisher rats were fed a 5% safflower oil diet (LF-SO) for 3 wks and then implanted with fibrosarcoma cells (subcutaneously, sc). Once tumors were palpable animals were switched onto one of two experimental diets (10% safflower [MF-SO] or 10% DHASCO™ [MF-DHA]) or maintained on the basal diet. One wk later animals were randomly assigned to receive saline injections or twice daily sc injections of 100 mg/kg arabinosycytosine (ara-C)for six days. Tumor and animal growth were monitored daily. Animals were euthenized and tissues examined after a 24 h drug-free period and 12 h fast. The largest tumors were observed in the saline injected MF-SO. Tumors were smaller in both the LF-SO and MF-DHA-saline groups and were not different between these two diets. Ara-C treatment resulted in substantial tumor regression for all three diets (no diet differences). Mucosal weights were highest in DHA fed animals both before and after ara-C treatment. The number of CFU-GM was higher and infection rates lowest in DHA-ara-C animals compared to MF-SO groups. In conclusion, DHA feeding promotes tumor regression compared to a moderate fat safflower-rich diet. Bone marrow suppression was less apparent in DHA fed animals and mucosa weight higher following ara-C treatment suggesting that DHA provides host benefits without compromising tumor responses.
Supported by the American Association for Cancer Research

Poster Abstract # 26

GROWTH AND EXPRESSION OF THE METASTATIC PHENOTYPE IN TWO HUMAN PROSTATE CANCER CELL LINES IN NUDE MICE FED DIFFERENT DIETARY FATTY ACIDS. Jeanne M. Connolly and David P. Rose, Division of Nutrition and Endocrinology, American Health Foundation, Valhalla, NY 10595

We have prevously shown that, compared to a diet rich in corn oil, high-fat diets rich in menhaden oil (MO) inhibit the growth of tumors derived from the human prostate cancer cell line DU145 injected subcutaneously into nude mice. To more closely model the human disease, we have now injected DU145 cells directly into the prostate gland of nude mice, and we report here the influence of these diets on the growth of the resulting intraprostatic tumors. In addition, in order to study the influence of these diets on metastatic progression, we have performed the same procedure using the PC3-M metastatic human prostate cancer cell line.

The weight of prostates at necropsy was significantly reduced by feeding MO diet ($p < 0.03$); no effect was observed on PC3-M tumors. Histologically, a high degree of vascularization was observed in DU145 cell prostatic tumors, which was not visible in PC3-M tumors; this was accompanied by a significantly higher level of vascular endothelial growth factor (VEGF), as detected by ELISA, in DU145 tumors than in PC3-M tumors (151 ± 72 vs. 39 ± 22 ng/mg protein, $p < 0.001$). Immunocytochemical examination showed that the expression of this angiogenesis-related growth factor occured in the cytoplasm of DU145 cells. These results suggest an association between neovascularization, an early step in tumor progression, and VEGF in DU145 cells, but not in the expression of the metastatic phenotype by PC3-M cells. No effect of the MO diet on VEGF levels was observed in the tumors studied.

Metastatic spread to locoregional lymph nodes occurred in a significantly higher proportion of PC3-M tumor-bearing mice compared with the DU145 tumor-bearing mice (43% vs. 16%, $p = 0.01$); lung metastasis was also more common in PC3-M mice, although this difference (17% vs. 10%) failed to achieve statistical significance. The greater expression of metastatic capacity by PC3-M cells was accompanied by a significantly higher level of urokinase-type plasminogen activator (uPA) in the PC3-M compared with the DU145 cell tumors (745 ± 598 vs.

64 ± 56 ng/mg protein, $p<0.001$). This proteolytic enzyme has been implicated in several steps of metastatic progression, as has been another proteolytic enzyme, matrix metalloproteinase-9 (MMP-9). MMP-9 activity, as determined by zymography, was also higher in PC3-M than in DU145 tumors; in contrast, the activity of another enzyme detected by this method, MMP-2, was no different in the two tumor types.

These results indicate that while growth of the poorly metastatic DU145 prostate tumors may be influenced by dietary MO, this process is not mediated by a reduction in VEGF-mediated vascularization. The metastatic phenotype, as exemplified by the PC3-M cell line, is associated with high levels of proteolytic enzymes including uPA and MMP-9, the activities of which are not reduced by dietary MO. Thus, while some early stages of prostate cancer progression seem responsive to dietary fatty acid manipulation, the fully metastatic phenotype appears to be resistant to such intervention.

This work was supported by grant No. 94B95 from the American Institute for Cancer Research.

Poster Abstract # 27

Efficacy of Vitamin D in Advanced Prostate Cancer: A Pilot Study. Van Veldhuizen PJ*, Drees B*, Sadasivan R**, Burns D*, Hamilton J*, Egbert A***, Taylor S*. Veterans Affairs Medical Center, Kansas City, MO* and Wichita, KS*** and the University of Kansas Medical Center, Kansas City KS**.

Bone is the most common site of metastases in prostatic cancer occurring in over ninety percent of patients with advanced disease. In these patients, bone pain is a common and often debilitating symptom. Small case series and case reports suggests that oncogenic osteomalacia may be a contributing factor to bone pain in this patient population. Oncogenic osteomalacia is characterized by low serum 1,25-dihydroxyvitamin D_3 levels, hypophosphatemia and phosphaturia as well as bone pain. Vitamin D deficiency can also result in muscle weakness and muscle pain, as well as unusual pain syndromes such as reflex sympathetic dystrophy and painful hyperesthesias. This patient population is susceptible to vitamin D deficiency because of decreased exposure to sunlight, advanced age and compromised nutritional intake. The objective of this study is to determine if the pain associated with advanced prostate cancer responds to treatment with vitamin D and to determine whether this response is secondary to a tumor response, to treatment of a vitamin D deficiency state or to treatment of oncogenic osteomalacia. Laboratory parameters of bone turnover are being monitored in additional to vitamin D levels. A strength assessment is also performed on a monthly basis. Patients are treated with 2000 U of ergocalciferol daily and 500mg of calcium twice daily. Five patients have been enrolled in this preliminary study and two have completed the planned treatment protocol. One patient had initial improvement of his pain score, stabilization of his prostate specific antigen and improved strength. One patient had disease and symptom progression while on treatment. ****Supported by a grant from the AICR.

Poster Abstract # 28

ROLE OF TRANSLATION IN THE DIFFERENTIATION-DEPENDENT STABILIZATION OF FATTY ACID SYNTHASE mRNA IN 3T3-F442A ADIPOCYTES. David Novick and Kathleen Sue Cook, Dept. of Biology, Tufts University, Medford, MA 02155

Regulated changes in mRNA turnover are known to play an important role in controlling the expression of a number of genes in adipocytes. For example, upon differentiation of preadipocytes into adipocytes, there is a 20-fold increase in the abundance of fatty acid synthase (FAS) mRNA. This increase is due to a nominal increase in transcription and a substantial stabilization of the message. The mechanism involved in this stabilization of FAS mRNA is not known. Since many elements involved in regulated turnover have been found to be dependent on protein synthesis, we investigated the role of protein synthesis in the differentiation-dependent stabilization of FAS mRNA. 3T3-F442A preadipocytes and adipocytes were treated with the protein synthesis inhibitors puromycin (100 uM) or cycloheximide (5 ug/ml), and the turnover of FAS mRNA and a control message, glyceraldehyde phosphate dehydrogenase (GAPDH), were measured after inhibiting transcription with actinomycin D. Our results show that puromycin and cycloheximide substantially stabilize FAS mRNA in adipocytes, but have little or no effect on the turnover of FAS mRNA in preadipocytes or GAPDH in either cell type. These results suggest that FAS mRNA enters a rapid decay pathway in preadipocytes that is independent of translation, and upon differentiation, FAS mRNA is subjected to a different decay pathway that is slower and dependent on translation. Since FAS plays a crucial role in fatty acid synthesis, it is important to understand the mechanism behind its regulation in adipocytes. The fact that recent findings have shown elevated levels of FAS in certain forms of human cancer suggests that a general understanding of how this gene is regulated could prove to be of clinical importance.

Poster Abstract # 29

Dietary Intervention Can Increase Omega-3/Omega-6 Polyunsaturated Fatty Acid Ratios in Serum and Breast Fat in Patients with Breast Cancer

Stefanie Capone, Dilprit Bagga, David Heber, Michael Lill and John Glaspy
Division of Hematology-Oncology, Department of Medicine, UCLA School of Medicine, Los Angeles, California

While there is interest in the role of polyunsaturated fatty acids (PUFA) as mediators of the nutritional component of breast cancer risk in Western countries, studies have not demonstrated that dietary intervention can result in significant changes in the PUFA composition of the breast microenvironment.

We are conducting a dietary intervention study in patients with high risk breast cancer, who have completed high dose chemotherapy with autologous progenitor cell support, to study the feasibility of nutritional intervention for the control of minimal residual disease. Patients receive a 15% calories from fat, fish oil and soy supplemented, isocaloric diet for three months. Subjects undergo extensive nutritional counseling weekly and are monitored with food diaries. Tolerance and compliance have been excellent. Breast and gluteal fat biopsies are done before and after dietary intervention.

Data are available for the first 15 patients and are expressed as means with standard errors; p values refer to results of paired t-tests. Serum and breast fat total fatty acids were analyzed by gas chromatography. In the serum, there were significant decreases in linoleic acid (LA, 18:2) (3228+/-405 μmol/L before, 2475+/-237 after, p=.06), arachidonic acid (AA, 20:4) (707+/-67 μmol/L before, 540+/54 after; p=.006) and in total omega-6 (ω-6) PUFA (4020+/-472 μmol/L before, 3015+/-283; p=.04). There was a significant increase in serum eicosapentanoic acid (EPA, 20:5) (38+/-26 μmol/L before, 497+/-109 after, p=.001), docosahexanoic acid (DHA, 22:6) (166+/-30 μmol/L before, 448+/-84 after, p=.001) and total ω-3 PUFA (232+/-55 μmol/L before, 960+/-181 after, p=.0006). The mean ω-3/ ω-6 PUFA ratio increased more than five fold from .064 to .34 (p=.0004).

In extracted breast fat, no significant decreases in LA, AA, or total ω-6 PUFA were observed. Significant increases were observed in EPA (0 μmol/gm before, 3.1+/-1 after, p=.01) and in DHA (.6 +/-.3 μmol/gm before, 4+/-.8 after, p=.001). The mean ω-3/ω-6 PUFA ratio increased from .046 to .064 (p=.01).

We conclude that dietary interventions aimed at altering PUFA levels in the microenvironment of the breast are safe and feasible in a highly motivated patient population. These interventions can lead to significant changes in the ω-3/ω-6 PUFA content of serum and breast fat. Longer follow up will be required to assess the full effects of this intervention, especially with respect to the composition of breast fat.

Poster Abstract # 30

NADPH-CYTOCHROME P450 REDUCTASE, CYTOCHROME P450 2C11, P450 1A1 AND THE ARYL HYDROCARBON RECEPTOR IN LIVERS OF RATS FED METHYL/FOLATE DEFICIENT DIETS. J.Zhang, S.M.Henning, D. Heber, M.E.Swendseid, V.L.W.Go, Clin. Nutr. Res. Unit, Dept. of Med., UCLA, L.A., CA 90095.

It is well established that rats chronically fed methyl-deficient diets develop hepatocellular carcinomas without exposure to known carcinogens. To determine whether the cytochrome P450s play a role in this model, we investigated several hepatic cytochrome P450 isozymes and the aryl hydrocarbon receptor (AHR) in rats fed the methyl-deficient diet. Rats in three groups were fed: 1) Control diet containing the AIN vitamin mixture; 2) control diet devoid of choline and folate (CFD); and 3) CFD diet devoid of niacin (CFND) for 15 months. Weight gain of rats in both deficient groups was decreased but food intake expressed per kg body weight was increased. The liver/body (wt) ratio increased 2-3 times in both deficient groups. Hepatic tumors were found in all rats fed deficient diets, but not in rats fed the control diet. Western blot analyses of hepatic tissue showed that compared to the control group NADPH-cytochrome P450 reductase increased significantly in both CFD and CFND groups, possiblly to compensate for a decreased reducing power. Cytochrome P4502C11 (CYP2C11) was not detected in rats fed either the deficient diets. In addition, AHR and cytochrome P4501A1 (CYP1A1) were detected in higher amounts in livers of rats fed the deficient diets. To our knowledge this is the first time it has been shown that the AHR and CYP1A1 are affected by dietary deficiencies. From our limited data it cannot be determined whether the increase in CYP1A1, promotes, impedes or has no effect on the development of hepatic tumors in methyl/folate deficient rats. Additional cytochrome P450 isozymes should be investigated including the lipid-related P450 2C series since CYP2C11 seems to be particularly sensitive to methyl/folate deficiency and because the increased hepatic lipid levels present in methyl deficiency are a prime source of free radicals. It is important to identify the endogenous ligand(s) stimulating the expression of the aryl hydrocarbon receptor, an issue that can be addressed in both cell cultures and aimal models. (Supported by NIH grants 5T32DK07688 and CA42710-10).

Poster Abstract # 31

TRANSFORMING GROWTH FACTOR-BETA MEDIATES RETINOIC ACID INHIBITION OF GROWTH AND HPV16 EARLY GENE EXPRESSION IN HPV16-IMMORTALIZED HUMAN KERATINOCYTES

Alfredo J. Canhoto[1], Lucia A. Pirisi[2], and Kim E. Creek[2,3]. Department of Chemistry and Biochemistry[1], University of South Carolina, and Department of Pathology[2], and Children's Cancer Research Laboratory, Department of Pediatrics[3], University of South Carolina School of Medicine, Columbia, South Carolina 29208.

Human papillomaviruses (HPV), in particular HPV16 and HPV18, are initiators of cervical cancer *in vivo*, and immortalize cultured human keratinocytes (HKc). Using a model system of HPV16-immortalized human keratinocytes (HKc/HPV16), we are investigating the role of all-*trans*-retinoic acid (RA), an active metabolite of vitamin A, in the chemoprevention of cervical cancer. We have previously shown that early passage HKc/HPV16 are about 100-fold more sensitive than normal HKc to the antiproliferative effects of RA and that treatment of HKc/HPV16 with 1 μM RA for 72 h decreases steady-state mRNA levels of the HPV16 oncogenes E6/E7 by 75%. In addition, we have also found that RA treatment of normal HKc and HKc/HPV16 increases transforming growth factor-beta1 (TGF-ß) and TGF-ß2 secretion 2- and 5-fold, respectively. Since TGF-ß inhibits the proliferation of early passage HKc/HPV16 and is also a potent inhibitor of E6/E7 mRNA expression, we reasoned that TGF-ß may mediate, at least in part, the inhibition of growth and HPV16 early gene expression by RA. Consistent with this possibility is the finding that polyclonal anti-TGF-ß antibodies blocked (up to 80%) RA inhibition of [^{3}H]thymidine incorporation by early passage HKc/HPV16 and prevented RA from decreasing steady-state E6/E7 mRNA levels. To study the effects of RA and TGF-ß on the HPV16 promoter, we have cloned the HPV16 upstream regulatory region (URR), including the promoter P97, upstream of a firefly luciferase reporter gene (p16URR-Luc). p16URR-Luc was transiently transfected into HKc/HPV16 and the effects of RA and TGF-ß on luciferase activity determined. Surprisingly, short term RA (1 μM) treatment (3 to 24 h) resulted in a modest increase (up to about 3-fold) in luciferase activity. As expected, TGF-ß1 treatment of HKc/HPV16 transfected with p16URR-Luc resulted in a time- and dose-dependent decrease of luciferase activity. Maximal inhibition of luciferase activity (84%) was observed after 48 h of treatment with 40 pM TGF-ß1. Current studies are aimed at identifying the RA and TGF-ß responsive elements within the HPV16 URR.

Supported by grant 95A18 from the American Institute for Cancer Research.

Poster Abstract #32

Dietary polyunsaturated fatty acids (PUFA) and antioxidant (α-tocopherol) in experimental mammary carcinogenesis.

C. Lhuillery[1], S. Cognault[1], E. Germain[1-2], P. Bougnoux[2].
[1]LNSA, INRA, 78352 Fr-Jouy en Josas; [2]UPRES 313, Clinique d'Oncologie, CHU, Université de Tours, 37044 Tours, France.

Whereas dietary lipid antioxidants are protective in early stages of cancer, recent data indicate that, in presence of peroxidable PUFA, they stimulate the growth of established tumors (1). To understand how lipid antioxidants could stimulate mammary tumor growth, we have examined the effects of dietary α-tocopherol (vitamin E) on tumor growth parameters in a rodent model of mammary carcinogenesis with rats receiving a high fat (15 w% fat) diet. Half of rats received a diet with no added vitamin E (-Vit E), the other half received 50 UI vit E/kg diet (+Vit E). All rats received an injection of N-methyl nitroso-urea to initiate mammary tumors. Tumor growth was followed by weekly palpation of the animals and by the measure of total tumor mass and number. At sacrifice, tumors were excised in both groups and we measured clinical and potential doubling times, cell cycle phase distribution and proportion of apoptotic cells.

Results:

1. Tumor appearance was significantly delayed in -Vit E group. At the end of the experiment, average tumor number and mass were significantly lower in the -Vit E group than in the +Vit E group.

2. Proportions of tumor cells in S or G2M phases were similar in both groups whereas apoptotic cell number was significantly decreased in +Vit E group.

These data show that the effects of dietary PUFA on mammary tumor growth are modulated by diet oxidative status and they suggest that lipoperoxides are involved in tumor cell loss by apoptosis in mammary carcinoma.

(1) Gonzalez et al., 1993, *Lipids,* 28, 827-832.

Poster Abstract # 33

Activation of a Tumor Suppressor Gene by a Nutrient Derivative
Richard A. Steinman,* Steven J. Shiff,[+] Jianping Huang,* Beatrice Yaroslavskiy,* and Michael Nalesnik[#]. University of Pittsburgh Cancer Institute* and Department of Pathology#, University of Pittsburgh School of Medicine, Pittsburgh, Pennsylvania, and [+]The Rockefeller University Hospital, New York, NY

We evaluated the hypothesis that high-fiber/low-fat diets protect against colon cancer by altering the topographic expression of the cell cycle inhibitor p21(WAF1) in colonic mucosa. We postulate that high fiber intake increases the level of butyrate in the colon, which in turn protects the gut from neoplasia by activating p21(WAF1). It should be noted that colonic neoplasms have been reported to lose normal distribution of p21(WAF1) within the crypt (El-Deiry, W.S., et. al, Cancer Research 55, 2910-2919 (1995)). We have established a plausible mechanism for such an effect by demonstrating in colon cancer cell lines that butyrate directly induces p21(WAF1) concurrent with growth arrest and differentiation. The concentration of butyrate which induces p21(WAF1) is comparable to levels of colonic butyrate which arise from colonic bacterial fermentation of dietary fiber.We have also demonstrated a requirement for p21(WAF1) for normal cell cycle control in the HT-29 cell line, and have identified a candidate element within the p21(WAF1) promoter through which butyrate upregulates p21(WAF1) expression. In order to determine whether the relationship observed between butyate levels and p21(WAF1) expression is observed in vivo, normal volunteers have been maintained for four weeks on high fat/low fiber or high fiber/low fat diets in an inpatient GCRC. In the one patient studied to date, increased fiber consumption was associated with alterations in stool butyrate levels and with an altered topographical distribution of p21(WAF1) expression within colonic crypts analyzed in biopsy specimens. A complete analysis of cell cycle gene expression patterns associated with these diametrically opposed diets is underway.

Poster Abstract # 34

Cloning and Identification of Rat Deoxyuridine Triphosphatase (dUTPase) as an Inhibitor of Peroxisome Proliferator-Activated Receptor α*

Ruiyin Chu, Yulian Lin, M. Sambasiva Rao, and Janardan K. Reddy[1]

Department of Pathology, Northwestern University Medical School, Chicago, IL 60611

Peroxisome proliferator-activated receptors (PPARs) are members of the nuclear receptor superfamily which transcriptionally regulate responsive genes by binding to the peroxisome proliferator response elements (PPREs). Protein(s) interacting with PPAR isoforms (α,δ, and γ) may modulate the PPAR-mediated transcriptional activation. Using a yeast two-hybrid system to screen a rat liver cDNA library, we have identified rat deoxyuridine triphosphatase (dUTPase; EC 3.6.1.23) as a PPARα-interacting protein. This cDNA encodes a polypeptide of 203 amino acids; the C-terminal 141 amino acid segment of this protein corresponds to the full-length human enzyme, which exhibits 92% identity with human dUTPase; the N-terminal extra 62 amino acid residue region is arginine-rich. *In vitro* binding assays indicate that rat dUTPase interacts with all three isoforms of mouse PPAR, but not with retinoid X receptor (RXR) and thyroid hormone receptor (TR). Interaction of PPARα with dUTPase is with the N-terminal 62 amino acid segment of rat dUTPase. Full-length rat dUTPase prevents PPAR/RXR heterodimerization resulting in an inhibition of PPAR activity in a ligand-independent manner. Immunostaining of human kidney tsA201 cells, transiently expressing dUTPase showed that this protein is present predominantly in the cytoplasm, but translocates into the nucleus with PPARα, when PPARα is coexpressed with dUTPase. Northern blot hybridization shows that rat dUTPase is encoded by an abundant 1 kb mRNA species present in all rat tissues. The identification of dUTPase as a PPAR-interacting protein suggests a possible link between tumorigenic peroxisome proliferators and the enzyme system involved in the maintenance of DNA fidelity.

Poster Abstract # 35

The Effect of Lipoxygenase Products of Fatty Acids on the Expression of Keratin 1 mRNA in Mouse Keratinocytes

Mary F. Locniskar, Susan M. Fischer[*] and Ruth A. Hagerman

Division of Nutritional Sciences, The University of Texas, Austin, TX, and
[*]The University of Texas M. D. Anderson Cancer Center, Science Park, Smithville, TX

This study was designed to investigate the effects of lipoxygenase products on keratinocyte differentiation. Differentiation of cultured keratinocytes is identified by the expression of differentiation specific keratins and is mediated by the calcium concentration of the culture medium. Treatment with the phorbol ester 12-O-tetradecanoylphorbol-13-acetate (TPA) alters the normal differentiation program by inhibiting Keratin 1 expression. We hypothesized that the lipoxygenase product of arachidonic acid, 12-hydroxyeicosatetraenoic acid (12(S)-HETE) would modulate Keratin 1 expression in cultured keratinocytes, based on evidence from other laboratories demonstrating that TPA and 12(S)-HETE have similar effects on cell adhesion. We found that 100 nM 12(S)-HETE mimics the effect of 500 nM TPA in suppressing Keratin 1 expression within 24 hours of calcium-induced differentiation. However, pretreatment with the lipoxygenase product of linoleic acid, 13-hydroxyoctadecadienoic acid (13(S)-HODE), blocked the 12(S)-HETE response but not that of TPA. Protein kinase C (PKC) was implicated on the basis of inhibitor studies where treatment of cultured keratinocytes with RO-31-8220 or bryostatin-1 abrogated the 12(S)-HETE and TPA effects. Furthermore, TPA and 12(S)-HETE stimulated PKC activity in the keratinocytes. Western analysis demonstrated that TPA caused a rapid but partial translocation of the PKCα isoform to the membrane, whereas PKCδ was not affected. On the other hand, 12(S)-HETE had no effect on PKCα, but did cause the translocation of PKCδ from the membrane to the cytosol. These data suggest that although 12(S)-HETE mimics the effects of TPA, different mechanisms are responsible.

This work was supported by AICR and NIH (CA46886).

Poster Abstract # 36

Molecular action of a phytoantiestrogen Indole-3-carbinol : Sidhanta A, Arvind P, Bradlow HL[1], Osborne MP[1] and Tiwari RK. Department of Microbiology and Immunology, New York Medical College, Valhalla, NY and [1]Strang Cancer Research Laboratory, New York, NY.

Indole-3-carbinol (I3C), a constituent of compounds present in cruciferous vegetables such as cabbage, broccoli, and brussels sprouts, has anticancer properties. Regression of spontaneous mammary tumors in C3H/HeJ in a dose dependent manner was noted when I3C was incorporated in the diet. Examination of the cellular and molecular mechanism of action of I3C in human breast cancer cells revealed that I3C preferentially affected the growth of estrogen receptor (ER) positive cells as compared with ER-negative cells. This effect was presumably mediated by the ability of I3C to selectively modulate estradiol metabolism and induction of associated cytochrome P-4501A1. Facilitative formation of 2-hydroxyestrone that is known to have antiestrogenic properties could partially explain the observed selective action of I3C. However, the most pronounced antiproliferative effect of I3C was observed on estrogen stimulated growth in estrogen responsive human breast cancer cell lines and MCF-7 cell line derived clones. Further examination of this observation showed that I3C inhibited the phosphorylation of ER without affecting the steady state level of the receptor. This inhibition of phosphorylation was not observed with 2-hydroxyestrone. We conclude the I3C inhibits key biochemical processes that affect functioning of ER without binding to the receptor. Phytoantiestrogens, such as I3C, provide a natural alternative to synthetic antiestrogens for human breast cancer prevention and an invaluable tool to dissect the molecular processes of estrogen mediated cell proliferation. (Supported by the American Institute of Cancer Research , Grant # 94B66)

Poster Abstract # 37

ROLE OF CORTICOSTERONE AND ENTEROSTATIN IN VOLUNTARY FAT INTAKE,

Chandan Prasad, PhD, Stanley S. Scott Cancer Center, Department of Medicine, Louisiana State

University Medical Center, 1542 Tulane Ave, New Orleans, LA 70112

Because excessive consumption of fat in diet is associated with many types of cancer, including breast cancer, it is important to understand the mechanisms controlling fat intake. Enterostatin is an endogenous inhibitor of fat intake. We recently observed a strong correlation between fat preference and the efficacy of enterostatin to decrease fat preference, which led us to wonder how a diet rich in fat could increase enterostatin sensitivity. The levels of two hormones, corticosterone and glucagon, consistently increase after chronic consumption of a fat-rich diet. Therefore, we explored whether chronic hypercorticosteronemia could lead to increased enterostatin sensitivity. To this end, rats were screened for fat preference, and 16 exhibiting low fat preference were retested on 2 consecutive days: on day one after saline and on day two after 0.25 mg/Kg enterostatin. After this they were treated with corticosterone-water ($50\mu g/ml$ as the sole drinking fluid) for 15 days. On the 14th and 15th days, fat preference was evaluated again after rats were given saline and enterostatin, respectively. Corticosterone treatment of rats led to a significant increase in total calorie intake which was largely due to increase in fat intake. Enterostatin did not decrease total calorie or fat calorie intake of low-fat preferring rats before corticosterone treatment. In contrast, after corticosterone treatment, enterostatin led to a pronounced decrease in total calorie intake, which was due exclusively to reduction in fat intake. Based on the results of these studies, it is tempting to speculate that chronic hypercorticosteronemia may lead to supersensitization of site(s) where enterostatin may act to reduce fat preference.

Poster Abstract # 38

DIFFERENTIAL EXPRESSION AND REGULATION OF CYCLOOXYGENASES IN TWO HUMAN BREAST CANCER CELL LINES. Xin-Hua Liu, David P. Rose. Division of Nutrition and Endocrinology, American Health Foundation, Valhalla, NY

Dietary linoleic acid provides the source of arachidonate which is further metabolized to prostaglandins and leukotrienes. Some human breast cancers synthesize large quantities of prostaglandin E_2 (PGE_2), but the regulatory mechanisms controlling its production are unclear. Two isoforms of prostaglandin endoperoxide synthase, also referred to as cyclooxygenases (COXs) have been identified. COX is a key enzyme in the conversion of arachidonic acid to PGs and other eicosanoids. We examined COX-1 and COX-2 expression and their regulation by tetradecanoyl phorbol acetate (TPA), as well as the corresponding PGE_2 production, in two breast cancer cell lines with different biological phenotypes. Estrogen-dependent MCF-7 cells exhibited a relatively high expression of COX-1; COX-2 was barely detectable, but was transiently induced by treatment with TPA (10nM). In contrast, the estrogen-independent, highly invasive and metastatic, MDA-MB-231 cell line showed a low expression of COX-1 but a high constitutive level of COX-2. This high COX-2 expression applied to both the protein and mRNA, and was increased further, and over a relatively long period of time, in the presence of TPA. The extent of PGE_2 production in both cell lines correlated well with the level of COX-2.

These results suggest that COX-2 is required for both constitutive and mitogen-induced PGE_2 synthesis in these human breast cancer cell lines. Overexpression and persistent expression of COX-2 may be influenced by breast tumor hormone status, and appear to be a feature of the aggressive, metastatic, phenotype.

The work was supported by Grant 94A12 from the American Institute for Cancer Research.

Poster Abstract # 39

DIFFERENTIAL GROWTH-INHIBITION OF HUMAN MAMMARY CARCINOMA
CELLS BY CURCUMIN

Mahitosh Mandal, Laxminarayanan Korutla, <u>Rakesh Kumar</u>

Departments of Medicine and Cellular and Molecular Physiology

Pennsylvania State University College of Medicine, Hershey, PA 17033

To further understand the growth inhibitory action of curcumin, we have began to explore the effect of curcumin on the growth of human breast cancer BT-20, ZR-75R and T-47D cells. Our results demonstrated that BT-20 cells were most-sensitive and T-47D cells were least-sensitive to growth inhibitory effect of curcumin. Since curcumin has been shown to inhibit the kinase activities of the cellular proteins, we examined the effect of curcumin of the state of phosphorylation of tumor-suppressor retinoblastoma protein, using a pair of curcumin-sensitive BT-20 cells and curcumin-insensitive T-47D cells. Results indicated that treatment of BT-20 cells (curcumin growth-sensitive) with curcumin specifically reduced the expression of slowly migrating phosphorylated RB protein. There was no effect of curcumin on the expression of RB protein in T-47D cells (curcumin growth-insensitive). The observed effect of curcumin on the expression of slowly migrating RB protein in BT-20 cells was dependent on the duration of curcumin treatment, and also on the doses of curcumin used. There was no effect of solvent DMSO on the expression of phosphorylated RB protein. Studies are in-progress to further delineate the influence of curcumin on cell cycle proteins (supported by the American Institute for Cancer Research grant #94B93).

Poster Abstract # 40

BCL-2 DEREGULATION LEADS TO INHIBITION OF SODIUM BYTYRATE-INDUCED APOPTOSIS IN HUMAN COLORECTAL CARCINOMA CELLS

Mahitosh Mandal, and Rakesh Kumar

Departments of Medicine and Cellular and Molecular Physiology

Pennsylvania State University College of Medicine, Hershey, PA 17033

Epidemiological and experimental studies have linked dietary fibre to the prevention of colorectal cancer, and suggest that short chain fatty acids such as butyric acid, the major short-chain fatty acid produced by fermentation of dietary fibre in the large intestine, may be an important mediator of protective effects of fibre. Homeostasis in colonic epithelial cells is regulated by a balance between proliferative activity and cell loss by apoptosis. We investigated the role of deregulation of anti-apoptotic gene product Bcl-2 on the sensitivity of colorectal carcinoma cells to undergo butyrate-induced apoptosis. Here we report an inverse relationship between the levels of Bcl-2 and the sensitivity of colorectal carcinoma cell lines to undergo apoptosis in response to butyrate. Overexpression of Bcl-2 in colorectal carcinoma DiFi cells resulted in suppression of butyrate-induced apoptosis and enhanced cell survival in response to butyrate. Butyrate-induced apoptosis was accompanied with the inhibition of expression of a 30 kD protein (p30 that immuno-recognized by anti-Bcl-2 mAb), and this cellular effect of butyrate was inhibited by Bcl-2 overexpression. These findings suggest that deregulation of Bcl-2 in human colorectal carcinoma cells confers resistance to induction of apoptosis by butyrate, a dietary micronutrient (supported by the American Institute for Cancer Research grant #96A077) .

Poster Abstract # 41

Attenuation of Intestinal Tumor Load by the Non-Steroidal Antiinflammatory Drug Sulindac in Mice with a Defect in the APC Gene. C.-H. Chiu, M. McEntee and J. Whelan. Department of Nutrition and Department of Pathology, University of Tennessee, Knoxville, TN 37996.

The heterozygous Min/+ mouse carries a dominant mutation in one allele of the tumor suppressor adenomatous polyposis coli (APC) gene and loss of the second allele results in spontaneous development of intestinal adenomas (100% incidence). The Min/+ mouse model is being studied because of close similarities to an inherited form of human intestinal cancer, familial adenomatous polyposis (FAP). Sixteen male Min/+ mice were randomly divided into 4 groups. At 77, 101 and 115 days of age, a group of animals were sacrificed at each time point. The fourth group was treated with sulindac (320 ppm) for 75 days and was sacrificed at 115 days of age. Sixteen male control mice (wild type) were placed on an identical protocol. Body weights in Min/+ mice plateaued at 63 days of age, an effect not observed in the sulindac-treated mice. Tumors associated with the Min/+ mice produced significantly higher levels of PGE_2 and LTB_4 as compared to the intestinal mucosa of controls at all ages. Intestinal eicosanoid formation in sulindac-treated Min/+ mice were not significantly different from control mice. Arachidonic acid content of intestinal phospholipids were higher in the Min/+ mice (independent of sulindac) as compared to control mice and these data are consistent with elevated eicosanoid formation; however, no differences were observed in the fatty acid composition of hepatic phospholipids. Min/+ mice produced an average of 42 ± 5 tumors per mouse. The average number of tumors did not significantly increase after 77 days of age. Sulindac-treated Min/+ mice (115 days of age) had an average of 60% fewer tumors that were 30-50% smaller in diameter as compared to untreated age-matched Min/+ mice. We conclude that sulindac, a pro-PGH synthase inhibitor which is activated in the liver and intestines, significantly decreases the number and size of intestinal tumors in Min/+ mice and eicosanoid metabolism may be related to the growth and/or maintenance of intestinal tumors in this model.

Poster Abstract # 42

American Institute for Cancer Research, 1996.

Dietary fat and Cancer: Genetic and Molecular Interactions

A high-fat diet during pregnancy increases estrogen receptors in the mammary gland and breast cancer risk. L. Hilakivi-Clarke, A. Stoica, D. Antonetti, I. Onojafe, M.-B. Martin, and R. Clarke. Lombardi Cancer Center, Georgetown University, Washington, DC.

High pregnancy estrogen levels appear to increase and low levels to reduce breast cancer risk. Since a high-fat diet increases serum estrogens and may increase breast cancer risk, we hypothesized that consumption of a diet high in n-6 polyunsaturated fatty acids (PUFA) during pregnancy increases the risk to develop this disease. In our study, pregnant Sprague-Dawley rats that were previously exposed to 7,12-dimethylbenz(a)anthracene (DMBA), were assigned to one of three isocaloric diets containing 16% calories from fat (low-fat), 31% calories from fat (medium-fat) or 43% calories from fat (high-fat) for the length of pregnancy (21 days). The fat source was corn oil, which is high in n-6 PUFA, primarily linoleic acid. The number of rats developing mammary tumors was significantly higher in the group that was exposed to a high-fat diet, than that exposed to a low-fat diet during pregnancy (Log Rank test: $z=2.15$, $p<.03$). Tumor multiplicity, latency to tumor appearance, and size of tumors upon first detection were similar among the dietary groups.

Our previous studies show an elevation in serum estradiol levels in pregnant rats consuming a high-fat diet. Thus, we measured the level of estrogen receptor (ER) protein in the mammary gland. Virgin female mice fed with the high-fat diet, exhibited a six-fold increase in ER protein in the gland, when compared with the low-fat group ($t=3.4$, $df=8$, $p<.01$). During pregnancy, mammary ER levels were slightly, but non-significantly increased in the high-fat group. However, four weeks after pregnancy, ER levels were higher in the female mice fed a high-fat diet during pregnancy than in the group fed a low-fat diet ($t=2.2$, $df=8$, $p<.06$). Our findings indicate that consumption of a high-fat diet during pregnancy increases the risk of developing carcinogen-induced mammary tumors, possibly through permanently altering ER levels in the mammary gland. Thus, ER may an important mediator between an exposure to a high-fat diet during pregnancy and breast cancer.

Poster Abstract # 43

American Institute for Cancer Research, 1996.

Dietary fat and Cancer: Genetic and Molecular Interactions

A high-fat diet during pregnancy reduces serum insulin levels and insulin resistance, and insulin expression in the mammary gland. L. Hilakivi-Clarke, E. Cho, and M. Raygada. Lombardi Cancer Center, Georgetown University, Washington, DC.

Accumulating evidence suggests that high pregnancy insulin levels and insulin resistance are causally related to cardiovascular diseases among the offspring. Insulin also plays a critical role in normal and malignant growth of the mammary tissue. We have recently shown that insulin mRNA is expressed in the human and rodent mammary gland and breast tumors. Since fat stimulates insulin release from the pancreas, and fat intake and obesity-linked insulin resistance is associated with breast cancer, we studied whether consumption of a high-fat diet during pregnancy alters circulating insulin and glucose levels, or expression of insulin mRNA in the mammary gland. Female Balb/c mice and Sprague-Dawley rats were assigned to two isocaloric diets containing 16% calories from fat (low-fat) or 43% calories from fat (high-fat), and bred. The fat source was corn oil, which is high in n-6 polyunsaturated fatty acids (PUFA), primarily linoleic acid. Consistent with earlier studies, we found that serum insulin levels and insulin resistance were significantly elevated in pregnant vs. non-pregnant animals. During pregnancy, the serum insulin levels (t=2.9, df=8, p<.02) and insulin/glucose ratio (an index of insulin resistance) (2.6, df=7, p<.03) in the animals fed the high-fat diet were significantly *lower* than in the low-fat fed animals. No differences in serum glucose levels were noted between the groups. Expression of insulin mRNA, as determined using RNase protection assay, was reduced in the pregnant mammary gland of animals exposed to a high-fat diet.

The significance of our findings remains unclear. Our previous data indicate that carcinogen-induced breast cancer risk is elevated in female rats and their offspring that were exposed to a high-fat diet during pregnancy (or fetal life). It also is true that women exhibiting a reduced subsequent breast cancer risk, have increased blood insulin levels and insulin resistance, and reduced serum levels of PUFA during pregnancy. Thus, low levels of linoleic acid and high levels of insulin during pregnancy appears to protect from breast cancer, while the opposite may be true for high levels of linoleic acid and reduced circulating levels of insulin.

Poster Abstract #44

SPHINGOMYELIN CONSUMPTION INFLUENCES TUMOR DEVELOPMENT IN CF1 MICE TREATED WITH 1,2-DIMETHYLHYDRAZINE: IMPLICATIONS FOR DIETARY SPHINGOLIPIDS AND COLON CARCINOGENESIS. E.-M. Schmelz, D. L. Dillehay,*# S. K. Webb#, A. Reiter+, J. Adams+ and A. H. Merrill, Jr. Departments of Biochemistry; * Pathology; +Chemistry and the Emory Mass Spectrometry Center; and #Division of Animal Resources, Emory University, Atlanta, Ga 30322.

Colon cancer is the second leading cause of cancer mortality in the United States. Although numerous epidemiological studies found a relationship between dietary factors and colon carcinogenesis, only a few potentially beneficial compounds have been evaluated until now. Prominent among these compounds are sphingolipids which modulate processes that are important in carcinogenesis. Ceramide, Ceramide-1-phosphate, sphingosine, sphingosine-1-phosphate, and sphingosyl-phosphorylcholine are able to stimulate or inhibit cell growth, enhance or inhibit differentiation, and induce cell death, in some systems, via apoptosis.

Complex sphingolipids are hydrolyzed in the gastrointestinal tract to ceramide, sphingosine, and other bioactive metabolites. To characterize the effects of dietary sphingolipids on colon carcinogenesis, female CF1 mice were administered 1,2-dimethylhydrazine to induce colonic tumors. They were then fed an essentially sphingolipid-free AIN 76A diet supplemented with 0 to 0.1% (w/w) sphingomyelin (SM) purified from powdered skim milk and buttermilk. As was found in a previous pilot study (Dillehay, D.L. et al., J. Nutr. 124: 615-620, 1996), feeding SM (@ 0.1%) for 4 weeks significantly reduced the number of aberrant colonic crypt foci and aberrant crypts per focus, which are early indicators of colon carcinogenesis. In longer term studies (34 weeks), SM had no effect on the colon tumor incidence nor multiplicity; however, up to 31% of the tumors of mice fed SM were adenomas, while all of the tumors of mice fed the diet without SM were adenocarcinomas.

These findings demonstrate that milk SM suppresses the appearance of more advanced, malignant tumors as well as early markers of colon carcinogenesis. Although the sphingolipid content of foods has not been widely studied, several foods (e.g., milk and soybeans) contain the sphingolipid levels used in these investigations; therefore, this class of compounds could be significant contributors to the cancer preventive effects of some foods.

Supported by funds from the National Dairy Board administered in cooperation with the National Dairy Council and NCI grant CA61820.

Poster Abstract # 45

Possible Role of Leptin in Cancer Anorexia. W.T. Chance, S. Sheriff, R. Dayal, J. Moore, A. Balasubramaniam and J.E. Fischer. VA Medical Center and Department of Surgery, University of Cincinnati Medical Center.

Anorexia is a common problem of neoplastic disease that limits therapy and increases cancer morbidity and mortality. Although many hypotheses of cancer anorexia have been advanced, few have adressed the problem at the molecular neurochemical level. Recent investigations have demonstrated an association of the obese gene product, leptin (LEP), with hypophagia, adipose tissue and lipid synthesis/metabolism. This protein affects hypothalamic control of food intake by reducing the synthesis and release of neuropeptide Y (NPY), a potent stimulator of feeding. Tumor-bearing (TB) organisms exhibit disturbances in lipid metabolism that are characterized by elevated lipolysis, hyperlipidemia and perhaps lipid recycling. Hypothalamic concentration and release of NPY are decreased in TB rats. In addition, intracellular cyclic AMP second messenger system activity for NPY-induced feeding is also reduced in anorectic TB rats. This suggested decreased in biological response to NPY is manifest as decreased feeding to NPY even prior to the development of overt anorexia in TB rats. Therefore, we hypothesized that alterations in circulating and hypothalamic LEP might be involved in the development of cancer anorexia.

Nine male, Fischer 344 rats were inoculated with approximately 50 mg fresh methycholanthrene sarcoma, while 14 additional rats received sham inoculations to form freely-feeding (FF) and pair-fed (PF) control groups. After the development of significant anorexia in TB rats, all rats were euthanized for the determination of plasma LEP levels by RIA and LEP receptor binding to hypothalamic membranes.

Group	BW (g)	Food (g/100 g BW)	Plasma LEP (ng/ml)	LEP Binding (fmole/mg prot)	Triglycerides (mg%)
FF	294 + 10	6.5 + 0.3	3.03 + 0.18	1.32 + 0.12	216 + 29
PF	209 + 4	3.0 + 0.7	1.24 + 0.19	2.05 + 0.38	71 + 10
TB	279 + 12	2.4 + 0.5	0.37 + 0.04	3.75 + 0.56	746 + 82

These results suggest major alterations in circulating levels of LEP in anorectic TB rats. Since TB organisms mobilize adipose tissue early in cachexia, adipocyte production of LEP may be decreased. It appears that the hypothalamus may respond to this decrease in circulating LEP by increasing LEP receptor density (B max). This increase in biological response to LEP in the hypothalamus may in turn lead to the decreases in synthesis and release of NPY observed in TB rats. Since NPY may represent the final common pathway for feeding, reduced biological activity of NPY feeding systems may result in the anorexia observed in TB organisms.

Poster Abstract # 46

Role of Arachidonic Acid Metabolism in Prostate Cancer Progression

Ghosh, J. and Myers, C. E. Jr., University of Virginia Cancer Center, Charlottesville, VA 22908.

Arachidonic acid [AA], an ω-6 poly-unsaturated fatty acid, was found to be a potent stimulator of invasion and growth of prostate cancer cells. Other than its direct effect on the activation of PKC and ras-GAP, AA is known to be metabolized through a variety of metabolic pathways to produce a host of active intermediates. To understand the nature of its effect we used metabolic blockers to selectively interrupt its conversion through different metabolic pathways. In prostate cancer cells the stimulatory effect of AA was completely abolished when the cells were treated with NDGA, which blocks the conversion of AA through the lipoxygenase pathways, whereas inhibitors of the cyclooxygenase pathway (ibuprofen, indomethacin etc.) and cytochrome P-450 mediated epoxygenase pathway (SKF-525A) were ineffective. These observations suggest that the effect of AA potentiating the malignant behavior of prostate cancer cells is mediated by its metabolic conversion through the lipoxygenase pathway. Moreover, further study revealed that in prostate cancer cells the 5-lipoxygenase pathway plays a critical role in mediating the effect of AA in these processes. In cellular environment AA can be made available from membrane phospholipids by the activity of the enzyme phospholipase A2 (PLA2). Of the many variants of PLA2 the cytosolic PLA2 (cPLA2) was recently found to be regulated by growth factor receptor mediated signals via MAP kinase. Uteroglobin an endogenous inhibitor of cPLA2 was found to have negligible expression in DU145 and PC3M prostate cancer cell lines. We observed that in PC3 cells cPLA2 is heavily phosphorylated and membrane localized which did not show any further increase after serum stimulation or treatment with calcium ionophore, A23187. Inhibition of PLA2 activity by quinacrine killed PC3 cells which could be prevented by extraneous addition of AA, a product of PLA2. Moreover, prostate cancer cells were observed to undergo apoptotic cell death when the metabolic conversion of AA through 5-lipoxygenase pathway was interrupted. Treatment of nude mice with 5-lipoxygenase inhibitor slowed growth of implanted prostate tumors and increased survival of the treated animals. These observations strongly suggest that metabolism of AA is very important for the survival of prostate cancer cells and opens up new target for prostate cancer chemotherapy.

Poster Abstract # 47

COMBINATION EFFECT OF SELECTED NUTRIENTS AND NON-NUTRIENTS ON AZOXYMETHANE-INDUCED ABERRANT CRYPT FOCI IN RAT COLON

D. RAMKISHAN RAO and A. Challa. Dept. of Food Science & Animal Industries, Alabama A&M University, Normal, AL 35762.

The association between diet and cancer has long been established by epidemiological and experimental studies. However, there are very few studies dealing with the effect of combining nutrients and non-nutrients, as generally consumed in diets, on colon tumorigenesis. Thus, the present study was aimed at determining the effect of combining green tea (GT) and phytic acid (PA) or *Bifidobacterium longum* (Bl; BB536, gift of Morinaga Milk Ind., Japan) and lactulose (L) with isoflavones (IF) on azoxymethane (AOM) -induced colonic aberrant crypt foci (ACF) in rat colon.

In each of the two experiments, 75 male Fisher 344 weanling rats were divided into 5 groups of 15 rats each and fed the experimental diets for 13 weeks. Diets in experiment 1 were: Control (AIN76A), 0.5% Bl, 2.5% L, 0.5% Bl+2.5% L, Bl+2.5% L+0.075% IF. Diets in experiment 2 were: Control (AIN76A), 1% GT, 1% PA, 1% GT+1% PA, 1% GT+1% PA+0.075% IF. Animals were injected twice with AOM @16mg/kg body wt. at 7 and 8 weeks of age and sacrificed at 17 weeks of age. Colons of 10 rats in each group were assayed for ACF and the liver and colonic mucosa of the remaining 5 rats in each group were used to assay glutathione-s-transferase(GST) activity.

Feeding B+L had an additive effect while addition of IF to Bl+L did not have any further effect ($p=0.0001$). Feeding GT and PA at a 1% level did not significantly reduce the number of ACF. However, combining the two reduced the incidence of ACF by 15.7% and addition of IF to GT+PA reduced the ACF by 33% compared to the control ($p=0.0005$). In all the dietary treatments, the number of ACF with 2 crypts/focus was significantly higher in number compared to number of ACF with 1,3,4 or 5+ crypts/focus. The diets had no significant effect on weight gains of the animals. Feeding 1% GT alone increased the colonic mucosal GST specific activity compared to the 1%PA, GT+PA and GT+PA+IF groups. Addition of IF increased the liver GST specific activity significantly ($p<0.05$). Feeding of L increased the liver GST total activity significantly compared to

B+L. Feeding of L,Bl+L, Bl+L+IF significantly increased the cecal wt. of the animals compared to the control($p<0.05$), with the IF group having the highest cecal wt. ($p=0.0001$). Addition of L reduced the cecal pH significantly ($p=0.0001$). Results indicate an additive effect of the nutrients and non-nutrients in suppressing the ACF. Since the mechanism of anticarcinogenic activity of the nutrients is different, combining some of them seems to augment their beneficial effect, thereby stressing the need to combine nutrients and non-nutrients in the diet to protect against colon tumorigenesis.

Key words: glutathione-s-transferase, colon, aberrant crypt foci, nutrients, non-nutrients.

Poster Abstract # 48

INDUCTION OF HEPATIC HEME OXYGENASE-1 AND FERRITIN IN RATS BY CANCER CHEMOPREVENTIVE DITHIOLETHIONES LEADS TO DECREASED INTRACELLULAR FREE IRON. Thomas Primiano[1*], Thomas R. Sutter[1], Periannan Kuppusamy[2], Jay L. Zweier[2], and Thomas W. Kensler[1*]. [1]The Department of Environmental Health Sciences, The Johns Hopkins School of Public Health, and [2]The Johns Hopkins Electron Paramagnetic Resonance Center, Baltimore, MD 21205, U.S.A.

Dithiolethiones including oltipraz, anethole dithiolethione and 1,2-dithiole-3-thione (D3T) inhibit experimental tumorigenesis elicited by many structurally diverse carcinogens in numerous target tissues. The knowledge of the mechanisms of how these compounds inhibit tumor formation would provide significant insight into the processes of carcinogenesis and anticarcinogenesis. Treatment of rats with the cancer chemopreventive agent 1,2-dithiole-3-thione (D3T) resulted in a significant increase of hepatic heme oxygenase (HO) activity, which corresponded to increased protein levels of HO-1. The levels of HO-1 protein were dramatically elevated over those of untreated controls at 6 h and remained elevated up to 24 h after treatment with D3T;whereas, HO-2 protein levels failed to increase in response to administration of D3T, indicating that enhanced heme oxygenase activities are the result of inductions in HO-1 levels alone. Additionally, the level of ferritin, the major iron storage protein in liver was elevated 3.7- and 5.0-fold in the hepatic S9 fractions of rats 6 and 24 h after treatment with D3T, respectively. These enhancements in protein levels were associated with increased steady-state RNA and a direct relationship between enhanced rates of gene transcription and elevated levels of HO-1 and ferritin RNA was found. To determine if increased ferritin synthesis affected intracellular iron concentrations, the levels of desferrioxamine-chelatable iron in rat liver tissue were quantitated by monitoring its specific paramagnetic signal. The steady-state levels of free iron in liver tissue decreased approximately 40 % 24 h after a single dose of D3T as measured by electron paramagnetic resonance spectroscopy (EPR). Furthermore, time-dependent reductions in the EPR signal for the steady-state levels of iron were found from 18-30 h after a single dose of dithiolethione. This decrease in intracellular iron coincided with the induction of ferritin protein. Therefore, we hypothesize that intracellular iron is sequestered into newly synthesized ferritin which may effectively reduce iron-mediated reactive oxygen generation. Thus, protective actions of D3T against the cytotoxic and carcinogenic consequences of chemicals that exert electrophilic or oxidative stresses may be mediated, in part, by the induction of HO-1, FL, and FH. Supported by NIH Grants CA 39416, ES 03819, ES 07141 and AICR Grant 95A119.

Poster Abstract #49

Photocarcinogenesis and Immune Suppression: preliminary studies on the effects of L-histidine on trans-urocanic acid levels in murine skin

Edward C. De Fabo & Lindsay J. Webber

Laboratory of Photoimmunology and Photobiology, Department of Dermatology,

The George Washington University, Washington, D.C.

Abstract

Dietary factors have been shown to affect UV-carcinogenesis. One of our previous studies showed that increased dietary L-histidine leads to increased levels of skin *trans*-urocanic acid (*trans*-UCA), an immune-regulating skin photoreceptor which, in turn, leads to increased sensitivity to UVB-induced immune suppression. In a two-phased photocarcinogenesis study, we wished to determine if changing the natural levels of *trans*-UCA in the skin, by increasing or decreasing L-histidine in the diet, would lead to increased levels of skin tumors. Preliminary results from phase I indicate that changing the histidine levels in the diet of BALB/c mice alters skin *trans*-UCA in a way suggesting an unusual type of control on its *in situ* formation. Preliminary data also suggest an increase in *trans*-UCA levels with low dietary intake of histidine. This may be the result of a type of protein malnourishment with histidine becoming available through self-digestion of protein. Such a condition could explain the observed changes in *trans*-UCA levels and lack of significant weight gain in those animals fed low levels of histidine. Precedent for this may be seen in human situations, for example kwashiorkor disease. Phase II of our study will determine how changes in dietary histidine affect UVB-induced skin cancer. Such experiments are currently underway in our laboratory.

Poster Abstract # 50

Genistein may inhibit the growth of human mammary epithelial (HME) cells by augmenting transforming growth factor beta (TGFβ)-signaling. T.G. Peterson, H. Kim and S. Barnes. Department of Pharmacology and Toxicology, University of Alabama at Birmingham, Birmingham, AL 35294, USA.

Mechanisms that have been proposed to account for the chemopreventive action of genistein mostly center on inhibition of membrane bound and intracellular protein tyrosine kinases (PTK). However, although genistein inhibits the epidermal growth factor (EGF)-stimulated growth of human breast cancer cell lines (Peterson and Barnes, Cell Growth & Differentiation, in press) and normal human mammary epithelial (HME) cells, it does not inhibit EGF receptor tyrosine autophosphorylation, suggesting that a mechanism other than PTK inhibition is involved. Since genistein inhibits cell cycle progression at G_1/S, we investigated the role of genistein on the regulation of cellular proteins which block cell growth at the G_1/S boundary, such as TGFβ. Addition of TGFβ (IC_{50} 2.4 ng/ml) or genistein (IC_{50} 5 μM) inhibited the EGF-stimulated growth of HME cells. The growth inhibitory effects of both genistein and TGFβ were reversed by the addition of anti-TGFβ$_{1,2,3}$ or anti-TGFβ$_1$ antibodies to the culture medium, suggesting that genistein-induced synthesis of TGFβ$_1$ was responsible for the growth inhibition in HME cells. The anti-TGFβ antibodies had no effect on the EGF-stimulated growth of untreated HME cells in the absence of TGFβ$_1$ or genistein. SELISA analysis revealed a 4.7-fold increase in the amount of TGFβ$_1$ secreted into the medium by the HME cells treated with genistein and EGF compared to the amount secreted by cells treated with EGF alone. We conclude that the mechanism of genistein's inhibition of the growth of HME cells is based on regulation of TGFβ synthesis, either through increased synthesis or decreased degradation, thereby modulating TGFβ-signaling. Support for this concept comes from clinical studies in which consumption of a soy-based beverage (that includes genistein) has been shown to be effective in the treatment of the human nosebleed disorder, hereditary hemorrhaegic telangiectasia (HHT) (Korzenik et al., 1996). HHT is a genetic disorder involving mutations in endoglin, a TGFβ-binding protein. This mechanism based on TGFβ function may not only underlie the action of genistein in the treatment of HHT, but also in the prevention of cancer and cardiovascular diseases.

Supported by grants from AICR and NCI (CA-61668).

Poster Abstract # 51

CIS-POLYUNSATURATED FATTY ACIDS INDUCE INTEGRIN- AND PROTEIN KINASE C-MEDIATED HUMAN BREAST CARCINOMA CELL ADHESION TO TYPE IV COLLAGEN. Palmantier, R., Roberts, J.D., George, M.D., and Olden, K., Lab of Molecular Carcinogenesis, Research Triangle Park, NC 27709.

Polyunsaturated fatty acids (PUFAs) have been shown to alter the outcome of breast cancer and to influence the formation of metastases *in vivo*. Tumor cell adhesion to basement membrane proteins, such as type IV collagen, is a crucial step in the metastatic cascade. In this study, we investigated the effect of FAs on the adhesion of a metastatic human breast cancer cell line, MDA-MB-435, to human type IV collagen. Arachidonic acid, linoleic acid, gamma-linoleic acid, eicosapentaenoic acid, and linolenic acid induced a dose dependent increase in cell adhesion to collagen following 30 min incubation. No significant increase in nonspecific adhesion to polylysine and BSA was observed. In contrast, oleic acid (monounsaturated), linoelaidic acid (*trans,trans*-linoleic acid) did not significanctly alter cell adhesion to collagen. Similarly, TPA, but not the phorbol ester 4-α-PDD which does not activate protein kinase C (PKC), induced cell adhesion to collagen, suggesting that PKC activation is involved in the modulation of adhesion. Furthermore, PUFA-stimulated adhesion to collagen was inhibited by calphostin C, a specific PKC inhibitor. Function-blocking antibodies against $\beta 1$ and $\alpha 2$ integrins, but not against $\alpha 3$, inhibited stimulated-cell adhesion to collagen, without affecting nonspecific cell adhesion to polylysine. We did not observe an enhancement in $\beta 1$ or $\alpha 2$ expression in PUsFA- or PMA-treated cells by flow-cytometry analysis, suggesting that the increase in adhesion was due to an activation of integrin function rather than an increase in integrin expression. Our results suggest that *cis*-PUFA specifically induced an integrin-mediated cell adhesion to collagen dependent on PKC activity. The lack of effect of other FAs than *cis*-PUFA raises the possibility that cyclooxygenase or lipoxygenase metabolites are involved in this cellular response rather than a direct effect of the *cis*-PUFAs on PKC activity and subsequent integrin function.

Poster Abstract # 52

Dietary Regulation of the Stearoyl-CoA Desaturase Gene in Liver: Characterization of a Novel DNA:Protein Complex

Robert J. Christy, Bihong Zhao and James P. Fitzgerald. University of Texas Health Science Center at San Antonio, Institute of Biotechnology, Center for Molecular Medicine, San Antonio, Texas 78250

In liver, the transcriptional activity of the Stearoyl-CoA Desaturase 1 (*Scd1*) gene is regulated in part by fasting and refeeding a fat-free, high carbohydrate diet (Ntambi, 1992). In order to understand the mechanism of this regulation, we examined the proximal 530 base pairs of the SCD1 5' flanking sequence using DNaseI protection and electrophoretic mobility shift assays. DNaseI protection assays revealed four regions of DNA/protein interaction in liver nuclear extracts. Two of these DNA/protein complexes remained unchanged in response to fasting and refeeding, and two were altered in response to fasting and refeeding. The first altered complex is at a site where complex formation has been demonstrated to change during 3T3-L1 adipocyte differentiation, and involves members of the CCAAT/Enhancer Binding Protein (C/EBP) family of transcriptional regulators. We demonstrate that the composition of C/EBP isoforms binding to the SCD1 5' flanking sequence is transiently altered upon refeeding. The other complex involves the binding of a protein(s), designated the Stearoyl-CoA Desaturase Binding Protein (SCDBP) (Christy et al., 1989), at a sequence adjacent to the C/EBP site. We show that SCDBP binding activity is disrupted by fasting, and reappears within two hours after refeeding. The inhibition of new protein synthesis by treatment with cycloheximide prevented both the reappearance of SCDBP binding and the accumulation of SCD1 mRNA following refeeding. The patterns of SCDBP binding activity and SCD1 RNA expression parallel each other in five tissues. SCDBP binding is present in liver and kidney, correlating with *Scd1* gene expression, while neither SCDBP binding nor SCD1 mRNA is detected in testis, muscle or intestine. We postulate that SCDBP may be required for *Scd1* transcriptional activation and is a central factor regulating the expression of *Scd1* in liver in response to fasting and refeeding.

Poster Abstract # 53

THE VITAMIN D ANALOG EB1089 INDUCES APOPTOSIS OF MCF-7 XENOGRAFTS
IN NUDE MICE. Kathryn Van Weelden, Martin Tenniswood and JoEllen Welsh, W. Alton Jones
Cell Science Center, Lake Placid, NY. 12946

We have demonstrated that $1,25(OH)_2D_3$, the physiological metabolite of vitamin D, and

its synthetic analog, EB1089, induce characteristic features of apoptosis such as chromatin

condensation, nuclear matrix degradation and DNA fragmentation of MCF-7 cells *in vitro*. These

morphological changes co-incide with up-regulation of clusterin and cathepsin B, proteins linked

to apoptosis in mammary gland, and down regulation of bcl-2, an anti-apoptotic protein. To

determine whether vitamin D compounds could mediate apoptotic tumor regression *in vivo*, we

treated nude mice carrying established MCF-7 xenografts with 60pmol EB1089 daily for up to 5

weeks. Compared to vehicle treated mice, tumors from EB1089 treated mice grew at a slower rate

from week 1 on. Tumor doubling times, calculated between week 1 and week 2 were 10.2 days in

control vs. 73.7 days in EB1089 treated mice. Tumor volume after 2 weeks of treatment was

significantly lower in EB1089 treated mice (382.5 ± 85 mm^3, n = 10) than in controls (612.1 mm^3

$\pm$ 54, n = 7). Preliminary histological analysis of formalin fixed sections indicates that tumors from

EB1089 treated mice exhibited large areas of apoptotic regression (detected by *in situ* end labeling

of DNA and Hoechst fluorescence) which were infrequent in tumors from vehicle treated mice. The

expression of tissue transglutaminase, an enzyme involved in apoptosis, was enhanced in tumors

derived from EB1089 treated mice compared to control treated mice. These studies demonstrate that

the synthetic vitamin D analog EB1089 induces regression of human breast tumors *in vivo* by a

process which involves apoptosis. [Supported by AICR]

Poster Abstract # 54

Vitamin D Mediated Apoptosis in an Estrogen Independent Cell Line (SUM-159PT).
Louise Flanagan and JoEllen Welsh, W. Alton Jones Cell Science Center, Lake Placid, NY. 12946;
Dept. Of Botany, University College Dublin, Belfield, Ireland.

1,25 dihydroxyvitamin D_3, $(1,25(OH)_2D_3)$, the active metabolite of vitamin D, and the 'negative growth regulator' TGFβ exert potent effects on cell growth and differentiation. Previous work has demonstrated that $1,25(OH)_2D_3$ induces apoptosis in MCF-7, estrogen dependent (ER+) human breast cancer cells. We are currently characterizing the effect of $1,25(OH)_2D_3$ and TGFβ on an estrogen independent (ER-) breast cancer cell line, SUM-159PT (kindly received from the University of Michigan Cell/Tissue Bank). To date we have demonstrated by western blotting that SUM-159PT cells express the vitamin D receptor (VDR). The functionality of the VDR was determined by analyzing its ability to bind ligand; to bind Vitamin D response elements (VDRE's) (gel shifts); and to modulate cell growth. Ligand binding assays have demonstrated that the number of VDR expressed in SUM-159PT is 27 pmole/mg which is less than that seen in MCF-7 cells, however the affinity of the VDR for $1,25(OH)_2D_3$ is approximately the same for both cell lines (K_d value = 9.4×10^{-11}M). Cell growth studies have demonstrated that treatment of SUM-159PT cells with 100nM $1,25(OH)_2D_3$ reduces viable cell numbers by 70% within 7 days. We have demonstrated that SUM-159PT cells are extremely sensitive to $1,25(OH)_2D_3$ mediated apoptosis (detected by *in situ* end labeling of DNA and Hoechst fluorescence). The role of TGFβ is being investigated with respect to the apoptotic pathway in this ER- model. Cell growth studies have shown that SUM-159PT are relatively sensitive to TGFβ1. Furthermore, when TGFβ1 is used in conjunction with $1,25(OH)_2D_3$, an additive effect on cell viability is observed ≈90% reduction of viable cell number compared to controls when treated for 7 days. We are currently testing the hypothesis that $1,25(OH)_2D_3$ exerts its effects on SUM-159PT cell growth through the TGFβ pathway. *In vivo* experiments carried out to date show that this estrogen independent cell line grows in ovarectomized nude mice in the absence of estradiol supplementation. The effect of EB 1089, a vitamin D analog, on SUM-159PT tumor growth and morphology is being investigated. (Supported by the AICR).

Poster Abstract # 55

AGN193109 IS A HIGHLY EFFECTIVE ANTAGONIST OF RETINOID ACTION IN HUMAN ECTOCERVICAL EPITHELIAL CELLS Chapla Agarwal, Roshantha A.S. Chandraratna, Alan T. Johnson, Ellen A. Rorke and **Richard L. Eckert**, Departments of Physiology and Biophysics, Dermatology, Reproductive Biology, Biochemistry and Environmental Health Sciences, Case Western Reserve University School of Medicine, Cleveland, Ohio 44106-4970 and Retinoid Research, Allergan Pharmaceuticals, Irvine, CA 92713

Retinoids are important physiological agents that regulate epithelial cell differentiation and proliferation. The importance of these agents in regulating growth, development and differentiation has led to a search for new retinoid agonists and antagonists. In the present manuscript we show that AGN193109, a retinoid analog, is an efficient antagonist of retinoid action in human cervical epithelial cells. Treatment of ECE16-1 cells with natural or synthetic retinoids reduces cytokeratin K5, K6, K14, K16 and K17 level, increases cytokeratin K7, K8 and K19 level, increases retinoic acid receptor-b (RARβ) mRNA levels, suppresses proliferation and alters cell morphology. Co-treatment with AGN193109 prevents these responses. Half-maximal and maximal antagonism is observed at a molar ratio of AGN193109:retinoid agonist of 1:1 and 10:1. When administered alone AGN193109 has no agonist activity. Thus, AGN193109 which binds to RARα, β and γ with Kd's = 2, 2 and 3 nM, respectively, but is unable to bind to the RXR receptors, is a highly active antagonist of retinoid action in ECE16-1 cells.

Poster Abstract # 56

RXR-SPECIFIC RETINOIDS INHIBIT THE ABILITY OF RAR-SPECIFIC RETINOIDS TO INCREASE THE LEVEL OF INSULIN-LIKE GROWTH FACTOR BINDING PROTEIN-3 (IGFBP-3) IN HUMAN ECTOCERVICAL EPITHELIAL CELLS   Joan R. Hembree, Chapla Agarwal, Richard L. Beard, Roshantha A.S. Chandraratna, and **Richard L. Eckert**, Departments of Physiology and Biophysics, Dermatology, Reproductive Biology, and Biochemistry, Case Western Reserve University School of Medicine, Cleveland, Ohio 44106-4970 and Retinoid Research, Allergan Incorporated, Irvine, CA 92713

The hormones derived from vitamin A and related synthetic ligands (retinoids) are important regulators of differentiation and development and have been shown to be therapeutically useful in the treatment of cervical cancer.   All-trans retinoic acid exerts its effects by activation of retinoic acid receptor (RAR) and retinoid X receptor (RXR) heterodimers. These heterodimers bind to the retinoic acid response elements (RAREs) of target genes to regulate gene expression. RXR ligands act through RXR homodimers to regulate gene expression. In the present manuscript we describe the effects of RAR- and RXR-specific ligands on regulation of IGFBP-3 production and cell proliferation in human ectocervical epithelial cell lines.   Treatment of ECE16-1 cells with an RAR-specific ligand (TTNPB) or a ligand that interacts with both RAR and RXR receptors (9-cis-retinoic acid) increases IGFBP-3 levels and suppresses cell proliferation. In contrast, RXR-specific ligands (AGN191701, SR11217 and SR11237) do not regulate proliferation and slightly suppress IGFBP-3 level. Cotreatment with increasing concentrations (0.01 - 1000 nm) of RXR-specific ligand antagonizes the growth suppressive and IGFBP-3 increasing effects of 1000 nM TTNPB. Similar results are observed in two other ectocervical epithelial cell lines, ECE16-D1 and ECD16-D2. These results indicate that RXR-specific ligands can antagonize RAR responses in these cell lines and suggests that an RAR-specific retinoid may be superior to one with mixed RAR/RXR binding activity for inhibiting cervical cancer cell proliferation. Moreover, the antagonism of RAR-dependent responses by RXR-specific ligands is consistent with a squelching model in which the RXR-specific ligand drives formation of RXR/RXR homodimers at the expense of the more active RAR/RXR heterodimers.

DIFFERENTIAL REGULATION OF HUMAN ECTOCERVICAL EPITHELIAL CELL LINE
PROLIFERATION AND DIFFERENTIATION BY RXR- AND RAR-SPECIFIC RETINOIDS   Chapla
Agarwal, Roshantha A.S. Chandraratna, Min Teng, Sunil Nagpal , Ellen A. Rorke and **Richard L.
Eckert**, Departments of Physiology and Biophysics, Dermatology, Reproductive Biology, Biochemistry
and Environmental Health Sciences, Case Western Reserve University School of Medicine, Cleveland,
Ohio 44106-4970 and Retinoid Research, Allergan Incorporated, Irvine, CA 92713

Retinoids are important regulators of human papillomavirus (HPV)-immortalized cervical epithelial
cell differentiation and have been successfully utilized in the treatment of HPV-involved cervical cancer.
In the present manuscript, we examine the effects of a series of natural and synthetic retinoids on
differentiation and proliferation of HPV16-positive lines, ECE16-1 and CaSki. RARα, RARγ and RXRα
are the major retinoid receptor subtypes expressed when ECE16-1 cells are grown in retinoid-free
medium. Our results indicate that ligands which interact with RARs only or both RARs and RXRs,
including all-trans-retinoic acid (all-trans-RA), 9-cis-retinoic acid (9-cis-RA), 13-cis-retinoic acid (13-cis-
RA) and several synthetic retinoids, suppress ECE16-1 cell proliferation, regulate expression of the
retinoid-responsive differentiation marker, cytokeratin K5 and increase RARβ mRNA levels. In contrast,
ligands which specifically interact with RXRs do not suppress proliferation and are less efficient
regulators of gene expression. CaSki cells express greatly reduced RAR and RXR levels compared to
ECE16-1 cells. However, both RAR- and RXR-specific ligands increase CaSki number by $\geq$20%. In
addition, RXR-specific ligands suppress cytokeratin K5 mRNA levels slightly, compared to RAR-specific
ligands which strongly suppress K5 mRNA levels. We also compare the effects of these agents on the
proliferation of other cervical cell lines including ECE16-D2, ME180 and SiHa cells. ECE16-D2 and
ME180 cells are growth suppressed by RAR-specific, but not RXR-specific retinoids. SiHa cells are not
responsive to either class of retinoid. Our results indicate that (i) the response of different human
cervical cell lines varies following treatment with receptor-type specific retinoids, and (ii) that the
relationship between retinoid regulation of proliferation and differentiation can be uncoupled.

INDEX

Acetate, 137
Actinomycin D, 77
Acute promyelocytic leukemia (APL), 206
Acyl-CoA dehydrogenase, 112
Acyl-CoA oxidase: *see* Peroxisomal acyl-CoA oxidase
Adenomatous polyposis coli (APC): *see* APC gene
Adipocyte lipid-binding protein (ALBP), 145–155
Adipocyte lipid-binding protein (ALBP) (aP2) knock-
    out mice, 149, 150–152, 154
Adipocytes
  gene expression regulation in, 145–155
  translation effects on fatty acid synthase in, 211
Aflatoxin, 98
AGN193109, 241
AIN 93G diets, 42–43
Alaskan natives, 5, 86
ALBP: *see* Adipocyte lipid-binding protein
α-Oxidation, 128
American Indians, 4–5
American Institute for Cancer Research (AICR), 13
AMF: *see* Autocrine motility factor
2-Amino-1-methyl-6-phenylimidazo (4,5-b)pyridine
    (PhIP), 98
Angiogenesis, 28–29
Angiotensin II, 77
Animal models
  of breast cancer, 40–43, 57
  of colon cancer, 87
  dietary lipids, energy intake, and cancer in, 29–32
  of PPARα-induced hepatocarcinogenesis, 127–133
Anorexia
  leptin and, 230
  ventromedial nuclei neurotransmitters and, 183, 184
AOM: *see* Azoxymethane
APC gene, 26, 77, 225
aP2 gene, 111
  polyunsaturated fatty acid regulation of, 145–155
aP2 null mice, 147, 150–152, 154
Apo genes, 113, 114
Apoptosis, 25–28
  Bcl-2 inhibition of butyrate-induced, 224
  butyrate effects on epithelial cell, 140
  choline deficiency and, 97–104

Apoptosis (*cont.*)
  1,25 dihydroxyvitamin D_3 induction of, 240
  EB1089 induction of, 239
  PPARα and, 115
Arabinosylcytosine, 207
Arachidonic acid (AA), 3
  breast cancer and, 48, 49–51, 52
  colon cancer and, 89
  metastasis and, 49–51, 72
  PPAR and, 146
  prostate cancer and, 231
ARP-1, 113, 173
Aryl hydrocarbon receptors, 213
Asians, 5, 11
Autocrine motility factor (AMF), 72, 75–76
Azoxymethane (AOM), 87, 88, 90, 232–233

B16 amelanotic melanoma cells, 73, 74, 75, 76, 78
BAPTA, 75
Bcl-1 gene, 26
Bcl-2 gene, 26, 27, 140, 224
B-CLL cells, 205
Benzene, 89
Benzofibrate, 110
N-Benzyl-N-hydroxy-5-phenylpentanamide (BHPP),
    74, 77
β-Adrenergic agonists, 146
Betaine, 99
β-Oxidation, 109–110, 112, 117, 168
  eicosapentaenoic acid and, 169
  fat catabolism and, 128
  hepatocarcinogenesis and, 114
  short-chain fatty acids and, 138
BHPP: *see* N-Benzyl-N-hydroxy-5-phenylpentanamide
Bifenazole, 110
Blacks, 9–10, 11
Bladder cancer, 12, 24
BRCA1 gene, 23, 24–25, 26, 177
BRCA2 gene, 23, 24–25, 26
Breast cancer, 29, 30–31, 32, 98, 177, 221–222
  cell growth in, 47–52
  cell invasion in, 51
  cell proliferation in, 57–65

Breast cancer (*cont.*)
   corn oil and, 202
   curcumin inhibition of, 223
   cyclooxygenase and: see Cyclooxygenase
   1,25 dihydroxyvitamin D_3 and, 240
   EB1089 and, 239
   epidemiologic research on, 6–8
   estrogen-dependent, 52
   estrogen-independent, 240
   estrogen receptor increase during pregnancy and,
      226
   flavonoid and, 189
   gene expression in, 193
   genetic factors in, 23, 24–25
   high-carbohydrate diet and, 187
   insulin levels during pregnancy and, 227
   integrin and protein kinase C-mediated, 237
   metastatic, 29, 49–51
   molecular studies on role of cholesterol in, 39–40,
      43–44
   molecular studies on role of dietary fat in, 39–43
   omega-3/omega-6 ratio and, 212
   palm oil and, 59, 189
   tamoxifen and, 189
   vitamin E and, 215
   weight-cycling and, 203
Bryostatins, 89
Butyrate
   Bcl-2 and, 140, 224
   fiber and, 216
   growth inhibition and differentation by, 140–142
   intestinal gene expression and, 137–142
   molecular and cell biology of, 138–140
   Nb2-SFJCD1 and B-CLL cells affected by, 205

Caco-2 colon carcinoma cells, 139, 140
Caffeic acid, 50
Calcium, 89, 182
Calphostin C, 75, 77
cAMP: *see* Cyclic adenosine monophosphate
Cancer cascade, 21–34; *see also* Dietary lipids
Carbohydrate ancillary factor (CAF), 170
Carbohydrate diet, 187
Carbon tetrachloride, 89
Carnitine palmitoyltransferase I, 112
Carotene, 185
CAT gene, 139, 170, 172
Cathepsin B, 72, 76
CCAAT/enhancer binding protein, 146, 153
Cell function, 168
Cellular proliferation, 25–28
   in breast cancer, 57–65
Ceramide, 87
c-erbB-1 gene, 64; *see also* Epidermal growth factor
   receptor
c-erbB-2 gene: *see neu* (c-erb-2/HER-2) gene
c-fms gene, 26
c-*fos* gene, 119, 139, 141–142

c-Ha-ras gene, 177
Charcoal-broiled foods, 25–26
Chinese people, 5, 6, 10
Cholesterol
   molecular studies on role in breast cancer, 39–40,
      43–44
   PPARα and, 130–131
Choline deficiency, 97–104
Chymotrypsin, 79
cis-Polyunsaturated fatty acids, 1, 237
9-*cis* Retinoic acid, 111, 113, 146
c-*jun* gene, 119, 141
Clofibrate, 110, 116, 117
Clofibric, 111
c-*myc* gene, 139–140, 141–142, 170
Coconut oil, 2
Collagen IV, 237
Collagenase IV, 29, 51, 72
Colon cancer, 29, 98–99, 177
   fiber and, 85, 216
   gene expression in, 137–142
   12(S)-HETE and, 77
   modulation of intracellular second messengers in,
      85–91
   nutrient and non-nutrient combination effect on,
      232–233
   PCNA and, 182
   sphingomyelin and, 228–229
   well-browned beef and, 188
Colorectal cancer
   Bcl-2 inhibition of butyrate-induced apoptosis in,
      224
   epidemiologic research on fat and, 4–6
Competition assays, 150
Corn oil
   breast cancer and, 202
   colon cancer and, 86, 88, 98–99
   prostate cancer and, 30, 208–209
Corticosterone, 220
COUP-TF, 173
COX: *see* Cyclooxygenase
CSF-1 gene, 72
c-*sis* gene, 139
Curcumin
   breast cancer growth inhibition by, 223
   inflammatory mediators inhibited by, 195–196
   skin tumorigenesis and, 186
CYC1 gene, 159
Cyclic adenosine monophosphate (cAMP), 77, 171
Cycloheximide, 142
Cyclooxygenase (COX), 47, 48
   differential expression and regulation of, 221–222
   metastasis and, 49–51, 72–73
Cyclooxygenase 1 (COX 1), 42, 43, 49, 52, 221–222
Cyclooxygenase 2 (COX 2), 42, 43, 49–51, 52,
   221–222
CYP4A family, 110, 117
CYP4A1 gene, 112

CYP4A6 gene, 113, 116
Cytochrome P-450, 72, 110, 112, 116, 117, 130
Cytochrome P-450 1A1, 213
Cytochrome P-450 2C11, 213

DCC gene, 77
DEHP: *see* Diethylhexylphthalate
5,6-Dehydroarachidonate, 50
Dehydroepiandrosterone (DHEA), 110, 117
Deoxyuridine triphosphate (dUTPase), 217
DHA: *see* Docosahexaenoic acid
DHEA: *see* Dehydroepiandrosterone
Diabetes, 171
Diacylglycerol (DAG), 72, 78–79, 80
    colon cancer and, 87–91
Dietary lipids, 21–34
    angiogenesis, metastasis and, 28–29
    cellular proliferation, apoptosis, immortality and,
        25–28
    energy intake, cancer and, 29–32
    genetic factors and, 23–25
Dietary supplements
    fatty acids as, 65
    ras expression and, 197
Diethylhexylphthalate (DEHP), 110, 115
1,25 Dihydroxyvitamin D_3 ($1,25(OH)_2D_3$), 240
7,12-Dimethylbenz(α)anthracene (DMBA), 31, 32,
        40, 44
1,2-Dimethylhydrazine, 228–229
Dithiolethiones, 234
DMBA: *see* 7,12-Dimethylbenz(α)anthracene
DNA damage
    corn oil and, 202
    lipid peroxidation and, 198
    p53 and, 27
    peroxisome proliferation and, 158
    PPARα and, 114
    weight-cycling and, 203
DNA replication, 114–115
Docosahexaenoic acid (DHA), 2, 49, 86, 207
Dopamine, 183
DR-1 motif, 113, 173
Dunning prostate cancer cells, 30, 73
DU 145 prostate cancer cells, 79, 208–209
dUTPase: *see* Deoxyuridine triphosphate

EB1089, 239
Ectocervical epithelial cells, 241, 242, 243
EETs: *see* Expoxyeicosatrienoic acids
EGF: *see* Epidermal growth factor
Eicosanoids, 49–51, 72–73
Eicosapentaenoic acid (EPA), 2, 49, 86, 169
Endometrium cancer, 12
Energy intake, 29–32
Enoyl-CoA hydratase/3-hydroxyacyl-CoA dehydro-
        genase, 110
Enterostatin, 220
EPA: *see* Eicosapentaenoic acid

Epidemiologic research, 1–13
    classification of fat in, 1–3
    issues related to fat and cancer in, 3–4
    specific cancer types in, 4–12
Epidermal growth factor (EGF)
    breast cancer and, 48, 51, 60–63
    choline deficiency and, 101–102
    metastasis and, 72–73
    PPARα and, 115
Epidermal growth factor receptor (EGFR), 59, 60, 61,
        62–63, 64–65
Epithelial cells
    apoptosis in, 140
    ectocervical, 241, 242, 243
erb-A gene, 26
erb-B gene, 26
erg-1 gene, 119
Esculetin, 50, 51
Eskimos: *see* Alaskan natives
Essential fatty acids, 2–3
Estrogen-dependent breast cancer, 52
Estrogen-independent breast cancer, 240
Estrogen receptors (ERs), 47, 48, 52
    indole-3-carbinol and, 194, 219
    pregnancy and, 226
Ethacrynic acid, 50
Etofibrate, 110
ETYA, 111
Experimental animal models: *see* Animal models
Experimental cell models, 57
Epoxeicosatrienoic acids (EETs), 72
Extracellular matrix (ECM), 71, 74–75

FABP: *see* Fatty acid-binding protein
FAK: *see* Focal adhesion kinase
Familial adenomatous polyposis, 225
Farnesoid X-activated receptor (FXR), 131–132
Farnesol, 131–132
Farnesylpyrophosphate (FPP), 130–131
Fat catabolism, 128–130
Fatty acid-binding protein (FABP), 146, 147, 154, 155
Fatty acids
    breast cancer cell growth and, 47–52
    breast cancer cell proliferation and, 57–65
    breast cancer metastasis and, 49–51
    as dietary supplements, 65
    essential, 2–3
    long-chain saturated, 59, 60, 65
    neurofibromin and GAP regulation by, 204
    non-essential, 2
    polyunsaturated: *see* Polyunsaturated fatty acids
    short-chain, 137–142
Fatty acid synthase, 211
Fecal mRNA, 91
Fenofibrate, 110, 113
Ferritin, 234
Fiber, 85, 216
Fibrate drugs, 113–114

Fish oil, 2
  breast cancer and, 59
  colon cancer and, 86, 88, 91
  hepatic lipid synthesis and metabolism and, 169
Flavonoids, 189
Focal adhesion kinase (FAK), 74
Folate deficiency, 99, 191, 213
fos gene, 26
Friend leukemia virus, 191
FXR: see Farnesoid X-activated receptor

GAGG sequence, 40
β-Galactosidase, 159
Gallbladder cancer, 12
GAPS: see GTPase activating proteins
Gastric cancer, 192
Gemfibrozil, 110, 113
Gene activation, 112–114
Gene expression
  in adipocytes, 145–155
  in breast cancer, 193
  hepatic, 169–173
  intestinal, 137–142
Genetic factors, 23–25
Gene transcription
  L-pyruvate kinase, 169–171
  S14, 171–172
Glioblastoma, 79
GLUT4, 154
Glutathione-S-transferase M1, 24
gp78, 75–76
G1 phase of cell cycle, 27, 140–141
G proteins, 44
  EGF and, 60, 62–63, 64
  12(S)-HETE and, 78, 79–80
Grilled meat, 98
Growth factor, 205
GSTM1 gene, 24
GTPase activating protein (GAP), 59, 204

H7, 171
Ha-ras gene, 40–42, 44
Helicobacter pylori, 192
Hemoglobin SS disease, 138
Hepatic gene expression, 169–173
Hepatic heme oxygenase-1, 234
Hepatic lipid synthesis and metabolism, 168–169
Hepatocarcinogenesis
  linoleic acid and, 200
  PPARα and, 114–115, 119, 127–133
Hepatocellular carcinoma
  choline deficiency and, 97–104
  methyl/folate deficiency and, 213
HER-2 gene: see Neu (c-erb-2/HER-2) gene
HETE: see Hydroxyeicosatetraenoic acid
Heterocyclic amines, 25, 188
High-carbohydrate diet, 187
High-density lipoprotein (HDL), 65

L-Histidine, 235
Histone deacetylase, 138
HMG-CoA: see 3-Hydroxy-3-methylglutaryl-CoA reductase/synthase
HNF-1, 170
HNF-4, 113, 170, 171, 173
HODE: see 13-Hydroxyoctadecadienoic acid
HPETE: see 5-/12-Hydroperoxy-2-trans,4-cis-pentadiene
HPV16: see Human papillomavirus
H-ras gene, 87
HT-29 colon adenocarcinoma cells, 137, 140–142
Human papillomavirus 16 (HPV16), 214
Hydrogen peroxide, 109, 114, 128
5-Hydroperoxy-2-trans,4-cis-pentadiene (HPETE), 50, 77
12-Hydroperoxy-2-trans,4-cis-pentadiene (HPETE), 50
8-Hydroxydeoxyguanosine, 114
12-Hydroxydodecanoic acid (12-OH-DDA), 128, 129
Hydroxyeicosatetraenoic acid (HETE), 72
5-Hydroxyeicosatetraenoic acid (HETE), 50, 51
8(S)-Hydroxyeicosatetraenoic acid (HETE), 118
12-Hydroxyeicosatetraenoic acid (HETE), 29, 50, 51, 52
12(S)-Hydroxyeicosatetraenoic acid (HETE)
  lysosomal enzyme release and, 76
  in multiple steps of metastasis, 73–74
  role in metastasis, 79–80
  as a signaling molecule, 77–79
  tumor cell–extracellular matrix interactions and, 74–75
  tumor cell infrastructure and cytoskeleton and, 76–77
  tumor cell motility and, 75–76
15-Hydroxyeicosatetraenoic acid (HETE), 50, 78
15(S)-Hydroxyeicosatetraenoic acid (HETE), 79
3-Hydroxy-3-methylglutaryl-CoA (HMG-CoA) reductase, 44
3-Hydroxy-3-methylglutaryl-CoA (HMG-CoA) synthase, 112
13-Hydroxyoctadecadienoic acid (HODE), 72, 73, 79
13(S)-Hydroxyoctadecadienoic acid (HODE), 78, 80

I3C: see Indole-3-carbinol
IGFBP-3: see Insulin-like growth factor binding protein-3
Immortality, 25–28
Indole-3-carbinol (I3C), 194, 219
Indomethacin, 51
Inositol hexaphosphate, 199
Inositol 1,4,5-triphosphate (IP_3), 78, 80, 89
Insulin, 146, 168, 170, 171, 227
Insulin-like growth factor binding protein-3 (IGFBP-3), 242
Insulin-like growth factor 1 (IGF1), 61
Insulin resistance, 154, 227
Integrins, 74, 75, 77, 78, 80, 237
Interleukin-1 (IL-1), 73, 184
Interleukin-1β (IL-1β), 195–196
Intestinal cancer
  gene expression in, 137–142
  sulindac and, 225
Intracellular second messengers, 85–91

IP$_3$: *see* Inositol 1,4,5-triphosphate
Iron, 234
Isoprenoids, 130–131

Japanese people
    breast cancer in, 6, 7, 49
    colorectal cancer in, 5
    prostate cancer in, 29
jun gene, 26
*jun*B gene, 119

Keratin 1, 218
Keratinocyte lipid-binding protein (KLBP), 154
    ligand-binding purification of, 147–148
    ligand-binding studies of, 149–150
    upregulation in ALBP(aP2) knock-out mice,
        150–152
Kidney cancer, 12
KLBP: *see* Keratinocyte lipid-binding protein
K-1735 melanoma cells, 73, 75, 76
Knock-out mice
    ALBP, 149, 150–152, 154
    p53, 25
Koreans
    breast cancer in, 6
    colorectal cancer in, 5
    prostate cancer in, 9–10
K-ras gene, 87

lacZ gene, 159
Lard, 59
Laurate, 59
Lecithin cholesterol acyltransferase, 113
Leptin, 230
Leukemia
    acute promyelocytic, 206
    Friend, 191
Leukotriene A$_4$ (LTA$_4$), 50
Leukotriene B$_4$ (LTB$_4$), 50, 72, 77, 118
Leukotriene D$_4$ (LTD$_4$) antagonists, 110
Ligand-binding purification, 147–148
Ligand-binding studies, 149–150
Linoleic acid (LA), 2, 3, 32, 72–73
    breast cancer and, 29, 48–49, 50–51, 59
    12(S)-HETE and, 79
    ornithine decarboylase and peroxisome proliferation
        with, 200
    PPARα and, 111
Lipid metabolism, 117
Lipid peroxidation, 198
Lipids: *see* Dietary lipids
Lipoxygenase (LOX)
    breast cancer and, 47, 48, 49–51
    keratin 1 and, 218
    metabolites of in metastasis, 71–80
5-Lipoxygenase (LOX), 50, 72
12-Lipoxygenase (LOX), 73
    breast cancer and, 48, 50, 52
    12(S)-HETE and, 75–76, 77–78

15-Lipoxygenase (LOX), 50, 72, 79
LNCaP prostate cell line, 30
Long-chain saturated fatty acids (LCSFA), 59, 60, 65
Lovastatin, 130
Low-density lipoprotein (LDL), 44, 65
LOX: *see* Lipoxygenase
Lung cancer, 11–12, 13
LY-171883, 111, 114
Lysosomal enzyme release, 76

Major late transcription factor-like (MLTF-like) bind,
    170
Malic enzymes, 112
MAP kinase, 64, 80
MCC gene, 77
MCF10Aneo T human mammary carcinoma cells, 76
MCF-7 breast cancer cells, 47, 48, 140, 189, 221–222,
    239
MDA-468 breast cancer cells, 60
MDA-MB-231 breast cancer cells, 48–49, 50, 51, 52,
    221–222
MDA-MB-435 breast cancer cells, 48–49, 51, 52, 189
MDM-2 gene, 77
*mdr*1 gene, 139
Meat, 2
    grilled, 98
    well-browned, 188
Menhaden oil, 42–43, 208–209
Metastasis, 28–29
    breast cancer, 29, 49–51
    lipoxygenase metabolites in, 71–80
Methionine, 99
Methyl-deficient diets, 213
N-Methylnitrosourea (MNU), 40–41, 43–44
Mevalonate, 130
Mn11 restriction endonuclease, 40–41
MNU: *see* N-Methylnitrosourea
Motility factor (MF), 75
mRNA
    fatty acid synthase, 211
    fecal, in colon cancer, 91
    12(S)-HETE and, 77
    keratin 1, 218
myc gene, 26
Myristate, 59

NADPH-cytochrome P-450 reductase, 213
Nafenopin, 115
Nb2-SFJCD1 (pre-T lymphoma), 205
Neocuproine, 99
neu (c-erb-2/HER-2) gene, 26, 64
Neurofibromin, 204
NF1 gene, 26, 170, 172
NF2 gene, 26
NFκB, 195–196
Nitric oxide, 72, 195–196
Non-essential fatty acids, 2
Non-genotoxic carcinogens, 119

Nordihydroguaiaretic acid, 50
Northern analysis, 149
N-*ras* gene, 139
Nucleophosmin (NMP), 206
Null mice
    aP2, 147, 150–152, 154
    PPARα, 116–117, 118, 119
nup/475 gene, 119, 141

OAF1 gene, 161, 162, 163
Oaf1p: *see* Oleate-activated factor 1
Okadaic acid, 77
Oleate-activated factor 1 (Oaf1p), 159–161, 163–164
Oleate response element, 159
Oleic acid, 59, 89, 112, 129
Olive oil, 86
Omega-3 polyunsaturated fatty acids, 29, 32
    breast cancer and, 42, 49, 52, 59, 212
    cell function and, 168
    colon cancer and, 86–87, 90, 91
    epidemiologic research on, 1, 2
    hepatic lipid synthesis and metabolism and, 169
    omega-6 ratio and dietary intervention, 212
Omega-6 polyunsaturated fatty acids, 32
    breast cancer and, 42–43, 47, 48–49, 51, 52, 212
    cell function and, 168
    colon cancer and, 86, 98–99
    epidemiologic research on, 1, 2
    hepatic lipid synthesis and metabolism and, 169
    normal mammary epithelial growth and, 48–49
    omega-3 ratio and dietary intervention, 212
Omega-9 polyunsaturated fatty acids, 86
Oncogenes, 26, 190
Ornithine decarboxylase, 199, 200
Ovarian cancer, 12, 177
Oxidation
    α-, 128
    β-: *see* β-Oxidation
Oxidative stress hypothesis, 128, 129, 132

Palmitate, 59
Palmitic acid, 128, 129
Palm oil, 2, 59, 189
Pancreatic cancer, 12, 98
Papillomaviruses, 194
    human, 214
PC2-M prostate cancer cells, 208–209
PCNA: *see* Proliferating nuclear cell antigen
PC-3 prostate cancer cells, 79
PEPCK: *see* Phosphoenolpyruvate carboxykinase
Peroxisomal acyl-CoA oxidase, 109, 110, 114, 116, 169
    cis-Acting elements in, 159
    regulation in *Saccharomyces cerevisiae*, 157–164
Peroxisome proliferation, 109–110
    in mammals, 158
    linoleic acid induction of, 200
    PPARα role in DHEA induction of, 117

Peroxisome proliferation (*cont.*)
    species difference in, 119–120
    in yeast, 158–159
Peroxisome proliferator-activated receptor (PPAR), 164, 169, 171
    function of, 111–112
    gene expression regulation in adipocytes by, 146, 151, 153, 154
    in mammals, 158
    polyunsaturated fatty acid regulation of, 172–173
Peroxisome proliferator-activated receptor α (PPARα), 109–120
    dUTPase and, 217
    farnesol as activator of, 131–132
    fat catabolism and, 128–130
    function of, 111–112
    gene activation by, 112–114
    gene expression regulation in adipocytes by, 153
    growth control and, 118
    hepatocarcinogenesis and, 114–115, 119, 127–133
    isoprenoids and, 130–131
    in mammals, 158
    non-genotoxic carcinogens and, 119
    oxidative stress hypothesis and, 128, 129, 132
    species differences in, 116, 119–120
Peroxisome proliferator-activated receptor α (PPARα)-null mice, 116–117, 118, 119
Peroxisome proliferator-activated receptor β (PPARβ), 111, 112, 117, 118, 127, 158
Peroxisome proliferator-activated receptor δ (PPARδ), 153
Peroxisome proliferator-activated receptor γ (PPARγ), 111, 117, 118, 127, 153, 158
Peroxisome proliferator-activated response element (PPRE), 112–113, 114, 116
Peroxisomes, 109–110
Pertussis toxin, 60, 78
p21 gene, 44
p53 gene, 27
    apoptosis and, 97, 98, 100–101
    butyrate and, 139, 140
    choline deficiency and, 97, 98, 100–101
    12(S)-HETE and, 77
    knock-out mouse, 25
    mechanism of, 26
Phenylacetate, 110, 118
PhIP: *see* 2-Amino-1-methyl-6-phenylimidazo (4,5-b)pyridine
Phorbol ester-induced ornithine decarboxylase activation, 199
Phosphatidylinositol bis 4,5-phosphate (PIP$_2$), 78, 80
Phosphoenolpyruvate carboxykinase (PEPCK), 146
Phospholipase A (PLA), 89
Phospholipase A$_2$ (PLA$_2$), 72
Phospholipase C (PLC), 72, 78, 79, 80
Phospholipase C-γ (PLC-γ)
    breast cancer cell proliferation and, 59, 60, 64
    colon cancer and, 87–91

Phospholipase D (PLD), 72, 89
Phytanic acid, 128, 129
pim-1 gene, 205
PIP$_2$: *see* Phosphatidylinositol bis 4,5-phosphate
Piroxicam, 50, 51
Placenta-like alkaline phosphatase (PLAP) promoter, 139
Platelet-activating factor (PAF), 72
Pluripotent response region (PRR), 172
Polycyclic aromatic hydrocarbons, 26
Polyunsaturated fatty acid-regulatory factors (PUFA-RF), 170, 172
Polyunsaturated fatty acid-regulatory region (PUFA-RR), 172–173
Polyunsaturated fatty acids (PUFA), 1
   cis, 1, 237
   hepatic gene expression and, 169–173
   hepatic lipid synthesis and metabolism and, 168–169
   omega: *see* Omega-3 polyunsaturated fatty acids;
      Omega-6 polyunsaturated fatty acids
   PPAR regulation by, 172–173
   regulation of gene expression in adipocytes by, 145–155
   S14 gene transcription suppression and, 171–172
   trans, 1
   vitamin E effect on, 215
POX1 gene, 159–161, 163–164
PPAR: *see* Peroxisome proliferator-activated receptor
PPRE: *see* Peroxisome proliferator-activated response element
Pregnancy
   estrogen receptor increase and breast cancer risk in, 226
   insulin levels and breast cancer risk in, 227
Preinitiation complex, 172
Proliferating nuclear cell antigen (PCNA), 182
Propionate, 137
Prostacyclin (PGI$_2$), 72, 77
Prostaglandin (PG), 42
Prostaglandin A (PGA), 72
Prostaglandin D$_2$ (PGD$_2$), 72
Prostaglandin E$_2$ (PGE$_2$), 51, 72
Prostaglandin G2 (PGG2), 49
Prostaglandin H2 (PGH2), 49, 79
Prostaglandin J$_2$ (PGJ$_2$), 72, 146, 153
Prostate cancer, 29–32, 33
   arachidonic acid metabolism and, 231
   corn oil and, 30, 208–209
   epidemiologic research on, 9–11
   12(S)-HETE and, 79
   menhaden oil and, 208–209
   vitamin D and, 210
Protein kinase C (PKC)
   breast cancer and, 237
   colon cancer and, 87–88, 89–91, 98–99
   12(S)-HETE and, 75, 78, 79
   interactions with unsaturated fatty acids and selenium, 201

Proto-oncogenes, 25
PUFA: *see* Polyunsaturated fatty acids
L-Pyruvate kinase gene transcription, 169–171

raf gene, 26
ras gene, 26, 44, 88, 197
Rb gene, 26, 27, 77
RET gene, 26
Retinoic acid (RA), 111, 112, 146, 214
Retinoic acid receptor (RAR), 173, 206, 242, 243
Retinoic acid X receptor (RXR), 242, 243
Retinoic acid X receptor α (RXRα), 111, 113, 127, 146, 153, 164, 169, 171
   farnesol and, 131
   polyunsaturated fatty acid regulation of PPAR and, 172–173
Retinoids, 241, 242, 243
Riboflavin deficiency, 98
Rous sarcoma virus (RSV) promoter, 172
RXR: *see* Retinoic acid X receptor

Saccharomyces cerevisiae, 157–164
Safflower oil, 42–43, 86
Scatter factor (SF), 75
Scd1 gene, 238
Selenium, 201
Serotonin (5HT), 183, 184
S14 gene transcription, 171–172
Short-chain fatty acids (SCFA), 137–142
sis gene, 26
Skin cancer, 24
Skin tumorigenesis, 186, 235
Smoked foods, 25–26
Smoking, 24
Sodium butyrate: *see* Butyrate
Soybean oil, 43
Sphingolipids, 228–229
Sphingomyelin, 228–229
Squalestatin, 130
src gene, 26
Staurosporin, 171
Stearate inhibition, 60–61
Stearoyl-CoA desaturase, 238
Sulindac, 225
SUM-159PT breast cancer cells, 240
SV40 large T-antigen, 97, 99
SW837 colon carcinoma cells, 140

Tamoxifen, 189
TATA boxes, 170
Tetradecanoyl phorbol acetate (TPA), 49, 52, 75, 78
3T3-F442A adipocytes, 211
TGF: *see* Transforming growth factor α/β/β$_1$
Thiazolidinediones, 111
Thin layer chromatography, 147
Thiolase, 130
Thromboxane A$_2$ (TXA$_2$), 51, 72, 79
Thymelea toxin, 78
Thymidine kinase, 139

Thyroid hormone (T$_3$), 168, 171, 173
Thyroid hormone receptor (TR), 113
Thyroid hormone response DR-2 element (TRRE), 113
Thyroid hormone response elements (TRE), 172
Thyroid hormone response region (TRR), 172
Tissue-specific factors (TSF), 172
TNF: see Tumor necrosis factor
α-Tocopherol: see Vitamin E
TPA: see Tetradecanoyl phorbol acetate
trans-Fatty acids, 1
Transforming growth factor α (TGFα)
    breast cancer and, 59, 64–65
    choline deficiency and, 97, 101–102
    metastasis and, 73
Transforming growth factor β (TGFβ), 214
Transforming growth factor β1 (TGFβ1)
    apoptosis and, 115
    butyrate and, 140, 141
    choline deficiency and, 97, 102–103
Transforming growth factor β2 (TGFβ2), 140
Translation, 211
trans-Urocanic acid (trans-UCA), 235
Trichloroethylene, 110
Triglycerides, 148, 149
Trypsin, 79
Tumor cell
    cytoskeleton, 76–77
    growth, 47–52
    intrastructure, 76–77
    invasion, 51
    motility, 75–76
    proliferation see Cellular proliferation
Tumor initiation, 186
Tumor necrosis factor (TNF), 73
Tumor necrosis factor-α (TNF-α)
    aP2 null mice and, 147, 150–152, 154
    curcumin and, 195–196

Tumor promotion, 89, 186
Tumor suppressor genes, 26, 216

UAS: see Upstream activating sequence
Ultraviolet (UV)-induced tumors, 190
Unsaturated fats, 1–2
Unsaturated fatty acids, 201
U251P glioblastoma cells, 79
Upstream activating sequence (UAS), 159, 163–164
Upstream stimulatory factor (USF), 170
Urokinase-type plasminogen activator (uPA), 51

Ventromedial nuclei (VMN) neurotransmitters, 183, 184
VHL gene, 26
Vietnamese people, 5
Vitamin A derivative receptor, abnormal, 206
Vitamin B12, 99
Vitamin C, 185
Vitamin D, 210, 239, 240
Vitamin D receptor (VDR), 173
Vitamin E, 185, 215

WAF1, 216
Weight-cycling, 203
Well-browned beef, 188
Western analysis, 149
Whites, 11
World Cancer Research Fund (WCRF), 13
WT1 gene, 26
WY-14,643, 110, 114, 115, 117, 172

Yeast, 157–164

Zellweger syndrome, 158
zif 268 gene, 141